SSA-based Compiler Design

Fabrice Rastello • Florent Bouchez Tichadou
Editors

SSA-based Compiler Design

 Springer

Editors
Fabrice Rastello
Inria, Grenoble, France

Florent Bouchez Tichadou
University of Grenoble Alpes
Grenoble, France

ISBN 978-3-030-80517-3   ISBN 978-3-030-80515-9   (eBook)
https://doi.org/10.1007/978-3-030-80515-9

This Springer imprint is published by the registered company Springer Nature Switzerland AG
The registered company address is: Gewerbestrasse 11, 6330 Cham, Switzerland

Foreword

Static Single Assignment (SSA) form is now the dominant framework upon which program analysis and transformation tools are currently built. It is possible to produce a state-of-the-art optimizing compiler that transforms the incoming program into SSA form after the initial syntactic and semantic analysis, and maintain that representation throughout the entire optimization process only to leave SSA form when the final machine code is emitted. This book describes the state-of-the-art representations of SSA form and the algorithms that use it.

The landscape of techniques for program analysis and transformation was very different when the two of us started the initial development of SSA form. Many of the compilers being built at the time were very simple, similar to what a student might generate in a first compiler class. They produced code based on a template for each production in the grammar of the language. The initial optimizing compilers transformed the code based on the branch-free context of the surrounding code, perhaps removing computation if the results were not to be used, perhaps computing a result in a simpler way by using a previous result.

The first attempts at more ambitious optimization occurred at IBM, in a few small companies in the Boston area and in a small number of academic institutions. These people demonstrated that large gains in performance were available if the scope of the optimizations was increased beyond the local level. In general, an optimizing compiler can make transformations to an intermediate representation of a program provided that the preconditions exist for the transformation to preserve the semantics of the program. Before SSA form, the dominant provers would construct a summary of what must be true about the state of the program before and after every statement in the program. The proof constructed would proceed by understanding what parts of the state were modified by the statement. This entire process was known as dataflow analysis. The results of that analysis were then used to make transformations to the intermediate code.

This framework had several drawbacks:

- The represented state was not sparse. The analysis to prove the set of facts involved solving a series of simultaneous equations which proved the validity of every proposition needed for that phase, at every location in the program. Because the state of the entire program had to be represented at each point in the program, the analysis was slow.
- The transformations voided the validity of the analysis. While incrementally updating the proofs was, in theory, possible, such algorithms were impractical because the representations were not sparse. Thus, each pass of the compiler consisted of an analysis phase followed by a transformation phase.
- Each type of transformation had its own set predicates that it needed to prove. Since the analysis was discarded after each transformation, there was no reason to try to prove any fact that was not necessary for the subsequent transformation.

  This had the effect of making it more difficult to design a new transformation because the programmer had to first figure out exactly what information was needed and then figure out how to perform the proofs before performing the transformation.
- The transformations were fragile. The intermediate representations available in early compilers used the programmer's variable names to express the data dependencies between different parts of the program. Part of designing any transformation involved separating those names into the values that reached each statement.

While significant work by many people had been done to improve the efficiency of the analysis process, the cost of having to perform the analysis from scratch, coupled with the need to represent every fact at each point, was significant. Fundamentally, this process did not scale: at the time that we developed SSA form, programs were getting larger and there was a desire to make optimizing compilers much more aggressive.

Many people were aiming to fix these problems separately. Def-use chains joined the origin of a value and its use but attempted to do this independently of control flow. Shapiro and Saint refined def-use chains by describing the birthpoints of values. These birthpoints would become the locations where SSA form introduces phi-functions. John Reif used birthpoints for some interesting analyses and found better algorithms to determine them than Shapiro and Saint. But birthpoints were expressed as an analysis technique of the original program and transformations still required rerunning the analysis between phases.

Our initial motivation for SSA form was to build a sparse representation that avoided the cost of proving every fact at every location. Our primary idea was to encode the data and control dependencies into a naming scheme for the variables. We chose to solve the extended form of the constant propagation problem which had been well studied by Wegbreit. We presented the algorithm at the 12th POPL conference. The algorithm was much more efficient than Wegbreit's version, but it also had one very interesting and unique property that was not appreciated at the time: the transformation did not adversely effect the correctness of the representation.

The first work was incomplete. It didn't cover the semantics of many common storage classes, and the construction algorithm was not well thought through. But the surprising lesson was that by constructing a uniform representation in which both transformations and analysis could be represented, we could simplify and improve the efficiency of both.

Over the next few years, we teamed up with others within IBM to develop not only an efficient method to enter SSA form but also a suite of techniques that each used SSA form and left the representation intact when finished. These techniques included dead code elimination, value numbering, and invariant code motion.

The optimizations that we developed are in fact quite simple, but that simplicity is a result of pushing much of the complexity into the SSA representation. The dataflow counterparts to these optimizations are complex, and this difference in complexity was not lost on the community of researchers outside of IBM. Being able to perform optimizations on a mostly functional representation is much easier than performing them on a representation that had all of the warts of a real programming language.

By today's standards, our original work was not really that useful: there were only a handful of techniques and the SSA form only worked for unaliased variables. But that original work defined the framework for how programming language transformation was to be performed. This book is an expression of how far SSA form has come and how far it needs to go to fully displace the older techniques with more efficient and powerful replacements. We are grateful and humbled that our work has led to such a powerful set of directions by others.

Chappaqua, NY, USA   Kenneth Zadeck
Yorktown Heights, NY, USA   Mark N. Wegman
March 2021

Preface

This book could have been given several names: *"SSA-based compiler design,"* or *"Engineering a compiler using the SSA form,"* or even *"The Static Single Assignment form in practice."* But now, if anyone mentions *"The SSA book,"* then they are certainly referring to this book.[1]

Twelve years were necessary to give birth to a book composed of 24 chapters and written by 31 authors. Twelve years: We are not very proud of this counter performance but here is how it comes: one day, a young researcher (myself) was told by a smart researcher (Jens Palsberg): *"we should write a book about all this!"* *"We"* was including Florent Bouchez, Philip Brisk, Sebastian Hack, Fernando Pereira, Jens, and myself. *"All this"* was referring to all the cool stuff related to register allocation of SSA variables. Indeed, the discovery made independently by the Californian (UCLA), German (U. of Karlsruhe), and French (Inria) research groups (Alain Darte was the one at the origin of the discovery in the French group) was intriguing, as it was questioning the relevance of all the register allocation papers published during the last thirty years. The way those three research groups came to know each other is worth mentioning by itself as without this anecdote the present world would probably be different (maybe not a lot, but most certainly with no SSA book...): A missing pdf file of a CGO article [240] on my web page made Sebastian and Fernando independently contact me. *"Just out of curiosity, why are you interested in this paper?"* led to a long friendly and fruitful collaboration... The lesson to learn from that story is: do not spend too much time on your web page, it may pay off in unexpected ways!

But why restricting a book on SSA to just register allocation? SSA was starting to be widely adopted in mainstream compilers, such as LLVM, Hotspot, and GCC, and was motivating many developers and research groups to revisit the different

---

[1] Put differently, this book fulfils the criterion of referential transparency, which seems to be a minimum requirement for a book about Static Single Assignment form... If this bad joke does not make sense to you, then you definitely need to read the book. If it does make sense, I am pretty sure you can still perfect your knowledge by reading it.

compiler analyses and optimizations for this new form. I was myself lucky to take my first steps in compiler development from 2000 to 2002 in LAO [102], an SSA-based assembly level compiler developed at STMicroelectronics.[2] Thanks to SSA, it took me only four days (including the two days necessary to read the related work) to develop a partial redundancy elimination very similar to the GVN-PRE of Thomas VanDrunen [295] for predicated code! Given my very low expertise, implementing the SSAPRE of Fred Chow (which is *not* SSA-based—see Chap. 11) would probably have taken me several months of development and debugging, for an algorithm less effective in removing redundancies. In contrast, my pass contained only one bug: I was removing the redundancies of the stack pointer because I did not know what it was used for... There is a lesson to learn from that story: If you do not know what a stack pointer is, do not give up, with substantial efforts you can still expect a position in a compiler group one day.

---

[2] More precisely ψ-SSA—see Chap. 15.

I realized later that many questions were not addressed by any of the existing compiler books: If register allocation can take advantage the structural properties of SSA, what about other low-level optimizations such as post-pass scheduling or instruction selection? Are the extensions for supporting aliasing or predication as powerful as the original form? What are the advantages and disadvantages of SSA form? I believed that writing a book that would address those broader questions should involve the experts of the domain.

With the help of Sebastian Hack, I decided to organize the SSA seminar. It was held in Autrans near Grenoble in France (yes for those who are old enough, this is where the 1968 winter Olympics were held) and regrouped 55 people (from left to right in the picture—speakers reported in italic): *Philip Brisk*, Christoph Mallon, *Sebastian Hack*, Benoit Dupont de Dinechin, David Monniaux, *Christopher Gautier*, *Alan Mycroft*, Alex Turjan, Dmitri Cheresiz, *Michael Beck*, *Paul Biggar*, Daniel Grund, *Vivek Sarkar*, Verena Beckham, Jose Nelson Amaral, Donald Nguyen, *Kenneth Zadeck*, *James Stanier*, *Andreas Krall*, *Dibyendu Das*, Ramakrishna Upadrasta, *Jens Palsberg*, *Ondrej Lhotak*, Hervé Knochel, *Anton*

Ertl, Cameron Zwarich, *Diego Novillo*, Vincent Colin de Verdière, *Massimiliano Mantione*, *Albert Cohen*, Valerie Bertin, *Sebastian Pop*, Nicolas Halbwachs, Yves Janin, Boubacar Diouf, *Jeremy Singer*, Antoniu Pop, Christian Wimmer, *Francois de Ferrière*, *Benoit Boissinot*, *Markus Schordan*, *Jens Knoop*, *Christian Bruel*, Florent Bouchez, Laurent Guerby, Benoît Robillard, *Alain Darte*, Fabrice Rastello, Thomas Heinze, *Keshav Pingali*, Christophe Guillon, *Wolfram Amme*, Quentin Colombet, and Julien Le Guen.

The goal of this seminar was twofold: first, have some of the best-known compiler experts climb to the top of the local mountain peak; second, regroup potential authors of the book, and work together on finding who could contribute to the writing. At the end of the week, some of the participants were still wondering why they accepted this invitation from this crazy mountain guy (although everyone made it to the top—I have pictures!), but the main layout of the book was roughly set up with authors for each chapter. The agreed objective was to have a book similar in spirit to a textbook: one that would cover most of the phases of an optimizing compiler; one that would use consistent notations and terminology across all the chapters of the book; one that would have all the required notions defined (possibly in a different chapter) before they are used. But it was not meant to be literally a textbook as it was for advanced compiler developers only. For a good textbook, we suggest, for example, the Tiger book [10] by Andrew Appel instead. In this current book, some of the chapters were constrained by page limit and might be really tough to understand otherwise. Paul Biggar created a LaTeX infrastructure, and I started pressing authors to fulfil their duty as soon as possible, convinced that within a year the book would be available... Lesson learned from that seminar: if a mountain climber asks you to join him for a "short" hike, tell him you have to work on the writing of a book.

This is when the smart guy declined to participate in the book adventure, and I suspect his intuition was telling him that the project was a bit too ambitious... It indeed turned out to be a very difficult task, especially for a young researcher. Writing a book by myself would probably have been much easier, but also less interesting. Concatenating independent chapters as one would have done for a compilation of the state of the art would also have been infinitely easier but would have lacked coherence and brought little or no added value. So, each chapter required significant rewriting from the authors compared to the original publishings, and the main reason it took such a long time is that it turns out to be extremely difficult to impose deadlines when there are no strong constraints that motivate those deadlines, as opposed for instance to the pressure to submit conference articles. We thought initially that six months would be enough for the different authors to write their chapter, and I anticipated two full months without any critical duty right after that period so as to be able to orchestrate the reviewing process: I wanted to review all chapters myself, but also have the different authors to do cross-reviewing of other chapters. After the deadline, it was already obvious that there was no way we would have a book within the time limit initially expected: most of the chapters were missing, and some were a carbon copy from the corresponding paper... Despite this, the reviewing process started, and it was clear a very deep reviewing was required. Indeed, almost every chapter contained at least one important technical flaw; people have different definitions (with slight but still important differences) for the same notion; some spent pages developing a new formalism that is already covered by existing well-established ones. But a deep review of a single chapter roughly takes a week... Starts a long walk in a tunnel were you context switch from one chapter review to another, try to orchestrate the cross-reviews, ask for the unfinished chapters to be completed, pray that the missing chapters finally change their status to "work-in-progress," and harass colleagues with metre-long back and forth email threads trying to finally clarify this *one* 8-word sentence that may be interpreted in the wrong way. You then realize no one will be able to implement the agreed changes in time because they got themselves swamped by other duties (the ones from their real job), and when these are available that is when you get yourself swamped by other duties (the ones from your real job), and your 2-month window is already flying away, leaving yourself realizing that the next one will be next year. And time flies because, for everybody including yourself, there is always something else more urgent than the book itself, and messages in this era of near-instantaneous communication behave as if they were using strange paths around the solar system, experiencing the joys of time dilation with 6-month return trips around Saturn's rings. More prosaically, I will lay the blame on the incompatibility between the highly imposed constraints and the absence of a strong common deadline for everybody. And so, the final lesson to learn from that experience is that if a mountain climber proposes to join him for a "short" hike, go for it: forget about the book, it will be less painful!

A few years after we initiated this project, as I was depressed by the time all this was taking, I met Charles Glaser from Springer who told me: *"You have plenty of time: I have a book that took 15 years to be completed."* At the time I am writing those lines, I still have a long list of to-dos, but I do not want to be the person Charles mentions as *"Do not worry, you cannot do worse than this book they started writing 15 years ago and which is still under construction..."* Sure, you might still find some typos or inconsistencies, but there should not be many: the book regroups the knowledge of many talented compiler experts, including substantial unpublished materials, and I am proud to release it today.

Grenoble, France   Fabrice Rastello
February 2021

Contents

Part III   Extensions

Part IV   Machine Code Generation and Optimization

About the Editors

Fabrice Rastello is an Inria research director and the leader of the CORSE (Compiler Optimization and Runtime SystEms) Inria team. His expertise includes automatic parallelization (PhD thesis on tiling as a loop transformations) and compiler back-end optimizations (engineer at STMicroelectronics's compiler group + researcher at Inria). Among others, he has advised several PhD theses so as to fully revisit register allocation for JIT compilation in the light of Static Single Assignment (SSA) properties. He likes mixing theory (mostly graphs, algorithmic, and algebra) and practice (industrial transfer). His current research topics include: (1) combining runtime techniques with static compilation, hybrid compilation being an example of such approach he is trying to promote; (2) performance debugging through static and dynamic (binary instrumentation) analysis; and (3) revisiting compilers infrastructure for pattern-specific programs.

Florent Bouchez Tichadou received his PhD in computer science in 2009 from the ENS Lyon in France, working on program compilation. He was then a postdoctoral fellow at the Indian Institute of Science (IISc) in Bangalore, India. He worked for 3 years at Kalray, a startup company in the Grenoble area in France. Since 2013, he is an assistant professor at the Université Grenoble Alpes (UGA).

Part I
Vanilla SSA

Chapter 1
Introduction

Jeremy Singer

In computer programming, as in real life, names are useful handles for concrete entities. The key message of this book is that having *unique names* for *distinct entities* reduces uncertainty and imprecision.

For example, consider overhearing a conversation about "Homer." Without any more contextual clues, you cannot disambiguate between Homer Simpson and Homer the classical Greek poet; or indeed, any other people called Homer that you may know. As soon as the conversation mentions Springfield (rather than Smyrna), you are fairly sure that the Simpsons television series (rather than Greek poetry) is the subject. On the other hand, if everyone had a *unique* name, then there would be no possibility of confusing twentieth century American cartoon characters with ancient Greek literary figures.

This book is about the *Static Single Assignment form* (SSA), which is a naming convention for storage locations (variables) in low-level representations of computer programs. The term *static* indicates that SSA relates to properties and analysis of program text (code). The term *single* refers to the uniqueness property of variable names that SSA imposes. As illustrated above, this enables a greater degree of precision. The term *assignment* means variable definitions. For instance, in the code

$$x = y + 1;$$

the variable x is being assigned the value of expression $(y + 1)$. This is a definition, or assignment statement, for x. A compiler engineer would interpret the above assignment statement to mean that the lvalue of x (i.e., the memory location labeled as x) should be modified to store the value $(y + 1)$.

J. Singer (✉)
School of Computing Science, University of Glasgow, Glasgow, UK
e-mail: Jeremy.Singer@glasgow.ac.uk

© The Author(s), under exclusive license to Springer Nature Switzerland AG 2022
F. Rastello, F. Bouchez Tichadou (eds.), *SSA-based Compiler Design*,
https://doi.org/10.1007/978-3-030-80515-9_1

1.1 Definition of SSA

The simplest, least constrained, definition of SSA can be given using the following informal prose:

> A program is defined to be in SSA form if each variable is a target of exactly one assignment statement in the program text.

However, there are various, more specialized, varieties of SSA, which impose further constraints on programs. Such constraints may relate to graph-theoretic properties of variable definitions and uses, or the encapsulation of specific control-flow or data-flow information. Each distinct SSA variety has specific characteristics. Basic varieties of SSA are discussed in Chap. 2. Part III of this book presents more complex extensions.

One important property that holds for all varieties of SSA, including the simplest definition above, is *referential transparency*: i.e., since there is only a single definition for each variable in the program text, a variable's value is *independent of its position* in the program. We may refine our knowledge about a particular variable based on branching conditions, e.g., we know the value of x in the conditionally executed block following an `if` statement that begins with

$$\texttt{if}(x == 0).$$

However, the *underlying value* of x does not change at this `if` statement. Programs written in pure functional languages are referentially transparent. Such referentially transparent programs are more amenable to formal methods and mathematical reasoning, since the meaning of an expression depends only on the meaning of its subexpressions and not on the order of evaluation or side effects of other expressions. For a referentially opaque program, consider the following code fragment.

$$
\begin{aligned}
x &= 1; \\
y &= x + 1; \\
x &= 2; \\
z &= x + 1;
\end{aligned}
$$

A naive (and incorrect) analysis may assume that the values of y and z are equal, since they have identical definitions of $(x + 1)$. However, the value of variable x depends on whether the current code position is before or after the second definition of x, i.e., variable values depend on their *context*. When a compiler transforms this program fragment to SSA code, it becomes referentially transparent. The translation process involves renaming to eliminate multiple assignment statements for the same variable. Now it is apparent that y and z are equal if and only if x_1 and x_2 are equal.

$$x_1 = 1;$$
$$y = x_1 + 1;$$
$$x_2 = 2;$$
$$z = x_2 + 1;$$

1.2   Informal Semantics of SSA

In the previous section, we saw how straight-line sequences of code can be transformed to SSA by simple renaming of variable definitions. The *target* of the definition is the variable being defined, on the left-hand side of the assignment statement. In SSA, each definition target must be a unique variable name. Conversely variable names can be used multiple times on the right-hand side of any assignment statements, as *source* variables for definitions. Throughout this book, renaming is generally performed by adding integer subscripts to original variable names. In general this is an unimportant implementation feature, although it can prove useful for compiler debugging purposes.

The ϕ-function is the most important SSA concept to grasp. It is a special statement, known as a *pseudo-assignment* function. Some call it a "notational fiction."[1] The purpose of a ϕ-function is to merge values from different incoming paths, at control-flow merge points.

Consider the following code example and its corresponding control-flow graph (CFG) representation:

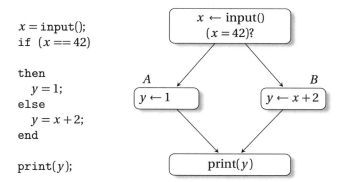

```
x = input();
if (x == 42)

then
    y = 1;
else
    y = x + 2;
end

print(y);
```

---

[1] Kenneth Zadeck reports that ϕ-functions were originally known as *phoney*-functions, during the development of SSA at IBM Research. Although this was an in-house joke, it did serve as the basis for the eventual name.

There is a distinct definition of y in each branch of the `if` statement. So multiple definitions of y reach the `print` statement at the control-flow merge point. When a compiler transforms this program to SSA, the multiple definitions of y are renamed as y_1 and y_2. However, the `print` statement could use either variable, dependent on the outcome of the `if` conditional test. A ϕ-function introduces a new variable y_3, which takes the value of either y_1 or y_2. Thus the SSA version of the program is

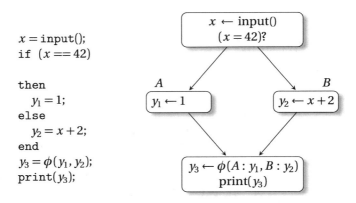

```
x = input();
if (x == 42)

then
    y₁ = 1;
else
    y₂ = x + 2;
end
y₃ = φ(y₁, y₂);
print(y₃);
```

In terms of their position, ϕ-functions are generally placed at control-flow merge points, i.e., at the heads of basic blocks that have multiple direct predecessors in control-flow graphs. A ϕ-function at block b has n parameters if there are n incoming control-flow paths to b. The behaviour of the ϕ-function is to select dynamically the value of the parameter associated with the actually executed control-flow path into b. This parameter value is assigned to the fresh variable name, on the left-hand side of the ϕ-function. Such pseudo-functions are required to maintain the SSA property of unique variable definitions, in the presence of branching control flow. Hence, in the above example, y_3 is set to y_1 if control flows from basic block A and set to y_2 if it flows from basic block B. Notice that the CFG representation here adopts a more expressive syntax for ϕ-functions than the standard one, as it associates direct predecessor basic block labels B_i with corresponding SSA variable names a_i, i.e., $a_0 = \phi(B_1 : a_1, \ldots, B_n : a_n)$. Throughout this book, basic block labels will be omitted from ϕ-function operands when the omission does not cause ambiguity.

It is important to note that, if there are multiple ϕ-functions at the head of a basic block, then these are executed in parallel, i.e., simultaneously *not* sequentially. This distinction becomes important if the target of a ϕ-function is the same as the source of another ϕ-function, perhaps after optimizations such as copy propagation (see Chap. 8). When ϕ-functions are eliminated in the SSA destruction phase, they are sequentialized using conventional copy operations, as described in Algorithm 21.6. This subtlety is particularly important in the context of register allocated code (see Chap. 22).

Strictly speaking, ϕ-functions are not directly executable in software, since the dynamic control-flow path leading to the ϕ-function is not explicitly encoded as an input to ϕ-function. This is tolerable, since ϕ-functions are generally only used during static analysis of the program. They are removed before any program interpretation or execution takes place. However, there are various executable extensions of ϕ-functions, such as ϕ_{if} or γ functions (see Chap. 14), which take an extra parameter to encode the implicit control dependence that dictates the argument the corresponding ϕ-function should select. Such extensions are useful for program interpretation (see Chap. 14), if conversion (see Chap. 20), or hardware synthesis (see Chap. 23).

We present one further example in this section, to illustrate how a loop control-flow structure appears in SSA. Here is the non-SSA version of the program and its corresponding control-flow graph SSA version:

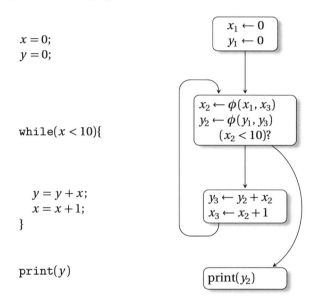

The SSA code features two ϕ-functions in the loop header; these merge incoming definitions from before the loop for the first iteration and from the loop body for subsequent iterations.

It is important to outline that SSA should not be confused with (*dynamic*) single assignment (DSA or simply SA) form used in automatic parallelization. *Static Single Assignment* does not prevent multiple assignments to a variable during program execution. For instance, in the SSA code fragment above, variables y_3 and x_3 in the loop body are redefined dynamically with fresh values at each loop iteration.

Full details of the SSA construction algorithm are given in Chap. 3. For now, it is sufficient to see that:

1. A ϕ-function has been inserted at the appropriate control-flow merge point where multiple reaching definitions of the same variable converged in the original program.
2. Integer subscripts have been used to rename variables x and y from the original program.

1.3 Comparison with Classical Data-Flow Analysis

As we will discover further in Chaps. 13 and 8, one of the major advantages of SSA form concerns data-flow analysis. Data-flow analysis collects information about programs at compile time in order to make optimizing code transformations. During actual program execution, information flows between variables. Static analysis captures this behaviour by propagating *abstract* information, or data-flow facts, using an operational representation of the program such as the control-flow graph (CFG). This is the approach used in classical data-flow analysis.

Often, data-flow information can be propagated more efficiently using a *functional*, or *sparse*, representation of the program such as SSA. When a program is translated into SSA form, variables are renamed at definition points. For certain data-flow problems (e.g., constant propagation) this is exactly the set of program points where data-flow facts may change. Thus it is possible to associate data-flow facts directly with variable names, rather than maintaining a vector of data-flow facts indexed over all variables, at each program point.

Figure 1.1 illustrates this point through an example of non-zero value analysis. For each variable in a program, the aim is to determine statically whether that variable can contain a zero integer value (i.e., null) at runtime. Here 0 represents the fact that the variable is null, Ø the fact that it is non-null, and ⊤ the fact that it is maybe-null. With classical dense data-flow analysis on the CFG in Fig. 1.1a, we would compute information about variables x and y for each of the entry and exit points of the six basic blocks in the CFG, using suitable data-flow equations. Using sparse SSA-based data-flow analysis on Fig. 1.1b, we compute information about each variable based on a simple analysis of its definition statement. This gives us seven data-flow facts, one for each SSA version of variables x and y.

For other data-flow problems, properties may change at points that are not variable definitions. These problems can be accommodated in a sparse analysis framework by inserting additional pseudo-definition functions at appropriate points to induce additional variable renaming. See Chap. 13 for one such instance. However, this example illustrates some key advantages of the SSA-based analysis.

1. Data-flow information *propagates directly* from definition statements to uses, via the def-use links implicit in the SSA naming scheme. In contrast, the classical

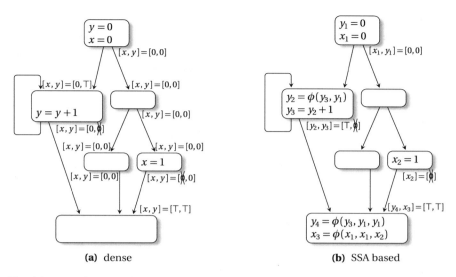

Fig. 1.1 Example control-flow graph for non-zero value analysis, only showing relevant definition statements for variables x and y

data-flow framework propagates information throughout the program, including points where the information does not change or is not relevant.

2. The results of the SSA data-flow analysis are *more succinct*. In the example, there are fewer data-flow facts associated with the sparse (SSA) analysis than with the dense (classical) analysis.

Part II of this textbook gives a comprehensive treatment of some SSA-based data-flow analysis.

1.4 SSA in Context

Historical Context Throughout the 1980s, as optimizing compiler technology became more mature, various intermediate representations (IRs) were proposed to encapsulate data dependence in a way that enabled fast and accurate data-flow analysis. The motivation behind the design of such IRs was the exposure of direct links between variable definitions and uses, known as *def-use chains*, enabling efficient propagation of data-flow information. Example IRs include the program dependence graph [118] and program dependence web [215]. Chapter 14 gives further details on dependence graph style IRs.

Static Single Assignment form was one such IR, which was developed at IBM Research and announced publicly in several research papers in the late 1980s [6, 89, 249]. SSA rapidly acquired popularity due to its intuitive nature and straightforward construction algorithm. The SSA property gives a standardized shape for variable def-use chains, which simplifies data-flow analysis techniques.

Current Usage The majority of current commercial and open source compilers, including GCC, LLVM, the HotSpot Java virtual machine, and the V8 JavaScript engine, use SSA as a key intermediate representation for program analysis. As optimizations in SSA are fast and powerful, SSA is increasingly used in just-in-time (JIT) compilers that operate on a high-level target-independent program representation such as Java byte-code, CLI byte-code (.NET MSIL), or LLVM bitcode.

Initially created to facilitate the development of high-level program transformations, SSA form has gained much interest due to its favourable properties that often enable the simplification of algorithms and reduction of computational complexity. Today, SSA form is even adopted for the final code generation phase (see Part IV), i.e., the back-end. Several industrial and academic compilers, static or just-in-time, use SSA in their back-ends, e.g., LLVM, HotSpot, LAO, libFirm, Mono. Many compilers that use SSA form perform SSA elimination before register allocation, including GCC, HotSpot, and LLVM. Recent research on register allocation (see Chap. 22) even allows the retention of SSA form until the very end of the code generation process.

SSA for High-Level Languages So far, we have presented SSA as a useful feature for compiler-based analysis of low-level programs. It is interesting to note that some high-level languages enforce the SSA property. The SISAL language is defined in such a way that programs automatically have referential transparency, since multiple assignments are not permitted to variables. Other languages allow the SSA property to be applied on a per-variable basis, using special annotations like `final` in Java, or `const` and `readonly` in C#.

The main motivation for allowing the programmer to enforce SSA in an explicit manner in high-level programs is that *immutability simplifies concurrent programming*. Read-only data can be shared freely between multiple threads, without any data dependence problems. This is becoming an increasingly important issue, with the shift to multi- and many-core processors.

High-level functional languages claim referential transparency as one of the cornerstones of their programming paradigm. Thus functional programming supports the SSA property implicitly. Chapter 6 explains the dualities between SSA and functional programming.

1.5   About the Rest of This Book

In this chapter, we have introduced the notion of SSA. The rest of this book presents various aspects of SSA, from the pragmatic perspective of compiler engineers and code analysts. The ultimate goals of this book are:

1. To demonstrate clearly the *benefits* of SSA-based analysis
2. To dispel the *fallacies* that prevent people from using SSA

This section gives pointers to later parts of the book that deal with specific topics.

1.5.1   Benefits of SSA

SSA imposes a strict discipline on variable naming in programs, so that each variable has a unique definition. Fresh variable names are introduced at assignment statements, and control-flow merge points. This serves to simplify the structure of variable *def-use* relationships (see Sect. 2.1) and *live ranges* (see Sect. 2.3), which underpin data-flow analysis. Part II of this book focuses on data-flow analysis using SSA. There are three major advantages to SSA:

Compile time benefit.   Certain compiler optimizations can be more efficient when operating on SSA programs, since referential transparency means that data-flow information can be associated directly with variables, rather than with variables at each program point. We have illustrated this simply with the non-zero value analysis in Sect. 1.3.

Compiler development benefit.   Program analyses and transformations can be easier to express in SSA. This means that compiler engineers can be more productive, in writing new compiler passes, and debugging existing passes. For example, the *dead code elimination* pass in GCC 4.x, which relies on an underlying SSA-based intermediate representation, takes only 40% as many lines of code as the equivalent pass in GCC 3.x, which does not use SSA. The SSA version of the pass is simpler, since it relies on the general-purpose, factored-out, data-flow propagation engine.

Program runtime benefit.   Conceptually, any analysis and optimization that can be done under SSA form can also be done identically out of SSA form. Because of the compiler development mentioned above, several compiler optimizations are shown to be more effective when operating on programs in SSA form. These include the class of *control-flow insensitive analyses*, e.g., [139].

1.5.2   Fallacies About SSA

Some people believe that SSA is too cumbersome to be an effective program representation. This book aims to convince the reader that such a concern is unnecessary, given the application of suitable techniques. The table below presents some common myths about SSA, and references in this first part of the book contain material to dispel these myths.

Myth	Reference
SSA greatly increases the number of variables	Chapter 2 reviews the main varieties of SSA, some of which introduce far fewer variables than the original SSA formulation.
SSA property is difficult to maintain	Chapters 3 and 5 discuss simple techniques for the repair of SSA invariants that have been broken by optimization rewrites.
SSA destruction generates many copy operations	Chapters 3 and 21 present efficient and effective SSA destruction algorithms.

Chapter 2
Properties and Flavours

Philip Brisk and Fabrice Rastello

Recall from the previous chapter that a procedure is in SSA form if every variable is defined only once, and every use of a variable refers to exactly one definition. Many variations, or flavours, of SSA form that satisfy these criteria can be defined, each offering its own considerations. For example, different flavours vary in terms of the number of ϕ-functions, which affects the size of the intermediate representation; some variations are more difficult to construct, maintain, and destruct than others. This chapter explores these SSA flavours and provides insights into their relative merits in certain contexts.

2.1 Def-Use and Use-Def Chains

Under SSA form, each variable is defined once. Def-use chains are data structures that provide, for the single definition of a variable, the set of all its uses. In turn, a use-def chain, which under SSA consists of a single name, uniquely specifies the definition that reaches the use. As we will illustrate further in the book (see Chap. 8), def-use chains are useful for forward data-flow analysis as they provide direct connections that shorten the propagation distance between nodes that generate and use data-flow information.

Because of its single definition per variable property, SSA form simplifies def-use and use-def chains in several ways. First, SSA form simplifies def-use chains

P. Brisk (✉)
UC Riverside, Riverside, CA, USA
e-mail: philip@cs.ucr.edu

F. Rastello
Inria, Grenoble, France
e-mail: fabrice.rastello@inria.fr

© The Author(s), under exclusive license to Springer Nature Switzerland AG 2022
F. Rastello, F. Bouchez Tichadou (eds.), *SSA-based Compiler Design*,
https://doi.org/10.1007/978-3-030-80515-9_2

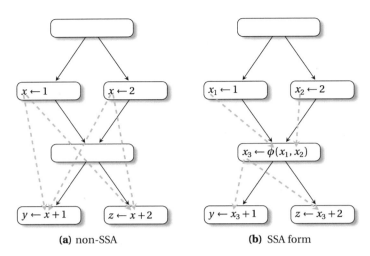

Fig. 2.1 Def-use chains (dashed) for non-SSA form and its corresponding SSA form program

as it combines the information as early as possible. This is illustrated by Fig. 2.1 where the def-use chain in the non-SSA program requires as many merges as there are uses of x, whereas the corresponding SSA form allows early and more efficient combination.

Second, as it is easy to associate each variable with its single defining operation, use-def chains can be represented and maintained almost for free. As this constitutes the skeleton of the so-called SSA graph (see Chap. 14), when considering a program under SSA form, use-def chains are implicitly considered as a given. The explicit representation of use-def chains simplifies backward propagation, which favours algorithms such as dead-code elimination.

For forward propagation, since def-use chains are precisely the reverse of use-def chains, computing them is also easy; maintaining them requires minimal effort. However, even without def-use chains, some lightweight forward propagation algorithms such as copy folding[1] are possible: Using a single pass that processes operations along a topological order traversal of a forward CFG,[2] most definitions are processed prior to their uses. When processing an operation, the use-def chain provides immediate access to the prior computed value of an argument. Conservative merging is performed when, at some loop headers, a ϕ-function encounters an unprocessed argument. Such a lightweight propagation engine proves to be fairly efficient.

---

[1] Copy folding removes any copy $a = b$ by renaming subsequent (dominated) uses of a by b. It is usually coupled with the SSA renaming phase.

[2] A forward control-flow graph is an acyclic reduction of the CFG obtained by removing back edges.

2.2   Minimality

SSA construction is a two-phase process: placement of ϕ-functions, followed by renaming. The goal of the first phase is to generate code that fulfils the single reaching-definition property, as already outlined. Minimality is an additional property relating to code that has ϕ-functions inserted, but prior to renaming; Chap. 3 describes the classical SSA construction algorithm in detail, while this section focuses primarily on describing the minimality property.

A definition D of variable v *reaches* a point p in the CFG if there exists a path from D to p that does not pass through another definition of v. We say that a code has the *single reaching-definition property* iff no program point can be reached by two definitions of the same variable. Under the assumption that the single reaching-definition property is fulfiled, the *minimality property* states the minimality of the number of inserted ϕ-functions.

This property can be characterized using the following notion of join sets. Let n_1 and n_2 be distinct basic blocks in a CFG. A basic block n_3, which may or may not be distinct from n_1 or n_2, is a *join node* of n_1 and n_2 if there exist at least two non-empty paths, i.e., paths containing at least one CFG edge, from n_1 to n_3 and from n_2 to n_3, respectively, such that n_3 is the only basic block that occurs on both of the paths. In other words, the two paths converge at n_3 and no other CFG node. Given a set S of basic blocks, n_3 is a join node of S if it is the join node of at least two basic blocks in S. The set of join nodes of set S is denoted by $\mathcal{J}(S)$.

Intuitively, a join set corresponds to the placement of ϕ-functions. In other words, if n_1 and n_2 are basic blocks that both contain a definition of variable v, then we ought to instantiate ϕ-functions for v at every basic block in $\mathcal{J}(\{n_1, n_2\})$. Generalizing this statement, if D_v is the set of basic blocks containing definitions of v, then ϕ-functions should be instantiated in every basic block in $\mathcal{J}(D_v)$. As inserted ϕ-functions are themselves definition points, some new ϕ-functions should be inserted at $\mathcal{J}(D_v \cup \mathcal{J}(D_v))$. Actually, it turns out that $\mathcal{J}(S \cup \mathcal{J}(S)) = \mathcal{J}(S)$, so the join set of the set of definition points of a variable in the original program characterizes exactly the minimum set of program points where ϕ-functions should be inserted.

We are not aware of any optimizations that require a strict enforcement of minimality property. However, placing ϕ-functions only at the join sets can be done easily using a simple topological traversal of the CFG as described in Chap. 4, Sect. 4.4. Classical techniques place ϕ-functions of a variable v at $\mathcal{J}(D_v \cup \{r\})$, with r the entry node of the CFG. There are good reasons for that, as we will explain further. Finally, as explained in Chap. 3, Sect. 3.3 for reducible flow graphs, some copy-propagation engines can easily turn a non-minimal SSA code into a minimal one.

2.3   Strict SSA Form and Dominance Property

A procedure is defined as *strict* if every variable is defined before it is used along every path from the entry to the exit point; otherwise, it is *non-strict*. Some languages, such as Java, impose strictness as part of the language definition; others, such as C/C++, impose no such restrictions. The code in Fig. 2.2a is non-strict as there exists a path from the entry to the use of a that does not go through the definition. If this path is taken through the CFG during the execution, then a will be used without ever being assigned a value. Although this may be permissible in some cases, it is usually indicative of a programmer error or poor software design.

Under SSA, because there is only a single (static) definition per variable, strictness is equivalent to the *dominance property*: Each use of a variable is dominated by its definition. In a CFG, basic block n_1 *dominates* basic block n_2 if every path in the CFG from the entry point to n_2 includes n_1. By convention, every basic block in a CFG dominates itself. Basic block n_1 *strictly dominates* n_2 if n_1 dominates n_2 and $n_1 \neq n_2$. We use the symbols n_1 dom n_2 and n_1 sdom n_2 to denote dominance and strict dominance, respectively.

Adding a (undefined) pseudo-definition of each variable to the entry point (root) of the procedure ensures strictness. The single reaching-definition property discussed previously mandates that each program point be reachable by exactly one definition (or pseudo-definition) of each variable. If a program point U is a use of variable v, then the reaching definition D of v will dominate U; otherwise, there would be a path from the CFG entry node to U that does not include D. If such a path existed, then the program would not be in strict SSA form, and a ϕ-function would

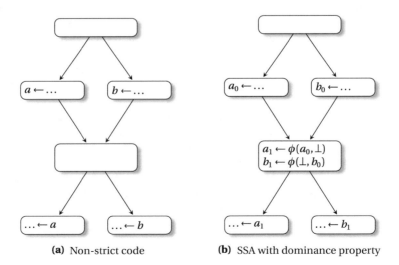

(a) Non-strict code   **(b)** SSA with dominance property

Fig. 2.2 A non-strict code and its corresponding strict SSA form. The presence of $\perp$ indicates a use of an undefined value

need to be inserted somewhere in $\mathscr{J}(r, D)$, as in our example of Fig. 2.2b where $\perp$ represents the undefined pseudo-definition. The so-called *minimal SSA form* is a variant of SSA form that satisfies both the minimality and dominance properties. As shall be seen in Chap. 3, minimal SSA form is obtained by placing the ϕ-functions of variable v at $\mathscr{J}(D_v, r)$ using the formalism of dominance frontier. If the original procedure is non-strict, conversion to minimal SSA will create a strict SSA-based representation. Here, strictness refers solely to the SSA representation; if the input program is non-strict, conversion to and from strict SSA form cannot address errors due to uninitialized variables. To finish with, the use of an implicit pseudo-definition in the CFG entry node to enforce strictness does not change the semantics of the program by any means.

SSA with dominance property is useful for many reasons that directly originate from the structural properties of the variable live ranges. The immediate dominator or "idom" of a node N is the unique node that strictly dominates N but does not strictly dominate any other node that strictly dominates N. All nodes but the entry node have immediate dominators. A dominator tree is a tree where the children of each node are those nodes it immediately dominates. Because the immediate dominator is unique, it is a tree with the entry node as root. For each variable, its live range, i.e., the set of program points where it is live, is a sub-tree of the dominator tree. Among other consequences of this property, we can cite the ability to design a fast and efficient method to query whether a variable is live at point q or an iteration-free algorithm to compute liveness sets (see Chap. 9). This property also allows efficient algorithms to test whether two variables *interfere* (see Chap. 21). Usually, we suppose that two variables interfere if their live ranges intersect (see Sect. 2.6 for further discussions about this hypothesis). Note that in the general case, a variable is considered to be live at a program point if there exists a definition of that variable that can reach this point (reaching-definition analysis), *and* if there exists a definition-free path to a use (upward-exposed use analysis). For strict programs, any program point from which you can reach a use without going through a definition is necessarily reachable from a definition.

Another elegant consequence is that the intersection graph of live ranges belongs to a special class of graphs called chordal graphs. Chordal graphs are significant because several problems that are NP-complete on general graphs have efficient linear-time solutions on chordal graphs, including graph colouring. Graph colouring plays an important role in register allocation, as the register assignment problem can be expressed as a colouring problem of the interference graph. In this graph, two variables are linked with an edge if they interfere, meaning they cannot be assigned the same physical location (usually, a machine register, or "colour"). The underlying chordal property highly simplifies the assignment problem otherwise considered NP-complete. In particular, a traversal of the dominator tree, i.e., a "tree scan," can colour all of the variables in the program, without requiring the explicit construction of an interference graph. The tree scan algorithm can be used for register allocation, which is discussed in greater detail in Chap. 22.

As we have already mentioned, most ϕ-function placement algorithms are based on the notion of dominance frontier (see Chaps. 3 and 4) and consequently do provide the dominance property. As we will see in Chap. 3, this property can be broken by copy propagation: In our example of Fig. 2.2b, the argument a_1 of the copy represented by $a_2 = \phi(a_1, \perp)$ can be propagated and every occurrence of a_2 can be safely replaced by a_1; the now identity ϕ-function can then be removed obtaining the initial code, that is, still SSA but not strict anymore. Making a non-strict SSA code strict is about the same complexity as SSA construction (actually we need a pruned version as described below). Still, the "strictification" usually concerns only a few variables and a restricted region of the CFG: The incremental update described in Chap. 5 will do the work with less effort.

2.4 Pruned SSA Form

One drawback of minimal SSA form is that it may place ϕ-functions for a variable at a point in the control-flow graph where the variable was not actually live prior to SSA. Many program analyses and optimizations, including register allocation, are only concerned with the region of a program where a given variable is live. The primary advantage of eliminating those dead ϕ-functions over minimal SSA form is that it has far fewer ϕ-functions in most cases. It is possible to construct such a form while still maintaining the minimality and dominance properties otherwise. The new constraint is that every *use point* for a given variable must be reached by exactly one definition, as opposed to all program points. Pruned SSA form satisfies these properties.

Under minimal SSA, ϕ-functions for variable v are placed at the entry points of basic blocks belonging to the set $\mathcal{J}(S, r)$. Under pruned SSA, we suppress the instantiation of a ϕ-function at the beginning of a basic block if v is not live at the entry point of that block. One possible way to do this is to perform liveness analysis prior to SSA construction, and then use the liveness information to suppress the placement of ϕ-functions as described above; another approach is to construct minimal SSA and then remove the dead ϕ-functions using dead-code elimination; details can be found in Chap. 3.

Figure 2.3a shows an example of minimal non-pruned SSA. The corresponding pruned SSA form would remove the dead ϕ-function that defines Y_3 since Y_1 and Y_2 are only used in their respective definition blocks.

2.5 Conventional and Transformed SSA Form

In many non-SSA and graph colouring based register allocation schemes, register assignment is done at the granularity of webs. In this context, a web is the maximum unions of def-use chains that have either a use or a def in common. As an example, the code in Fig. 2.4a leads to two separate webs for variable a. The conversion to

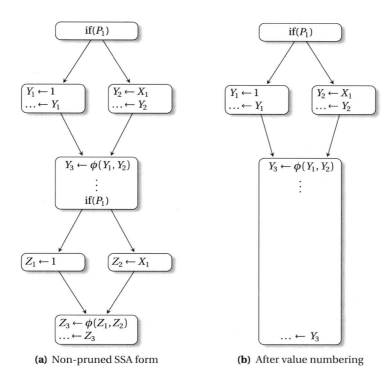

(a) Non-pruned SSA form   **(b)** After value numbering

Fig. 2.3 Non-pruned SSA form allows value numbering to determine that Y_3 and Z_3 have the same value

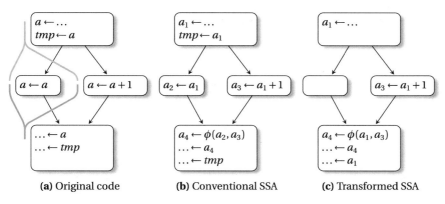

(a) Original code   **(b)** Conventional SSA   **(c)** Transformed SSA

Fig. 2.4 **(a)** Non-SSA register webs of variable a and **(b)** corresponding SSA ϕ-webs $\{a_1\}$ and $\{a_2, a_3, a_4\}$; **(b)** conventional and **(c)** corresponding transformed SSA form after copy propagation of a_1

minimal SSA form replaces each web of a variable v in the pre-SSA program with some variable names v_i. In pruned SSA, these variable names partition the live range of the web: At every point in the procedure where the web is live, *exactly* one variable v_i is also live; and none of the v_i is live at any point where the web is not.

Based on this observation, we can partition the variables in a program that has been converted to SSA form into ϕ-equivalence classes that we will refer as ϕ-webs. We say that x and y are ϕ-*related* to one another if they are referenced by the same ϕ-function, i.e., if x and y are either parameters or defined by the ϕ-function. The transitive closure of this relation defines an equivalence relation that partitions the variables defined locally in the procedure into equivalence classes, the ϕ-webs. Intuitively, the ϕ-equivalence class of a resource represents a set of resources "connected" via ϕ-functions. For any freshly constructed SSA code, the ϕ-webs exactly correspond to the register web of the original non-SSA code.

Conventional SSA form (C-SSA) is defined as SSA form for which each ϕ-web is interference-free. Many program optimizations such as copy propagation may transform a procedure from conventional to a non-conventional (T-SSA for Transformed SSA) form, in which some variables belonging to the same ϕ-web interfere with one another. Figure 2.4c shows the corresponding transformed SSA form of our previous example: Here variable a_1 interferes with variables a_2, a_3, and a_4, since it is defined at the top and used last.

Bringing back the conventional property of a T-SSA code is as "difficult" as translating out of SSA (also known as SSA "destruction," see Chap. 3) . Indeed, the destruction of conventional SSA form is straightforward: Each ϕ-web can be replaced with a single variable; all definitions and uses are renamed to use the new variable, and all ϕ-functions involving this equivalence class are removed. SSA destruction starting from non-conventional SSA form can be performed through a conversion to conventional SSA form as an intermediate step. This conversion is achieved by inserting copy operations that dissociate interfering variables from the connecting ϕ-functions. As those copy instructions will have to be inserted at some points to get rid of ϕ-functions, for machine-level transformations such as register allocation or scheduling, T-SSA provides an inaccurate view of the resource usage. Another motivation for sticking to C-SSA is that the names used in the original program might help capture some properties otherwise difficult to discover. Lexical partial redundancy elimination (PRE) as described in Chap. 11 illustrates this point.

Apart from those specific examples most current compilers choose not to maintain the conventional property. Still, we should outline that, as later described in Chap. 21, checking if a given ϕ-web is (and if necessary turning it back to) interference-free can be done in linear time (instead of the naive quadratic time algorithm) in the size of the ϕ-web.

2.6   A Stronger Definition of Interference

Throughout this chapter, two variables have been said to interfere if their live ranges intersect. Intuitively, two variables with overlapping lifetimes will require two distinct storage locations; otherwise, a write to one variable will overwrite the value of the other. In particular, this definition has applied to the discussion of interference graphs and the definition of conventional SSA form, as described above.

Although it suffices for correctness, this is a fairly restrictive definition of interference, based on static considerations. Consider for instance the case when two simultaneously live variables in fact contain the same value, then it would not be a problem to put both of them in the same register. The ultimate notion of interference, which is obviously undecidable because of a reduction to the halting problem, should decide for two distinct variables whether there exists an execution for which they simultaneously hold two different values. Several "static" extensions to our simple definition are still possible, in which, under very specific conditions, variables whose live ranges overlap one another may not interfere. We present two examples.

Firstly, consider the double-diamond graph of Fig. 2.2a again, which, although non-strict, is correct as soon as the two if conditions are the same. Even if a and b are unique variables with overlapping live ranges, the paths along which a and b are respectively used and defined are mutually exclusive with one another. In this case, the program will either pass through the definition of a and the use of a, or the definition of b and the use of b, since all statements involved are controlled by the same condition, albeit at different conditional statements in the program. Since only one of the two paths will ever execute, it suffices to allocate a single storage location that can be used for a or b. Thus, a and b do not actually interfere with one another. A simple way to refine the interference test is to check if one of the variables is live at the definition point of the other. This relaxed but correct notion of interference would not make a and b in Fig. 2.2a interfere while variables a_1 and b_1 of Fig. 2.2b would still interfere. This example illustrates the fact that live range splitting required here to make the code fulfil the dominance property may lead to less accurate analysis results. As far as the interference is concerned, for a SSA code with dominance property, the two notions are strictly equivalent: Two live ranges intersect iff one contains the definition of the other.

Secondly, consider two variables u and v, whose live ranges overlap. If we can prove that u and v will always hold the same value at every place where both are live, then they do not actually interfere with one another. Since they always have the same value, a single storage location can be allocated for both variables, because there is only one unique value between them. Of course, this new criterion is in general undecidable. Still, a technique such as global value numbering that is straightforward to implement under SSA (see Chap. 11.5.1) can do a fairly good job, especially in the presence of a code with many variable-to-variable copies, such as one obtained after a naive SSA destruction pass (see Chap. 3). In that case (see Chap. 21), the difference between the refined notion of interference and the non-value-based one is significant.

This refined notion of interference has significant implications if applied to SSA form. In particular, the interference graph of a procedure is no longer chordal, as any edge between two variables whose lifetimes overlap could be eliminated by this property.

2.7 Further Reading

The advantages of def-use and use-def chains provided almost for free under SSA are well illustrated in Chaps. 8 and 13.

The notion of minimal SSA and a corresponding efficient algorithm to compute it were introduced by Cytron et al. [90]. For this purpose they extensively develop the notion of dominance frontier of a node n, $\mathscr{DF}(n) = \mathscr{J}(n, r)$. The fact that $\mathscr{J}^+(S) = \mathscr{J}(S)$ was actually discovered later, with a simple proof by Wolfe [307]. More details about the theory on (iterated) dominance frontier can be found in Chaps. 3 and 4. The post-dominance frontier, which is its symmetric notion, also known as the control dependence graph, finds many applications. Further discussions on control dependence graph can be found in Chap. 14.

Most SSA papers implicitly consider the SSA form to fulfil the dominance property. The first technique that really exploits the structural properties of the strictness is the fast SSA destruction algorithm developed by Budimlić et al. [53] and revisited in Chap. 21.

The notion of pruned SSA has been introduced by Choi, Cytron and Ferrante [67]. The example of Fig. 2.3 to illustrate the difference between pruned and non-pruned SSA has been borrowed from Cytron et al. [90]. The notions of conventional and transformed SSA were introduced by Sreedhar et al. in their seminal paper [267] for destructing SSA form. The description of the existing techniques to turn a general SSA into either a minimal, a pruned, a conventional, or a strict SSA is provided in Chap. 3.

The ultimate notion of interference was first discussed by Chaitin in his seminal paper [60] that presents the graph colouring approach for register allocation. His interference test is similar to the refined test presented in this chapter. In the context of SSA destruction, Chap. 21 addresses the issue of taking advantage of the dominance property with this refined notion of interference.

Chapter 3
Standard Construction and Destruction Algorithms

Jeremy Singer and Fabrice Rastello

This chapter describes the standard algorithms for construction and destruction of SSA form. SSA *construction* refers to the process of translating a non-SSA program into one that satisfies the SSA constraints. In general, this transformation occurs as one of the earliest phases in the middle-end of an optimizing compiler, when the program has been converted to three-address intermediate code. SSA *destruction* is sometimes called out-of-SSA translation. This step generally takes place in an optimizing compiler after all SSA optimizations have been performed, and prior to code generation. Note however that there are specialized code generation techniques that can work directly on SSA-based intermediate representations such as instruction selection (see Chap. 19), if-conversion (see Chap. 20), and register allocation (see Chap. 22).

The algorithms presented in this chapter are based on material from the seminal research papers on SSA. These original algorithms are straightforward to implement and have acceptable efficiency. Therefore such algorithms are widely implemented in current compilers. Note that more efficient, albeit more complex, alternative algorithms have been devised. These are described further in Chaps. 4 and 21.

Figure 3.1 shows the control-flow graph (CFG) of an example program. The set of nodes is $\{r, A, B, C, D, E\}$, and the variables used are $\{x, y, tmp\}$. Note that the program shows the complete control-flow structure, denoted by directed edges between the nodes. However, the program only shows statements that define relevant variables, together with the unique `return` statement at the exit point of the CFG. All of the program variables are undefined on entry. On certain control-flow paths,

J. Singer (✉)
University of Glasgow, Glasgow, UK
e-mail: Jeremy.Singer@glasgow.ac.uk

F. Rastello
Inria, Grenoble, France
e-mail: fabrice.rastello@inria.fr

© The Author(s), under exclusive license to Springer Nature Switzerland AG 2022
F. Rastello, F. Bouchez Tichadou (eds.), *SSA-based Compiler Design*,
https://doi.org/10.1007/978-3-030-80515-9_3

Fig. 3.1 Example
control-flow graph, before
SSA construction occurs

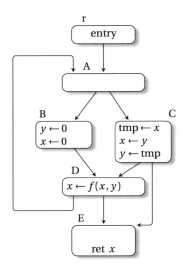

some variables may be used without being defined, e.g., x on the path $r \rightarrow A \rightarrow C$. We discuss this issue later in the chapter. We intend to use this program as a running example throughout the chapter, to demonstrate various aspects of SSA construction.

3.1 Construction

The original construction algorithm for SSA form consists of two distinct phases.

1. **ϕ-function insertion** performs *live range splitting* to ensure that any use of a given variable v is reached[1] by exactly one definition of v. The resulting live ranges exhibit the property of having a single definition, which occurs at the beginning of each live range.
2. **Variable renaming** assigns a unique variable name to each live range. This second phase rewrites variable names in program statements such that the program text contains only one definition of each variable, and every use refers to its corresponding unique reaching definition.

As already outlined in Chap. 2, there are different flavours of SSA with distinct properties. In this chapter, we focus on the *minimal* SSA form.

---

[1] A program point p is said to be *reachable* by a definition of v if there exists a path in the CFG from that definition to p that does not contain any other definition of v.

3.1.1   Join Sets and Dominance Frontiers

In order to explain how ϕ-function insertions occur, it will be helpful to review the related concepts of *join sets* and *dominance frontiers*.

For a given set of nodes S in a CFG, the join set $\mathscr{J}(S)$ is the set of *join nodes* of S, i.e., nodes in the CFG that can be reached by two (or more) distinct elements of S using disjoint paths. Join sets were introduced in Chap. 2, Sect. 2.2.

Let us consider some join set examples from the program in Fig. 3.1.

1. $\mathscr{J}(\{B, C\}) = \{D\}$, since it is possible to get from B to D and from C to D along different, non-overlapping, paths.
2. Again, $\mathscr{J}(\{r, A, B, C, D, E\}) = \{A, D, E\}$ (where r is the entry), since the nodes A, D, and E are the only nodes with multiple direct predecessors in the program.

The *dominance frontier* of a node n, DF(n), is the border of the CFG region that is dominated by n. More formally,

- Node x *strictly dominates* node y if x dominates y and $x \neq y$.
- The set of nodes DF(n) contains all nodes x such that n dominates a direct predecessor of x but n does not strictly dominate x.

For instance, in our Fig. 3.1, the dominance frontier of the y defined in block B is the first operation of D, while the DF of the y defined in block C would be the first operations of D and E.

Note that DF is defined over individual nodes, but for simplicity of presentation, we overload it to operate over sets of nodes too, i.e., $\text{DF}(S) = \bigcup_{s \in S} \text{DF}(s)$. The *iterated dominance frontier* $\text{DF}^+(S)$ is obtained by iterating the computation of DF until reaching a fixed point, i.e., it is the limit $DF_{i \to \infty}(S)$ of the sequence:

$$\begin{aligned} \text{DF}_1(S) &= \text{DF}(S) \\ \text{DF}_{i+1}(S) &= \text{DF}(S \cup \text{DF}_i(S)) \end{aligned}$$

Construction of minimal SSA requires for each variable v the insertion of ϕ-functions at $\mathscr{J}(\text{Defs}(v))$, where $\text{Defs}(v)$ is the set of nodes that contain definitions of v. The original construction algorithm for SSA form uses the iterated dominance frontier $\text{DF}^+(\text{Defs}(v))$. This is an over-approximation of join set, since $\text{DF}^+(S) = \mathscr{J}(S \cup \{r\})$, i.e., the original algorithm assumes an *implicit* definition of every variable at the entry node r.

3.1.2 ϕ-Function Insertion

This concept of dominance frontiers naturally leads to a straightforward approach that places ϕ-functions on a per-variable basis. For a given variable v, we place ϕ-

Fig. 3.2 Example
control-flow graph, including
inserted ϕ-functions for
variable x

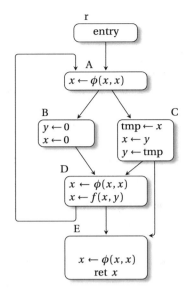

functions at the iterated dominance frontier $DF^+(\text{Defs}(v))$ where $\text{Defs}(v)$ is the set
of nodes containing definitions of v. This leads to the construction of SSA form that
has the dominance property, i.e., where each definition of each renamed variable
dominates its entire live range.

Consider again our running example from Fig. 3.1. The set of nodes containing
definitions of variable x is $\{B, C, D\}$. The iterated dominance frontier of this set is
$\{A, D, E\}$ (it is also the DF, no iteration needed here). Hence we need to insert ϕ-
functions for x at the beginning of nodes A, D, and E. Figure 3.2 shows the example
CFG program with ϕ-functions for x inserted.

As far as the actual algorithm for ϕ-functions insertion is concerned, we will
assume that the dominance frontier of each CFG node is pre-computed and that the
iterated dominance frontier is computed on the fly, as the algorithm proceeds. The
algorithm works by inserting ϕ-functions iteratively using a worklist of definition
points, and flags (to avoid multiple insertions). The corresponding pseudo-code for
ϕ-function insertion is given in Algorithm 3.1. The worklist of nodes W is used
to record definition points that the algorithm has not yet processed, i.e., it has not
yet inserted ϕ-functions at their dominance frontiers. Because a ϕ-function is itself a
definition, it may require further ϕ-functions to be inserted. This is the cause of node
insertions into the worklist W during iterations of the inner loop in Algorithm 3.1.
Effectively, we compute the iterated dominance frontier on the fly. The set F is used
to avoid repeated insertion of ϕ-functions on a single block. Dominance frontiers
of distinct nodes may intersect, e.g., in the example CFG in Fig. 3.1, $DF(B)$ and
$DF(C)$ both contain D, but once a ϕ-function for a particular variable has been
inserted at a node, there is no need to insert another, since a single ϕ-function per
variable handles all incoming definitions of that variable to that node.

Algorithm 3.1: Standard algorithm for inserting ϕ-functions

```
1   for v: variable names in original program do
2   │   F ← {}                              ▷ set of basic blocks where φ is added
3   │   W ← {}                              ▷ set of basic blocks that contain definitions of v
4   │   for d ∈ Defs(v) do
5   │   │   let B be the basic block containing d
6   │   │   W ← W ∪ {B}
7   │   while W ≠ {} do
8   │   │   remove a basic block X from W
9   │   │   for Y: basic block ∈ DF(X) do
10  │   │   │   if Y ∉ F then
11  │   │   │   │   add v ← φ(...) at entry of Y
12  │   │   │   │   F ← F ∪ {Y}
13  │   │   │   │   if Y ∉ Defs(v) then
14  │   │   │   │   │   W ← W ∪ {Y}
```

Table 3.1 Walkthrough of placement of ϕ-functions for variable x in example CFG

while loop #	X	DF(X)	F	W
–	–	–	{}	$\{B, C, D\}$
1	B	$\{D\}$	$\{D\}$	$\{C, D\}$
2	C	$\{D, E\}$	$\{D, E\}$	$\{D, E\}$
3	D	$\{E, A\}$	$\{D, E, A\}$	$\{E, A\}$
4	E	{}	$\{D, E, A\}$	$\{A\}$
5	A	$\{A\}$	$\{D, E, A\}$	{}

We give a walkthrough example of Algorithm 3.1 in Table 3.1. It shows the stages of execution for a single iteration of the outermost for loop, inserting ϕ-functions for variable x. Each row represents a single iteration of the while loop that iterates over the worklist W. The table shows the values of X, F, and W at the start of each while loop iteration. At the beginning, the CFG looks like Fig. 3.1. At the end, when all the ϕ-functions for x have been placed, then the CFG looks like Fig. 3.2.

Provided that the dominator tree is given, the computation of the dominance frontier is quite straightforward. As illustrated by Fig. 3.3, this can be understood using the DJ-graph notation. The skeleton of the DJ-graph is the dominator tree of the CFG that makes the D-edges (dominance edges). This is augmented with J-edges (join edges) that correspond to all edges of the CFG whose source does not strictly dominate its destination. A DF-edge (dominance frontier edge) is an edge whose destination is in the dominance frontier of its source. By definition, there is a DF-edge (a, b) between every CFG nodes a, b such that a dominates a direct predecessor of b, but does not strictly dominate b. In other words, for each J-edge (a, b), all ancestors of a (including a) that do not strictly dominate b have b in their dominance frontier. For example, in Fig. 3.3, (F, G) is a J-edge, so $\{(F, G), (E, G), (B, G)\}$ are DF-edges. This leads to the pseudo-code given in

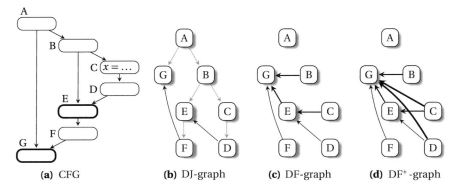

Fig. 3.3 An example CFG and its corresponding DJ-graph (*D*-edges are top-down), DF-graph, and DF$^+$-graph

Algorithm 3.2, where for every edge (a, b) we visit all ancestors of a to add b to their dominance frontier.

Since the iterated dominance frontier is simply the transitive closure of the dominance frontier, we can define the DF$^+$-graph as the transitive closure of the DF-graph. In our example, as $\{(C, E),\ (E, G)\}$ are DF-edges, (C, G) is a DF$^+$-edge. Hence, a definition of x in C will lead to inserting ϕ-functions in E and G. We can compute the iterated dominance frontier for each variable independently, as outlined in this chapter, or "cache" it to avoid repeated computation of the iterated dominance frontier of the same node. This leads to more sophisticated algorithms detailed in Chap. 4.

Algorithm 3.2: Algorithm for computing the dominance frontier of each CFG node

```
1  for (a, b) ∈ CFG edges do
2  │   x ← a
3  │   while x does not strictly dominate b do
4  │   │   DF(x) ← DF(x) ∪ b
5  │   └   x ← immediate dominator(x)
```

Once ϕ-functions have been inserted using this algorithm, the program usually still contains several definitions per variable; however, now there is a single definition statement in the CFG that reaches each use. For each variable use in a ϕ-function, it is conventional to treat them as if the use actually occurs on the corresponding incoming edge or at the end of the corresponding direct predecessor node. If we follow this convention, then def-use chains are aligned with the CFG dominator tree. In other words, *the single definition that reaches each use dominates that use*.

3.1.3 Variable Renaming

To obtain the desired property of a Static Single Assignment per variable, it is necessary to perform variable renaming, which is the second phase in the SSA construction process. ϕ-function insertions have the effect of splitting the live range(s) of each original variable into pieces. The variable renaming phase associates each individual live range with a new variable name, also called a *version*. The pseudo-code for this process is presented in Algorithm 3.3. Because of the dominance property outlined above, it is straightforward to rename variables using a depth-first traversal of the dominator tree. During the traversal, for each variable v, it is necessary to remember the version of its unique reaching definition at some point p in the graph. This corresponds to the closest definition that dominates p. In Algorithm 3.3, we compute and cache the reaching definition for v in the per-variable slot "v.reachingDef" that is updated as the algorithm traverses the dominator tree of the SSA graph. This per-variable slot stores the in-scope, "new" variable name (version) for the equivalent variable at the same point in the un-renamed program.

Algorithm 3.3: Renaming algorithm for second phase of SSA construction

---

    ▷ *rename variable definitions and uses to have one definition per variable name*
1 **foreach** v : variable **do**
2   v.reachingDef $\leftarrow \perp$

3 **foreach** *BB*: basic block in depth-first search preorder traversal of the dom. tree **do**
4   **foreach** i : instruction in linear code sequence of *BB* **do**
5   **foreach** v : variable used by non-ϕ-function i **do**
6   updateReachingDef(v, i)
7   replace this use of v by v.reachingDef in i

8   **foreach** v : variable defined by i (may be a ϕ-function) **do**
9   updateReachingDef(v, i)
10   create fresh variable v'
11   replace this definition of v by v' in i
12   v'.reachingDef $\leftarrow v$.reachingDef
13   v.reachingDef $\leftarrow v'$

14   **foreach** ϕ: ϕ-function in a direct successor of *BB* **do**
15   **let** v : variable used by ϕ coming from *BB*
16   updateReachingDef$(v,$ end of $BB)$
17   replace this use of v by v.reachingDef in ϕ

The variable renaming algorithm translates our running example from Fig. 3.1 into the SSA form of Fig. 3.4a. The table in Fig. 3.4b gives a walkthrough example of Algorithm 3.3, only considering variable x. The labels l_i mark instructions in the program that mention x, shown in Fig. 3.4a. The table records (1) when

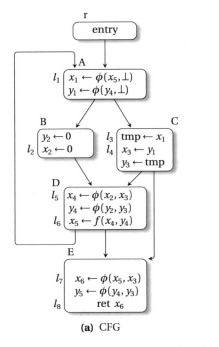

BB	x mention	x.reachingDef
r	l_1 use	$\perp$
A	def l_1	$\perp$ then x_1
B	def l_2	x_1 then x_2
B	l_5 use	x_2
C	l_3 use	x_2 updated into x_1
C	def l_4	x_1 then x_3
C	l_5 use	x_3
C	l_7 use	x_3
D	def l_5	x_3 updated into x_1 then x_4
D	l_6 use	x_4
D	def l_6	x_4 then x_5
D	l_1 use	x_5
D	l_7 use	x_5
E	def l_7	x_5 then x_6
E	l_8 use	x_6

(a) CFG   (b) Walk-through of renaming for variable x

Fig. 3.4 SSA form of the example of Fig. 3.1

x.reachingDef is updated from x_{old} into x_{new} due to a call of updateReachingDef, and (2) when x.reachingDef is x_{old} before a definition statement, then x_{new} afterwards.

Procedure updateReachingDef(v,p) Utility function for SSA renaming

Data: v : variable from program
Data: p : program point
▷ *search through chain of definitions for v until we find the closest definition that dominates p, then update v.reachingDef in-place with this definition*
1 $r \leftarrow v$
2 **repeat**
3 | $r \leftarrow r.$reachingDef
4 **until** $r == \perp$ or definition(r) dominates p
5 $v.$reachingDef $\leftarrow r$

The original presentation of the renaming algorithm uses a per-variable stack that stores all variable versions that dominate the current program point, rather than the slot-based approach outlined above. In the stack-based algorithm, a stack value is pushed when a variable definition is processed (while we explicitly update the reachingDef field at this point). The top stack value is peeked when a variable use is encountered (we read from the reachingDef field at this point). Multiple stack values may be popped when moving to a different node in the dominator tree (we always check whether we need to update the reachingDef field before we read from it). While the slot-based algorithm requires more memory, it can take advantage of an existing working field for a variable and be more efficient in practice.

3.1.4 Summary

Now let us review the flavour of SSA form that this simple construction algorithm produces. We refer back to several SSA properties that were introduced in Chap. 2.

- It is *minimal* (see Sect. 2.2). After the ϕ-function insertion phase, but before variable renaming, the CFG contains the minimal number of inserted ϕ-functions to achieve the property that exactly one definition of each variable v reaches every point in the graph.
- It is *not pruned* (see Sect. 2.4). Some of the inserted ϕ-functions may be dead, i.e., there is not always an explicit use of the variable subsequent to the ϕ-function (e.g., y_5 in Fig. 3.4a).
- It is *conventional* (see Sect. 2.5). The transformation that renames all ϕ-related variables into a unique representative name and then removes all ϕ-functions is a correct SSA destruction algorithm.
- Finally, it has the *dominance* property (that is, it is *strict*—see Sect. 2.3). Each variable use is dominated by its unique definition. This is due to the use of iterated dominance frontiers during the ϕ-placement phase, rather than join sets. Whenever the iterated dominance frontier of the set of definition points of a variable differs from its join set, then at least one program point can be reached both by r (the entry of the CFG) and one of the definition points. In other words, as in Fig. 3.1, one of the uses of the ϕ-function inserted in block A for x does not have any actual reaching definition that dominates it. This corresponds to the $\bot$ value used to initialize each reachingDef slot in Algorithm 3.3. Actual implementation code can use a NULL value, create a fake undefined variable at the entry of the CFG, or create undefined pseudo-operations on the fly just before the particular use.

3.2 Destruction

SSA form is a sparse representation of program information, which enables simple, efficient code analysis and optimization. Once we have completed SSA-based optimization passes, and certainly before code generation, it is necessary to eliminate ϕ-functions since these are not executable machine instructions. This elimination phase is known as *SSA destruction*.

When freshly constructed, an untransformed SSA code is conventional and its destruction is straightforward: One simply has to rename all ϕ-related variables (source and destination operands of the same ϕ-function) into a unique representative variable. Then, each ϕ-function should have syntactically identical names for all its operands, and thus can be removed to coalesce the related live ranges.

We refer to a set of ϕ-related variables as a ϕ-web. We recall from Chap. 2 that conventional SSA is defined as a flavour under which each ϕ-web is free from interferences. Hence, if all variables of a ϕ-web have non-overlapping live ranges, then the SSA form is conventional. The discovery of ϕ-webs can be performed efficiently using the classical *union-find* algorithm with a disjoint-set data structure, which keeps track of a set of elements partitioned into a number of disjoint (non-overlapping) subsets. The ϕ-webs discovery algorithm is presented in Algorithm 3.4.

Algorithm 3.4: The ϕ-webs discovery algorithm, based on the union-find pattern

---

1 **for** each variable v **do**
2      $\lfloor$ phiweb$(v) \leftarrow \{v\}$;
3 **for** each instruction of the form $a_{\text{dest}} = \phi(a_1, \ldots, a_n)$ **do**
4   **for** each source operand a_i in instruction **do**
5          $\lfloor$ union(phiweb(a_{dest}), phiweb(a_i))

While freshly constructed SSA code is conventional, this may not be the case after performing some optimizations such as copy propagation. Going back to conventional SSA form requires the insertion of copies. The simplest (although not the most efficient) way to destroy non-conventional SSA form is to split all *critical edges*, and then replace ϕ-functions by copies at the end of direct predecessor basic blocks. A critical edge is an edge from a node with several direct successors to a node with several direct predecessors. The process of splitting an edge, say (b_1, b_2), involves replacing edge (b_1, b_2) by (1) an edge from b_1 to a freshly created basic block and by (2) another edge from this fresh basic block to b_2. As ϕ-functions have a parallel semantic, i.e., have to be executed simultaneously not sequentially, the same holds for the corresponding copies inserted at the end of direct predecessor basic blocks. To this end, a pseudo instruction called a *parallel copy* is created to represent a set of copies that have to be executed in parallel. The replacement of parallel copies by sequences of simple copies is handled later on. Algorithm 3.5 presents the corresponding pseudo-code that makes non-conventional SSA conventional. As already mentioned, SSA destruction of such form is straightforward. However, Algorithm 3.5 can be slightly modified to directly destruct SSA by deleting line 13, replacing a_i' by a_0 in the following lines, and adding "remove the ϕ-function" after them.

Algorithm 3.5: Critical edge splitting algorithm for making non-conventional SSA form conventional

---

```
 1  foreach B: basic block of the CFG do
 2  │   let (E₁, ..., Eₙ) be the list of incoming edges of B
 3  │   foreach Eᵢ = (Bᵢ, B) do
 4  │   │   let PCᵢ be an empty parallel copy instruction
 5  │   │   if Bᵢ has several outgoing edges then
 6  │   │   │   create fresh empty basic block B'ᵢ
 7  │   │   │   replace edge Eᵢ by edges Bᵢ → B'ᵢ and B'ᵢ → B
 8  │   │   │   insert PCᵢ in B'ᵢ
 9  │   │   else
10  │   │   │   append PCᵢ at the end of Bᵢ
11  │   │   foreach φ-function at the entry of B of the form a₀ = φ(B₁ : a₁, ..., Bₙ : aₙ) do
12  │   │   │   foreach aᵢ (argument of the φ-function corresponding to Bᵢ) do
13  │   │   │   │   let a'ᵢ be a freshly created variable
14  │   │   │   │   add copy a'ᵢ ← aᵢ to PCᵢ
15  │   │   │   │   replace aᵢ by a'ᵢ in the φ-function
```

We stress that the above destruction technique has several drawbacks: first because of specific architectural constraints, region boundaries, or exception handling code, the compiler might not permit the splitting of a given edge; second, the resulting code contains many temporary-to-temporary copy operations. In theory, reducing the frequency of these copies is the role of the coalescing during the register allocation phase. A few memory- and time-consuming coalescing heuristics mentioned in Chap. 22 can handle the removal of these copies effectively. Coalescing can also, with less effort, be performed prior to the register allocation phase. As opposed to a (so-called conservative) coalescing during register allocation, this *aggressive* coalescing would not cope with the interference graph colourability. Further, the process of copy insertion itself might take a substantial amount of time and might not be suitable for dynamic compilation. The goal of Chap. 21 is to cope both with non-splittable edges and difficulties related to SSA destruction at machine code level, but also aggressive coalescing in the context of resource constrained compilation.

Once φ-functions have been replaced by parallel copies, we need to sequentialize the parallel copies, i.e., replace them by a sequence of simple copies. This phase can be performed immediately after SSA destruction or later on, perhaps even after register allocation (see Chap. 22). It might be useful to postpone the copy sequentialization since it introduces arbitrary interference between variables. As an example, $a_1 \leftarrow a_2 \parallel b_1 \leftarrow b_2$ (where $inst_1 \parallel inst_2$ represents two instructions $inst_1$ and $inst_2$ to be executed simultaneously) can be sequentialized into $a_1 \leftarrow a_2; \ b_1 \leftarrow b_2$, which would make b_2 interfere with a_1 while the other way round $b_1 \leftarrow b_2; \ a_1 \leftarrow a_2$ would make a_2 interfere with b_1 instead.

If we still decide to replace parallel copies into a sequence of simple copies immediately after SSA destruction, this can be done as shown in Algorithm 3.6. To see that this algorithm converges, one can visualize the parallel copy as a graph where nodes represent resources and edges represent transfers of values: the number of steps is exactly the number of cycles plus the number of non-self-edges of this graph. The correctness comes from the invariance of the behaviour of *seq*; *pcopy*. An optimized implementation of this algorithm will be presented in Chap. 21.

Algorithm 3.6: Replacement of parallel copies with sequences of sequential copy operations

---

1 **let** *pcopy* denote the parallel copy to be sequentialized
2 **let** *seq* = () denote the sequence of copies
3 **while** ¬ [∀(*b* ← *a*) ∈ *pcopy*, *a* = *b*] **do**
4   **if** ∃(*b* ← *a*) ∈ *pcopy* s.t. ∄(*c* ← *b*) ∈ *pcopy* **then**   ▷ *b is not live-in of pcopy*
5   append *b* ← *a* to *seq*
6   remove copy *b* ← *a* from *pcopy*

7   **else**   ▷ *pcopy is only made-up of cycles; Break one of them*
8   **let** *b* ← *a* ∈ *pcopy* s.t. *a* ≠ *b*
9   **let** *a'* be a freshly created variable
10   append *a'* ← *a* to *seq*
11   replace in *pcopy* *b* ← *a* into *b* ← *a'*

3.3 SSA Property Transformations

As discussed in Chap. 2, SSA comes in different flavours. This section describes algorithms that transform arbitrary SSA code into the desired flavour. Making SSA *conventional* corresponds exactly to the first phase of SSA destruction (described in Sect. 3.2) that splits critical edges and introduces parallel copies (sequentialized later in bulk or on-demand) around ϕ-functions. As already discussed, this straightforward algorithm has several drawbacks addressed in Chap. 21.

Making SSA *strict*, i.e., fulfil *the dominance property*, is as "hard" as constructing SSA. Of course, a pre-pass through the graph can detect the offending variables that have definitions that do not dominate their uses. Then there are several possible single-variable ϕ-function insertion algorithms (see Chap. 4) that can be used to patch up the SSA, by restricting attention to the set of non-conforming variables. The renaming phase can also be applied with the same filtering process. As the number of variables requiring repair might be a small proportion of all variables, a costly traversal of the whole program can be avoided by building the def-use chains (for non-conforming variables) during the detection pre-pass. Renaming can then be done on a per-variable basis or better (if pruned SSA is preferred) the reconstruction algorithm presented in Chap. 5 can be used for both ϕ-functions placement and renaming.

The construction algorithm described above does not build *pruned* SSA form. If available, liveness information can be used to filter out the insertion of ϕ-functions wherever the variable is not live: The resulting SSA form is pruned. Alternatively, pruning SSA form is equivalent to a dead code elimination pass after SSA construction. As use-def chains are implicitly provided by SSA form, dead-ϕ-function elimination simply relies on marking actual uses (non-ϕ-function ones) as *useful* and propagating *usefulness* backwards through ϕ-functions. Algorithm 3.7 presents the relevant pseudo-code for this operation. Here, *stack* is used to store useful and unprocessed variables defined by ϕ-functions.

Algorithm 3.7: ϕ-function pruning algorithm

```
1  stack ← ()
        ▷ — initial marking phase —
2  foreach I : instruction of the CFG in dominance order do
3       if I is φ-function defining variable a then mark a as useless
4
5       else I is not φ-function
6           foreach x : source operand of I do
7               if x is defined by some φ-function then
8                   mark x as useful; stack.push(x)

        ▷ — usefulness propagation phase —
9  while stack not empty do
10      a ← stack.pop()
11      let I be the φ-function that defines a
12      foreach x : source operand of I do
13          if x is marked as useless then
14              mark x as useful; stack.push(x)

        ▷ — final pruning phase —
15 foreach I : φ-function do
16      if destination operand of I marked as useless then delete I
```

To construct pruned SSA form via dead code elimination, it is generally much faster to first build *semi-pruned SSA form*, rather than minimal SSA form, and then apply dead code elimination. Semi-pruned SSA form is based on the observation that many variables are *local*, i.e., have a small live range that is within a single basic block. Consequently, pruned SSA would not instantiate any ϕ-functions for these variables. Such variables can be identified by a linear traversal over each basic block of the CFG. All of these variables can be filtered out: minimal SSA form restricted to the remaining variables gives rise to the so-called semi-pruned SSA form.

3.3.1   Pessimistic ϕ-Function Insertion

Construction of minimal SSA, as outlined in Sect. 3.1, comes at the price of a
sophisticated algorithm involving the computation of iterated dominance frontiers.
Alternatively, ϕ-functions may be inserted in a *pessimistic* fashion, as detailed
below. In the pessimistic approach, a ϕ-function is inserted at the start of each CFG
node (basic block) for each variable that is live at the start of that node. A less
sophisticated, or *crude*, strategy is to insert a ϕ-function for each variable at the
start of each CFG node; the resulting SSA will not be pruned. When a pessimistic
approach is used, many inserted ϕ-functions are redundant. Code transformations
such as code motion, or other CFG modifications, can also introduce redundant ϕ-
functions, i.e., make a minimal SSA program become non-minimal.

The application of *copy propagation* and rule-based ϕ-function rewriting can
remove many of these redundant ϕ-functions. As already mentioned in Chap. 2,
copy propagation can break the dominance property by propagating variable a_j
through ϕ-functions of the form $a_i = \phi(a_{x_1}, \ldots, a_{x_k})$ where all the source operands
are syntactically equal to either a_j or $\bot$. If we want to avoid breaking the dominance
property we simply have to avoid applying copy propagations that involve $\bot$. A
more interesting rule is the one that propagates a_j through a ϕ-function of the
form $a_i = \phi(a_{x_1}, \ldots, a_{x_k})$ where all the source operands are syntactically equal
to either a_i or a_j. These ϕ-functions turn out to be "identity" operations, where all
the source operands become syntactically identical and equivalent to the destination
operand. As such, they can be trivially eliminated from the program. Identity ϕ-
functions can be simplified this way from inner to outer loops. To be efficient,
variable def-use chains should be pre-computed—otherwise as many iterations as
the maximum loop depth might be necessary for copy propagation. When the CFG
is reducible,[2] this simplification produces minimal SSA. The underlying reason
is that the def-use chain of a given ϕ-web is, in this context, isomorphic with a
subgraph of the reducible CFG: All nodes except those that can be reached by
two different paths (from actual non-ϕ-function definitions) can be simplified by
iterative application of *T1* (removal of self-edge) and *T2* (merge of a node with
its unique direct predecessor) graph reductions. Figure 3.5 illustrates these graph
transformations. To be precise, we can simplify an SSA program as follows:

1. Remove any ϕ-node $a_i = \phi(a_{x_1}, \ldots, a_{x_k})$ where all a_{x_n} are a_j. Replace all
   occurrences of a_i by a_j. This corresponds to *T2* reduction.
2. Remove any ϕ-node $a_i = \phi(a_{x_1}, \ldots, a_{x_k})$ where all $a_{x_n} \in \{a_i, a_j\}$. Replace all
   occurrences of a_i by a_j. This corresponds to *T1* followed by *T2* reduction.

---

[2] A CFG is reducible if there are no jumps into the middle of loops from the outside, so the only
entry to a loop is through its header. Section 3.4 gives a fuller discussion of reducibility, with
pointers to further reading.

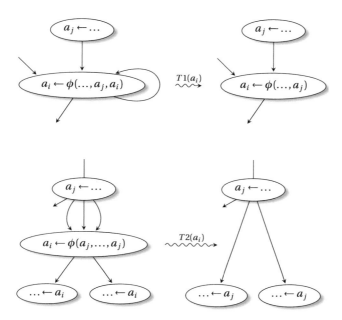

Fig. 3.5 *T1* and *T2* rewrite rules for SSA-graph reduction, applied to def-use relations between SSA variables

This approach can be implemented using a worklist, which stores the candidate nodes for simplification. Using the graph made up of def-use chains (see Chap. 14), the worklist can be initialized with ϕ-functions that are direct successors of non-ϕ-functions. However, for simplicity, we may initialize it with all ϕ-functions. Of course, if loop nesting forest information is available, the worklist can be avoided by traversing the CFG in a single pass from inner to outer loops, and in a topological order within each loop (header excluded). But since we believe the main motivation for this approach to be its simplicity, the pseudo-code shown in Algorithm 3.8 uses a work queue.

This algorithm is guaranteed to terminate in a fixed number of steps. At every iteration of the while loop, it removes a ϕ-function from the work queue W. Whenever it adds new ϕ-functions to W, it removes a ϕ-function from the program. The number of ϕ-functions in the program is bounded so the number of insertions to W is bounded. The queue could be replaced by a worklist, and the insertions/removals done at random. The algorithm would be less efficient, but the end result would be the same.

Algorithm 3.8: Removal of redundant ϕ-functions using rewriting rules and work queue

---

1   $W \leftarrow ()$
2   **foreach** I: ϕ-function in program in reverse post-order **do**
3   W.enqueue(I); mark I

4   **while** W not empty **do**
5      $I \leftarrow W$.dequeue(); unmark I
6   let I be of the form $a_i = \phi(a_{x_1}, \ldots, a_{x_k})$
7   **if** all source operands of the ϕ-function are $\in \{a_i, a_j\}$ **then**
8   Remove I from program
9   **foreach** I': instruction different from I that uses a_i **do**
10   replace a_i by a_j in I'
11   **if** I' is a ϕ-function and I' is not marked **then**
12   mark I'; W.enqueue(I')

3.4 Further Reading

The early literature on SSA form [89, 90] introduces the two phases of the construction algorithm we have outlined in this chapter and discusses algorithmic complexity on common and worst-case inputs. These initial presentations trace the ancestry of SSA form back to early work on data-flow representations by Shapiro and Saint [258].

Briggs et al. [50] discuss pragmatic refinements to the original algorithms for SSA construction and destruction, with the aim of reducing execution time. They introduce the notion of semi-pruned form, show how to improve the efficiency of the stack-based renaming algorithm, and describe how copy propagation must be constrained to preserve correct code during SSA destruction.

There are numerous published descriptions of alternative algorithms for SSA construction, in particular for the ϕ-function insertion phase. The pessimistic approach that first inserts ϕ-functions at all control-flow merge points and then removes unnecessary ones using simple *T1/T2* rewrite rules was proposed by Aycock and Horspool [14]. Brandis and Mössenböck [44] describe a simple, syntax-directed approach to SSA construction from well structured high-level source code. Throughout this textbook, we consider the more general case of SSA construction from arbitrary CFGs.

A *reducible* CFG is one that will collapse to a single node when it is transformed using repeated application of *T1/T2* rewrite rules. Aho et al. [2] describe the concept of reducibility and trace its history in early compilers literature.

Sreedhar and Gao [263] pioneer linear-time complexity ϕ-function insertion algorithms based on DJ-graphs. These approaches have been refined by other researchers. Chapter 4 explores these alternative construction algorithms in depth.

Blech et al. [30] formalize the semantics of SSA, in order to verify the correctness of SSA destruction algorithms. Boissinot et al. [35] review the history of SSA destruction approaches and highlight misunderstandings that led to incorrect destruction algorithms. Chapter 21 presents more details on alternative approaches to SSA destruction.

There are instructive dualisms between concepts in SSA form and functional programs, including construction, dominance, and copy propagation. Chapter 6 explores these issues in more detail.

Chapter 4
Advanced Construction Algorithms for SSA

Dibyendu Das, Upadrasta Ramakrishna, and Vugranam C. Sreedhar

The insertion of ϕ-functions is an important step in the construction of Static Single Assignment (SSA) form. In SSA form every variable is assigned only once. At control-flow merge points ϕ-functions are added to ensure that every use of a variable corresponds to exactly one definition. In the rest of this chapter we will present three different approaches for inserting ϕ-functions at appropriate merge nodes. Recall that SSA construction falls into two phases of ϕ-function insertion and variable renaming. Here we present different approaches for the first phase for minimal SSA form. We first present some properties of the DJ-graph that allow us to compute the iterated dominance frontier (DF^+) of a given set of nodes S by traversing the DJ-graph from leaves to root. ϕ-functions can then be placed using the DF^+ set. Based on the same properties, we then present an alternative scheme for computing DF^+-graph based on data-flow equations, this time using a traversal of the DJ-graph from root to leaves. Finally, we describe another approach for computing the iterated dominance frontier based on the loop nesting forest.

D. Das (✉)
Intel Corporation, Bangalore, India
e-mail: dibyendu.das@intel.com

U. Ramakrishna
IIT Hyderabad, Kandi, India
e-mail: ramakrishna@iith.ac.in

V. C. Sreedhar
IBM, Yorktown Heights, NY, USA
e-mail: vugranam@us.ibm.com

© The Author(s), under exclusive license to Springer Nature Switzerland AG 2022
F. Rastello, F. Bouchez Tichadou (eds.), *SSA-based Compiler Design*,
https://doi.org/10.1007/978-3-030-80515-9_4

4.1 Basic Algorithm

We start by recalling the basic algorithm already described in Chap. 3. The original algorithm for ϕ-functions is based on computing the dominance frontier (DF) set for the given control-flow graph. The dominance frontier $DF(x)$ of a node x is the set of all nodes z such that x dominates a direct predecessor of z, without strictly dominating z. For example, $DF(8) = \{6, 8\}$ in Fig. 4.1. The basic algorithm for the insertion of ϕ-functions consists in computing the iterated dominance frontier (DF^+) for a set of all definition points (or nodes where variables are defined). Let $Defs(v)$ be the set of nodes where variable v is defined. Given that the dominance frontier for a set of nodes is just the union of the DF set of each node, we can compute $DF^+(Defs(v))$ as a limit of the following recurrence equation (where S is initially $Defs(v)$):

$$DF^+{}_1(S) = DF(S)$$

$$DF^+{}_{i+1}(S) = DF(S \cup DF^+{}_i(S))$$

A ϕ-function is then inserted at each join node in the $DF^+(Defs(v))$ set.

4.2 Computation of $DF^+(S)$ Using DJ-Graphs

We now present a linear time algorithm for computing the $DF^+(S)$ set of a given set of nodes S without the need for explicitly pre-computing the full DF set. The algorithm uses the DJ-graph (see Chap. 3, Sect. 3.1.2 and Fig. 3.3b) representation

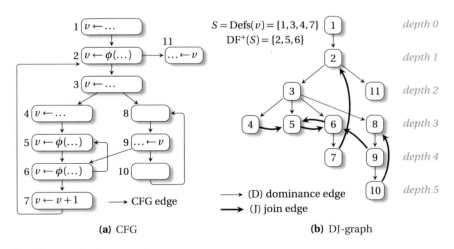

(a) CFG   **(b)** DJ-graph

Fig. 4.1 A motivating example

of a CFG. The DJ-graph for our example CFG is also shown in Fig. 4.1b. Rather than explicitly computing the DF set, this algorithm uses a DJ-graph to compute the $DF^+(Defs(v))$ on the fly.

4.2.1 Key Observations

Now let us try to understand how to compute the DF set for a *single node* using the DJ-graph. Consider the DJ-graph shown in Fig. 4.1b where the depth of a node is the distance from the root in the dominator tree. The first key observation is that a DF-edge never goes down to a greater depth. To give a raw intuition of why this property holds, suppose there was a DF-edge from 8 to 7, then there would be a path from 3 to 7 through 8 without flowing through 6, which contradicts the dominance of 7 by 6.

As a consequence, to compute DF(8) we can simply walk down the dominator (D) tree from node 8 and from each visited node y, identify all join (J) edges $y \twoheadrightarrow z$ such that $z.depth \leq 8.depth$. For our example the J-edges that satisfy this condition are $10 \twoheadrightarrow 8$ and $9 \twoheadrightarrow 6$. Therefore DF(8) $= \{6, 8\}$. To generalize the example, we can compute the DF of a node x using the following formula (see Fig. 4.2a for an illustration):

$$DF(x) = \{z \mid \exists \, y \in \text{dominated}(x) \ \wedge \ y \twoheadrightarrow z \in \text{J-edges} \ \wedge \ z.depth \leq x.depth\}$$

where

$$\text{dominated}(x) = \{y \mid x \text{ dom } y\}$$

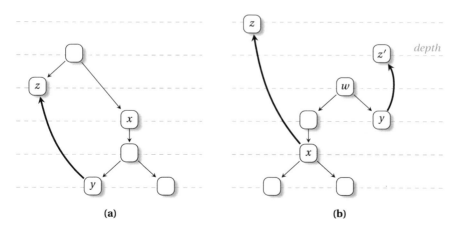

(a) (b)

Fig. 4.2 Illustrating examples to the DF formulas

Now we can extend the above idea to compute the DF^+ for *a set of nodes*, and hence the insertion of ϕ-functions. This algorithm does not precompute DF; given a set of initial nodes $S = \text{Defs}(v)$ for which we want to compute the relevant set of ϕ-functions, a key observation can be made. Let w be an ancestor node of a node x on the dominator tree. If $DF(x)$ has already been computed before the computation of $DF(w)$, the traversal of dominated(x) can be avoided and $DF(x)$ directly used for the computation of $DF(w)$. This is because nodes reachable from dominated(x) are already in $DF(x)$. However, the converse may not be true, and therefore the order of the computation of DF is crucial.

To illustrate the key observation consider the example DJ-graph in Fig. 4.1b, and let us compute $DF^+(\{3, 8\})$. It is clear from the recursive definition of DF^+ that we have to compute $DF(3)$ and $DF(8)$ as a first step. Now, suppose we start with node 3 and compute $DF(3)$. The resulting DF set is $DF(3) = \{2\}$. Now, suppose we next compute the DF set for node 8, and the resulting set is $DF(8) = \{6, 8\}$. Notice here that we have already visited node 8 and its subtree when visiting node 3. We can avoid such duplicate visits by ordering the computation of DF set so that we first compute $DF(8)$ and then during the computation of $DF(3)$ we avoid visiting the subtree of node 8 and use the result $DF(8)$ that was previously computed.

Thus, to compute $DF(w)$, where w is an ancestor of x in the DJ-graph, we do not need to compute it from scratch as we can re-use the information computed as part of $DF(x)$ as shown. For this, we need to compute the DF of deeper (based on depth) nodes (here, x), before computing the DF of a shallower node (here, w). The formula is as follows, with Fig. 4.2b illustrating the positions of nodes z and z'.

$$DF(w) = \{z \mid z \in DF(x) \wedge z.\text{depth} \leq w.\text{depth}\} \,\cup$$
$$\{z' \mid y \in \text{subtree}(w) \setminus \text{subtree}(x) \wedge y \xrightarrow{J} z' \wedge z'.\text{depth} \leq w.\text{depth}\}$$

4.2.2 Main Algorithm

In this section we present the algorithm for computing DF^+. Let, for node x, $x.\text{depth}$ be its depth from the root node r, with $r.\text{depth} = 0$. To ensure that the nodes are processed according to the above observation we use a simple array of sets *OrderedBucket*, and two functions defined over this array of sets: (1) InsertNode(n) that inserts the node n in the set *OrderedBucket*[n.depth], and (2) GetDeepestNode() that returns a node from the *OrderedBucket* with the deepest depth number.

In Algorithm 4.1, at first we insert all nodes belonging to S in the *OrderedBucket*. Then the nodes are processed in a bottom-up fashion over the DJ-graph from deepest node depth to least node depth by calling Visit(x). The procedure Visit(x) essentially walks top-down in the DJ-graph avoiding already visited nodes. During this traversal it also peeks at destination nodes of J-edges. Whenever it notices that the depth number of the destination node of a J-edge is less than or equal to the depth number of the *current_x*, the destination node is added to the DF^+ set (Line 4) if it is not

---

Algorithm 4.1: Algorithm for computing $DF^+(S)$ set

Input: DJ-graph representation of a program
Input: S: set of CFG nodes
Output: $DF^+(S)$

1   $DF^+ \leftarrow \emptyset$
2   **foreach** node $x \in S$ **do**
3   InsertNode(x)
4   x.visited $\leftarrow$ false
5   **while** $x \leftarrow$ GetDeepestNode() **do**   ▷ *Get node from the deepest depth*
6   current_x $\leftarrow$ x
7   x.visited $\leftarrow$ true
8   Visit(x)   ▷ *Find DF(x)*
9   **return** DF^+

Procedure Visit(y)

---

1   **foreach** J-edge $y \overset{J}{\dashrightarrow} z$ **do**
2   **if** z.depth $\leq$ current_x.depth **then**
3   **if** $z \notin DF^+$ **then**
4              $DF^+ \leftarrow DF^+ \cup \{z\}$
5   **if** $z \notin S$ **then** InsertNode(z)

6   **foreach** D-edge $y \overset{D}{\rightarrow} y'$ **do**
7   **if** y'.visited $=$ false **then**
8   y'.visited $\leftarrow$ true
          /* if y'.boundary = false */   ▷ *See the section on Further Reading for details*
9   Visit(y')

---

present in DF^+ already. Notice that at Line 5 the destination node is also inserted in the *OrderedBucket* if it was never inserted before. Finally, at Line 9 we continue to process the nodes in the subtree by visiting over the D-edges. When the algorithm terminates, the set DF^+ contains the iterated dominance frontier for the initial set S.

In Fig. 4.3, some of the phases of the algorithm are depicted for clarity. The *OrderedBucket* is populated with the nodes 1, 3, 4, and 7 corresponding to $S =$ Defs(v) $= \{1, 3, 4, 7\}$. The nodes are inserted in the buckets corresponding to the depths at which they appear. Hence, node 1 which appears at depth 0 is in the 0-th bucket, node 3 is in bucket 2 and so on. Since the nodes are processed bottom-up, the first node that is visited is node 7. The J-edge $7 \dashrightarrow 2$ is considered, and the DF^+ set is empty: the DF^+ set is updated to hold node 2 according to Line 4 of the Visit procedure. In addition, InsertNode(2) is invoked and node 2 is inserted in bucket 2. The next node visited is node 4. The J-edge $4 \dashrightarrow 5$ is considered, which results in the new $DF^+ = \{2, 5\}$. The final DF^+ set converges to $\{2, 5, 6\}$ when node 5 is visited. Subsequent visits of other nodes do not add anything to the DF^+ set. An interesting case arises when node 3 is visited. Node 3 finally causes nodes 8, 9, and 10 also to be visited (Line 9 during the down traversal of the D-graph). However, when node 10

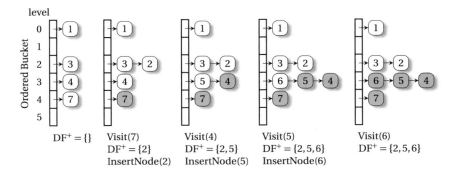

Fig. 4.3 Phases of Algorithm 4.1 for $S = \text{Defs}(v) = \{1, 3, 4, 7\}$

is visited, considering J-edge $10 \overset{J}{\nrightarrow} 8$ does not result in an update of the DF$^+$ set as the depth of node 8 is deeper than that of node 3.

4.3   Data-Flow Computation of DF$^+$-Graph Using DJ-Graph

In this section we describe a method to iteratively compute the DF$^+$ relation using a data-flow formulation. As already mentioned, the DF$^+$ relation can be computed using a transitive closure of the DF-graph, which in turn can be computed from the DJ-graph. In the algorithm proposed here, explicit DF-graph construction or the transitive closure formation are not necessary. Instead, the same result can be achieved by formulating the DF$^+$ relation as a data-flow problem and solving it iteratively. For several applications, this approach has been found to be a fast and effective method to construct DF$^+$(x) for each node x and the corresponding ϕ-function insertion using the DF$^+$(x) sets.

Data-flow equation:
Consider a J-edge $y \overset{J}{\nrightarrow} z$.
Then for all nodes x such that x dominates y and x.depth $\geq$ z.depth:

$$\text{DF}^+(x) = \text{DF}^+(x) \cup \text{DF}^+(z) \cup \{z\}$$

The set of data-flow equations for each node n in the DJ-graph can be solved iteratively using a top-down pass over the DJ-graph. To check whether multiple passes are required over the DJ-graph before a fixed point is reached for the data-flow equations, we devise an "inconsistency condition" stated as follows:

Inconsistency Condition:

For a J-edge, $y' \overset{J}{\nrightarrow} x$, if y' does not satisfy DF$^+$(y') $\supseteq$ DF$^+$(x),
then the node y' is said to be inconsistent.

The algorithm described in the next section is directly based on the method of building up the $DF^+(x)$ sets of the nodes as each J-edge is encountered in an iterative fashion by traversing the DJ-graph top-down. If no node is found to be *inconsistent* after a single top-down pass, all the nodes are assumed to have reached fixed-point solutions. If any node is found to be inconsistent, multiple passes are required until a fixed-point solution is reached.

4.3.1   Top-Down DF^+ Set Computation

Function TDMSC-Main(DJ-graph)

Input: A DJ-graph representation of a program.
Output: The DF^+ sets for the nodes.
1 **foreach** node $x \in$ DJ-graph **do**
2 $\quad$ $DF^+(x) \leftarrow \{\}$

3 **repeat** **until** TDMSC-I(DJ-graph)

Function TDMSC-I(DJ-graph)

---

1 *RequireAnotherPass* $\leftarrow$ false
2 **foreach** edge e **do** e.visited $\leftarrow$ false
3 **while** $z \leftarrow$ next node in B(readth) F(irst) S(earch) order of DJ-graph **do**
4 $\quad$ **foreach** incoming edge $e = y \overset{J}{\rightarrow} z$ **do**
5 $\quad\quad$ **if** not e.visited **then**
6 $\quad\quad\quad$ e.visited $\leftarrow$ true
7 $\quad\quad\quad$ $x \leftarrow y$
8 $\quad\quad\quad$ **while** $(x.\text{depth} \geq z.\text{depth})$ **do**
9 $\quad\quad\quad\quad$ $DF^+(x) \leftarrow DF^+(x) \cup DF^+(z) \cup \{z\}$
10 $\quad\quad\quad\quad$ $lx \leftarrow x$
11 $\quad\quad\quad\quad$ $x \leftarrow \text{parent}(x)$ $\qquad\qquad\qquad\qquad$ ▷ *dominator tree parent*
12 $\quad\quad\quad$ **foreach** incoming edge $e' = y' \overset{J}{\rightarrow} lx$ **do**
13 $\quad\quad\quad\quad$ **if** e'.visited **then**
14 $\quad\quad\quad\quad\quad$ **if** $DF^+(y') \not\supseteq DF^+(lx)$ **then** $\qquad$ ▷ *Check inconsistency*
15 $\quad\quad\quad\quad\quad\quad$ *RequireAnotherPass* $\leftarrow$ true;

16 **return** *RequireAnotherPass*

The first and direct variant of the approach laid out above is poetically termed TDMSC-I. This variant works by scanning the DJ-graph in a top-down fashion as shown in Line 3 of Function TDMSC-I. All $DF^+(x)$ sets are set to the empty set before the initial pass of TDMSC-I. The $DF^+(x)$ sets computed in a previous pass are carried over if a subsequent pass is required.

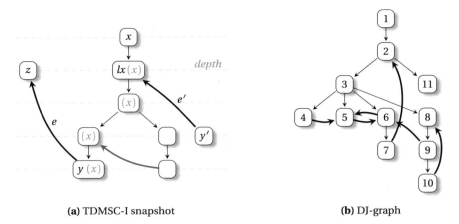

(a) TDMSC-I snapshot   (b) DJ-graph

Fig. 4.4 (**a**) Snapshot during execution of the TDMSC-I algorithm and (**b**) recall of running example

The DJ-graph is visited depth by depth. During this process, for each node z encountered, if there is an *incoming* J-edge $y \dashrightarrow z$ as in Line 4, then a separate bottom-up pass starts at Line 8 (see Fig. 4.4a for a snapshot of the variables during algorithm execution).

This bottom-up pass traverses all nodes x such that x dominates y and x.depth $\geq$ y.depth, updating the $DF^+(x)$ values using the aforementioned data-flow equation. Line 12 is used for the inconsistency check. *RequireAnotherPass* is set to true only if a fixed point is not reached and the inconsistency check succeeds for any node.

There are some subtleties in the algorithm that should be noted. Line 12 of the algorithm visits incoming edges to lx only when lx is at the same depth as z, which is the current depth of inspection and the incoming edges to lx's posterity are at a depth greater than that of node z and are unvisited yet.

Here, we will briefly walk through TDMSC-I using the DJ-graph of Fig. 4.1b (reprinted here as Fig. 4.4b). Moving top-down over the graph, the first J-edge encountered is when $z = 2$, i.e., $7 \dashrightarrow 2$. As a result, a bottom-up climbing of the nodes happens, starting at node 7 and ending at node 2 and the DF^+ sets of these nodes are updated so that $DF^+(7) = DF^+(6) = DF^+(3) = DF^+(2) = \{2\}$. The next J-edge to be visited can be any of $5 \dashrightarrow 6$, $9 \dashrightarrow 6$, $6 \dashrightarrow 5$, $4 \dashrightarrow 5$, or $10 \dashrightarrow 8$ at depth $= 3$. Assume node 6 is visited first, and thus it is $5 \dashrightarrow 6$, followed by $9 \dashrightarrow 6$. This results in $DF^+(5) = DF^+(5) \cup DF^+(6) \cup \{6\} = \{2, 6\}$, $DF^+(9) = DF^+(9) \cup DF^+(6) \cup \{6\} = \{2, 6\}$, and $DF^+(8) = DF^+(8) \cup DF^+(6) \cup \{6\} = \{2, 6\}$. Now, let $6 \dashrightarrow 5$ be visited. Hence, $DF^+(6) = DF^+(6) \cup DF^+(5) \cup \{5\} = \{2, 5, 6\}$. At this point, the *inconsistency check* comes into the picture for the edge $6 \dashrightarrow 5$, as $5 \dashrightarrow 6$ is another J-edge that is already visited and is an incoming edge of node 6. Checking for $DF^+(5) \supseteq DF^+(6)$ fails, implying that the $DF^+(5)$ needs to be computed again. This will be done in a succeeding pass as suggested by the *RequireAnotherPass* value of true. In a second iterative pass, the J-edges are visited in the same order.

Now, when $5 \dashrightarrow 6$ is visited, $DF^+(5) = DF^+(5) \cup DF^+(6) \cup \{6\} = \{2, 5, 6\}$ as this time $DF^+(5) = \{2, 6\}$ and $DF^+(6) = \{2, 5, 6\}$. On a subsequent visit of $6 \dashrightarrow 5$, $DF^+(6)$ is also set to $\{2, 5, 6\}$. The inconsistency no longer appears and the algorithm proceeds to handle the edges $4 \dashrightarrow 5$, $9 \dashrightarrow 6$ and $10 \dashrightarrow 8$ which have also been visited in the earlier pass. TDMSC-I is repeatedly invoked by a different function which calls it in a loop till *RequireAnotherPass* is returned as false, as shown in the procedure TDMSCMain.

Once the iterated dominance frontier relation is computed for the entire CFG, inserting the ϕ-functions is a straightforward application of the $DF^+(x)$ values for a given $Defs(x)$, as shown in Algorithm 4.2.

Algorithm 4.2: ϕ-function insertion for $Defs(x)$ using DF^+ sets

---

Input: A CFG with DF^+ sets computed and $Defs(x)$.
Output: Blocks augmented by ϕ-functions for the given $Defs(x)$.
1 **foreach** $n \in Defs(x)$ **do**
2   **foreach** $n' \in DF^+(n)$ **do**
3   Add a ϕ-function for n in n' if not inserted already

4.4   Computing Iterated Dominance Frontier Using Loop Nesting Forests

This section illustrates the use of *loop nesting forests* to construct the iterated dominance frontier (DF^+) of a set of vertices in a CFG. This method works with reducible as well as irreducible loops.

4.4.1   Loop Nesting Forest

A loop nesting forest is a data structure that represents the loops in a CFG and the containment relation between them. In the example shown in Fig. 4.5a the loops with back edges $11 \rightarrow 9$ and $12 \rightarrow 2$ are both reducible loops. The corresponding loop nesting forest is shown in Fig. 4.5b and consists of two loops whose header nodes are 2 and 9. The loop with header node 2 contains the loop with header node 9.

4.4.2   Main Algorithm

The idea is to use the forward CFG, an acyclic version of the control-flow graph (i.e., without back edges), and construct the DF^+ for a variable in this context: whenever

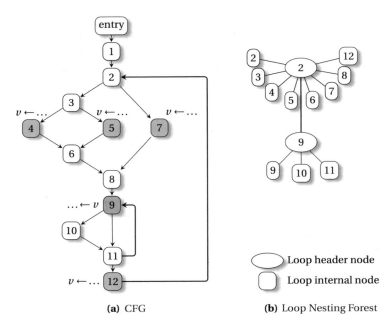

(a) CFG   **(b)** Loop Nesting Forest

Fig. 4.5 A sample CFG with four defs and one use of variable v and its loop nesting forest

two distinct definitions reach a join point, it belongs to the DF^+. Then, we take into account the back edges using the loop nesting forest: if a loop contains a definition, its header also belongs to the DF^+.

A definition node d "reaches" another node u if there is non-empty a path in the graph from d to u which does not contain any redefinition. If at least two definitions reach a node u, then u belongs to $DF^+(S)$ where $S = \text{Defs}(x)$ consists of these definition nodes. This suggests Algorithm 4.3 which works for *acyclic* graphs. For a given S, we can compute $DF^+(S)$ as follows:

- Initialize DF^+ to the empty set.
- Using a topological order, compute the subset of $DF^+(\text{Defs}(x))$ that can reach a node using forward data-flow analysis.
- Add a node to DF^+ if it is reachable from multiple nodes.

For Fig. 4.5, the forward CFG of the graph G, termed G_{fwd}, is formed by dropping the back edges $11 \rightarrow 9$ and $12 \rightarrow 2$. Also, r is a specially designated node that is the root of the CFG. For the definitions of x in nodes $4, 5, 7$, and 12 in Fig. 4.5, the subsequent nodes (forward) reached by multiple definitions are 6 and 8: node 6 can be reached by any one of the two definitions in nodes 4 or 5, and node 8 by either the definition from node 7 or one of 4 or 5. Note that the back edges do not exist in the forward CFG and hence node 2 is not part of the DF^+ set yet. We will see later how the DF^+ set for the entire graph is computed by considering the contribution of the back edges.

Algorithm 4.3: Algorithm for DF$^+$ in an acyclic graph

Input: G_{fwd}: forward CFG
Input: S: subset of nodes
Output: DF$^+(S)$

1   DF$^+ \leftarrow \emptyset$
2   **foreach** node n **do** *UniqueReachingDef*$(n) \leftarrow \{r\}$
3   **foreach** node $u \neq r$ in topological order **do**
4   *ReachingDefs* $\leftarrow \emptyset$
5   **foreach** v direct predecessor of u **do**
6   **if** $v \in$ DF$^+ \cup S \cup r$ **then**
7   *ReachingDefs* $\leftarrow$ *ReachingDefs* $\cup \{v\}$;
8   **else**
9   *ReachingDefs* $\leftarrow$ *ReachingDefs* $\cup$ *UniqueReachingDef*(v);
10   **if** $|ReachingDefs| = 1$ **then**
11   *UniqueReachingDef*$(u) \leftarrow$ *ReachingDefs*;
12   **else**
13   DF$^+ \leftarrow$ DF$^+ \cup \{u\}$;

14   **return** DF$^+$

Let us walk through this algorithm computing DF$^+$ for variable v, i.e., $S = \{4, 5, 7, 12\}$. The nodes in Fig. 4.5 are already numbered in topological order. Nodes 1 to 5 have only one direct predecessor, none of them being in S, so their *UniqueReachingDef* stays r, and DF$^+$ is still empty. For node 6, its two direct predecessors belong to S, hence *ReachingDefs* $= \{4, 5\}$, and 6 is added to DF$^+$. Nothing changes for 7, then for 8 its direct predecessors 6 and 7 are, respectively, in DF$^+$ and S: they are added to *ReachingDefs*, and 8 is then added to DF$^+$. Finally, for nodes 8 to 12, their *UniqueReachingDef* will be updated to node 8, but this will no longer change DF$^+$, which will end up being $\{6, 8\}$.

DF$^+$ on Reducible Graphs

A reducible graph can be decomposed into an acyclic graph and a set of back edges. The contribution of back edges to the iterated dominance frontier can be identified by using the loop nesting forest. If a vertex v is contained in a loop, then DF$^+(v)$ will contain the loop header, i.e., the unique entry of the reducible loop. For any vertex v, let HLC(v) denote the set of loop headers of the loops containing v. Given a set of vertices S, it turns out that DF$^+(S) =$ HLC$(S) \cup$ DF$^+_{fwd}(S \cup$ HLC$(S))$ where HLC$(S) = \bigcup_{v \in S}$ HLC(v), and where DF$^+_{fwd}$ denote the DF$^+$ restricted to the forward CFG G_{fwd}.

Reverting back to Fig. 4.5, we see that in order to find the DF$^+$ for the nodes defining x, we need to evaluate DF$^+_{fwd}(\{4, 5, 7, 12\} \cup$ HLC$(\{4, 5, 7, 12\}))$. As all these nodes are contained in a single loop with header 2, HLC$(\{4, 5, 7, 12\}) = \{2\}$. Computing DF$^+_{fwd}(\{4, 5, 7, 12\})$ gives us $\{6, 8\}$, and finally, DF$^+(\{4, 5, 7, 12\}) = \{2, 6, 8\}$.

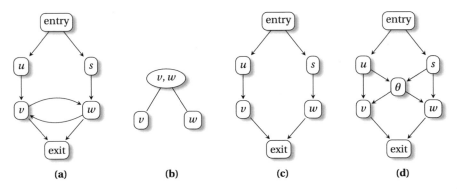

Fig. 4.6 (a) An irreducible graph. (b) The loop nesting forest. (c) The acyclic subgraph. (d) Transformed graph

DF$^+$ on Irreducible Graphs

We will now briefly explain how graphs containing irreducible loops can be handled. The insight behind the implementation is to transform the irreducible loop in such a way that an acyclic graph is created from the loop without changing the dominance properties of the nodes.

The loop in the graph of Fig. 4.6a, which is made up of nodes v and w, is irreducible as it has two entry nodes, v and w. We let the headers be those entry nodes. It can be transformed to the acyclic graph (c) by removing the back edges, i.e., the edges within the loop that point to the header nodes, in other words, edges $v \rightarrow w$ and $w \rightarrow v$. We create a dummy node, θ, to which we connect all direct predecessors of the headers (u and s), and that we connect to all the headers of the loop (v and w), creating graph (d).

Following this transformation, the graph is now acyclic, and computing the DF$^+$ for the nodes in the original irreducible graph translates to computing DF$^+$ using the transformed graph to get DF$^+_{fwd}$ and using the loop forest of the original graph (b).

The crucial observation that allows this transformation to create an equivalent acyclic graph is the fact that the dominator tree of the transformed graph remains identical to the original graph containing an irreducible cycle.

4.5   Concluding Remarks and Further Reading

Concluding Remarks

Although all these algorithms claim to be better than the original algorithm by Cytron et al., they are difficult to compare due to the unavailability of these algorithms in a common compiler framework.

In particular, while constructing the whole DF$^+$ set seems very costly in the classical construction algorithm, its cost is actually amortized as it will serve to insert ϕ-functions for many variables. It is, however, interesting not to pay this cost whenever we only have a few variables to consider, for instance, when repairing SSA as in the next chapter.

Note also that people have observed in production compilers that, during SSA construction, what seems to be the most expensive part is the renaming of the variables and not the insertion of ϕ-functions.

Further Reading

Cytron's approach for ϕ-function insertion involves a fully eager approach of constructing the entire DF-graph [90]. Sreedhar and Gao proposed the first algorithm for computing DF^+ sets without the need for explicitly computing the full DF set [263], producing the linear algorithm that uses DJ-graphs.

The lazy algorithm presented in this chapter that uses DJ-graph was introduced by Sreedhar and Gao and constructs DF on the fly only when a query is encountered. Pingali and Bilardi [29] suggested a middle-ground by combining both approaches. They proposed a new representation called ADT (Augmented Dominator Tree). The ADT representation can be thought of as a DJ-graph, where the DF sets are precomputed for certain nodes called "boundary nodes" using an eager approach. For the rest of the nodes, termed "interior nodes," the DF needs to be computed on the fly as in the Sreedhar-Gao algorithm. The nodes which act as "boundary nodes" are detected in a separate pass. A factor β is used to determine the partitioning of the nodes of a CFG into boundary or interior nodes by dividing the CFG into zones. β is a number that represents space/query-time tradeoff. $\beta \ll 1$ denotes a fully eager approach where storage requirement for DF is maximum but query time is faster while $\beta \gg 1$ denotes a fully lazy approach where storage requirement is zero but query is slower.

Given the ADT of a control-flow graph, it is straightforward to modify Sreedhar and Gao's algorithm for computing ϕ-functions in linear time. The only modification that is needed is to ensure that we need not visit all the nodes of a subtree rooted at a node y when y is a boundary node whose DF set is already known. This change is reflected in Line 8 of Algorithm 4.1, where a subtree rooted at y is visited or not visited based on whether it is a boundary node or not.

The algorithm computing DF^+ without the explicit DF-graph is from Das and Ramakrishna [94]. For iterative DF^+ set computation, they also exhibit TDMSC-II, an improvement to algorithm TDMSC-I. This improvement is fueled by the observation that for an inconsistent node u, the DF^+ sets of all nodes w such that w dominates u and $w.depth \geq u.depth$, can be locally corrected for some special cases. This heuristic works very well for certain classes of problems—especially for CFGs with DF-graphs having cycles consisting of a few edges. This eliminates extra passes as an inconsistent node is made consistent immediately on being detected.

Finally, the part on computing DF^+ sets using loop nesting forests is based on Ramalingam's work on loops, dominators, and dominance frontiers [236].

Chapter 5
SSA Reconstruction

Sebastian Hack

Some optimizations break the single-assignment property of the SSA form by inserting additional definitions for a single SSA value. A common example is live range splitting by inserting copy operations or inserting spill and reload code during register allocation. Other optimizations, such as loop unrolling or jump threading, might duplicate code, thus adding additional variable definitions, *and* modify the control flow of the program. We will first mention two examples before we present algorithms to properly repair SSA.

The first example is depicted in Fig. 5.1. Our spilling pass decided to spill a part of the live range of variable x_0 in the right block in (a), resulting in the code shown in (b), where it inserted a store and a load instruction. This is indicated by assigning to the memory location X. The load is now a second definition of x_0, violating the SSA form that has to be reconstructed as shown in (c). This example shows that maintaining SSA also involves placing new ϕ-functions.

Many optimizations perform such program modifications, and maintaining SSA is often one of the more complicated and error-prone parts in such optimizations, owing to the insertion of additional ϕ-functions and the correct redirection of the uses of the variable.

Another example for such a transformation is *path duplication* which is discussed in Chap. 20. Several popular compilers such as GCC and LLVM perform one or another variant of path duplication, for example, when *threading jumps*.

S. Hack (✉)
Saarland University, Saarbrücken, Germany
e-mail: hack@cs.uni-saarland.de

© The Author(s), under exclusive license to Springer Nature Switzerland AG 2022
F. Rastello, F. Bouchez Tichadou (eds.), *SSA-based Compiler Design*,
https://doi.org/10.1007/978-3-030-80515-9_5

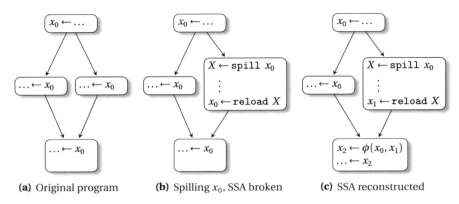

(a) Original program   **(b)** Spilling x_0, SSA broken   **(c)** SSA reconstructed

Fig. 5.1 Adding a second definition as a side-effect of spilling

5.1 General Considerations

In this chapter, we will discuss two algorithms. The first is an adoption of the classical dominance-frontier based algorithm. The second performs a search from the uses of the variables to the definition and places ϕ-functions on demand at appropriate places. In contrast to the first, the second algorithm might not construct minimal SSA form in general; however, it does not need to update its internal data structures when the CFG is modified.

We consider the following scenario: The program is represented as a control-flow graph (CFG) and is in SSA form with dominance property. For the sake of simplicity, we assume that each instruction in the program only writes to a single variable. An optimization or transformation violates SSA by inserting additional definitions for an existing SSA variable, like in the examples above. The original variable and the additional definitions can be seen as a single non-SSA variable that has multiple definitions and uses. Let in the following v be such a non-SSA variable.

When reconstructing SSA for v, we will first create fresh variables for every definition of v to establish the single-assignment property. What remains is associating every use of v with a suitable definition. In the algorithms, v.defs denotes the set of all instructions that define v. A use of a variable is a pair consisting of a program point (an instruction) and an integer denoting the index of the operand at this instruction.

Both algorithms presented in this chapter share the same driver routine described in Algorithm 5.1. First, we scan all definitions of v so that for every basic block b we have the list b.defs that contains all instructions in the block which define one of the variables in v.defs. It is best to sort this according to the schedule of the instructions in the block from back to front, making the latest definition the first in the list.

Then, all uses of the variable v are traversed to associate them with the proper definition. This can be done by using precomputed use-def chains if available or scanning all instructions in the dominance subtree of v's original SSA definition. For each use, we have to differentiate whether the use is in a ϕ-function or not. If so, the use occurs at the end of the direct predecessor block that corresponds to the position of the variable in the ϕ's argument list. In that case, we start looking for the reaching definition from the end of that block. Otherwise, we scan the instructions of the block backwards until we reach the first definition that is before the use (Line 14). If there is no such definition, we have to find one that reaches this block from *outside*.

We use two functions, FindDefFromTop and FindDefFromBottom that search the reaching definition, respectively, from the beginning or the end of a block. Find-DefFromBottom actually just returns the last definition in the block, or call Find-DefFromTop if there is none.

The two presented approaches to SSA repairing differ in the implementation of the function FindDefFromTop. The differences are described in the next two sections.

Algorithm 5.1: SSA reconstruction driver

Input: v, a variable that breaks SSA property

1 **foreach** $d \in v.$defs **do**
2   create fresh variable v'
3   rewrite def of v by v' in d
4       $b \leftarrow d.$block
5   insert d in $b.$defs

6 **foreach** use (inst, index) of v **do**
7   **if** inst is a ϕ-function **then**
8           $b \leftarrow$ inst.block.directpred(*index*)
9           $d \leftarrow$ FindDefFromBottom(v, b)
10   **else**
11          $d \leftarrow \bot$
12          $b \leftarrow$ inst.block
13   **foreach** $l \in b.$defs from bottom to top **do**
14   **if** l is before inst in b **then**
15                  $d \leftarrow l$
16   break
17   **if** $d = \bot$ **then**   ▷ *no local def. found, searching in the direct preds*
18              $d \leftarrow$ FindDefFromTop(v, b)

19      $v' \leftarrow$ version of v defined by d
20   rewrite use of v by v' in inst

Procedure FindDefFromBottom(v, b)

---
1   **if** b.defs $\neq \emptyset$ **then**
2   │   **return** latest instruction in b.defs

3   **else**
4   │   **return** FindDefFromTop(v, b)

5.2   Reconstruction Based on the Dominance Frontier

This algorithm follows the same principles as the classical SSA construction algorithm by Cytron at al. as described in Chap. 3. We first compute the iterated dominance frontier (DF^+) of v. This set is a sound approximation of the set where ϕ-functions must be placed—it might contain blocks where a ϕ-function would be dead. Then, we search for each use u the corresponding reaching definition. This search starts at the block of u. If that block b is in the DF^+ of v, a ϕ-function needs to be placed at its entrance. This ϕ-function becomes a new definition of v and has to be inserted in v.defs and in b.defs. The operands of the newly created ϕ-function will query their reaching definitions by recursive calls to FindDefFromBottom on direct predecessors of b. Because we inserted the ϕ-function into b.defs *before* searching for the arguments, no infinite recursion can occur (otherwise, it could happen, for instance, with a loop back edge).

If the block is not in the DF^+, the search continues in the immediate dominator of the block. This is because in SSA, every use of a variable must be dominated by its definition.[1] Therefore, the reaching definition is the same for all direct predecessors of the block, and hence for the immediate dominator of this block.

Procedure FindDefFromTop(v, b)

▷ *SSA Reconstruction based on Dominance Frontiers*
1   **if** $b \in DF^+(v.\text{defs})$ **then**
2   │ $v' \leftarrow$ fresh variable
3   │ $d \leftarrow$ new ϕ-function in b: $v' \leftarrow \phi(\dots)$
4   │   append d to b.defs
5   │   **foreach** $p \in b$.directpreds **do**
6   │   │ $o \leftarrow$ FindDefFromBottom(v, p)
7   │   │ $v' \leftarrow$ version of v defined by o
8   │   │   set corresponding operand of d to v'

9   **else**
10 │ $d \leftarrow$ FindDefFromBottom(v, b.idom)   ▷ *search in immediate dominator*
11 **return** d

---

[1] The definition of an operand of a ϕ-function has to dominate the corresponding direct predecessor block.

5.3   Search-Based Reconstruction

The second algorithm presented here is adapted from an algorithm designed to construct SSA from the abstract syntax tree, but it also works well on control-flow graphs. Its major advantage over the algorithm presented in the last section is that it requires neither dominance information nor dominance frontiers. Thus it is well suited to be used in transformations that change the control-flow graph. Its disadvantage is that potentially more blocks have to be visited during the reconstruction. The principal idea is to start a search from every use to find the corresponding definition, inserting ϕ-functions on the fly while caching the SSA variable alive at the beginning of basic blocks. As in the last section, we only consider the reconstruction for a single variable called v in the following. If multiple variables have to be reconstructed, the algorithm can be applied to each variable separately.

Search on an Acyclic CFG

The algorithm performs a backward depth-first search in the CFG to collect the reaching definitions of v in question at each block, recording the SSA variable that is alive at the beginning of a block in the "beg" field of this block. If the CFG is an acyclic graph (DAG), all predecessors of a block can be visited before the block itself is processed, as we are using a post-order traversal following edges backwards. Hence, we know all the definitions that reach a block b: if there is more than one definition, we need to place a ϕ-function in b, otherwise it is not necessary.

Search on a General CFG

If the CFG has loops, there are blocks for which not all reaching definitions can be computed before we can decide whether a ϕ-function has to be placed or not. In a loop, recursively computing the reaching definitions for a block b will end up at b itself. To avoid infinite recursion when we enter a block during the traversal, we first create a ϕ-function without arguments, "pending_ϕ." This creates a new definition v_ϕ for v which is the variable alive at the beginning of this block.

When we return to b after traversing the rest of the CFG, we decide whether a ϕ-function has to be placed in b by looking at the reaching definition for every direct predecessor. These reaching definitions can be either v_ϕ itself (loop in the CFG without a definition of v), or some other definitions of v. If there is only one such other definition, say w, then pending_ϕ is not necessary and we can remove it, propagating w downwards instead of v_ϕ. Note that in this case it will be necessary to "rewrite" all uses that referred to pending_ϕ to w. Otherwise, we keep pending_ϕ and fill its missing operands with the reaching definitions. In this version of function FindDefFromTop, this check is done by the function Phi-Necessary.

Removing More Unnecessary ϕ-Functions

In programs with loops, it can be the case that the local optimization performed when function FindDefFromTop calls Phi-Necessary does not remove all unnecessary ϕ-functions. This can happen in loops where ϕ-functions can become

---

Procedure FindDefFromTop(b)

▷ *Search-based SSA Reconstruction*
Input: b, a basic block
1 **if** b.top $\neq \perp$ **then**
2 | **return** b.top

3 pending_$\phi \leftarrow$ new ϕ-function in b
4 $v_\phi \leftarrow$ result of pending_ϕ
5 b.top $\leftarrow v_\phi$
6 reaching_defs $\leftarrow$ []
7 **foreach** $p \in b$.directpreds **do**
8 | reaching_defs $\leftarrow$ reaching_defs $\cup$ FindDefFromBottom(v, p)

9 $v_{\text{def}} \leftarrow$ Phi-Necessary(v_ϕ, reaching_defs)
10 **if** $v_{\text{def}} = v_\phi$ **then**
11 | set arguments of pending_ϕ to reaching_defs

12 **else**
13 | rewire all uses of pending_ϕ to v_{def}
14 | remove pending_ϕ
15 | b.top $\leftarrow v_{\text{def}}$

16 **return** v_{def}

Procedure Phi-Necessary(v_ϕ, reaching_defs)

▷ *Checks if the set reaching_defs makes pending_ϕ necessary. This is the case if it is a subset of $\{v_\phi, other\}$*
Input: v_ϕ, the variable defined by pending_ϕ
Input: reaching_defs, list of variables
Output: v_ϕ if pending_ϕ is necessary, the variable to use instead otherwise
1 other $\leftarrow \perp$
2 **foreach** $v' \in$ reaching_defs **do**
3 | **if** $v' = v_\phi$ **then** continue
4 | **if** other $= \perp$ **then**
5 | | other $\leftarrow v'$
6 | **else if** $v' \neq$ other **then**
7 | | **return** v_ϕ

▷ *this assertion is violated if reaching_defs contains only pending_ϕ which never can happen*
8 assert (other $\neq \perp$)
9 **return** other

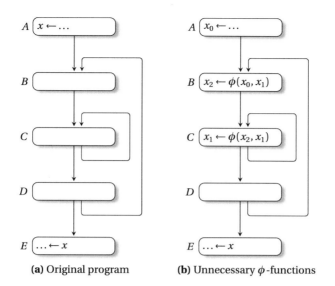

(a) Original program **(b)** Unnecessary ϕ-functions

Fig. 5.2 Removal of unnecessary ϕ-functions

Fig. 5.3 The irreducible
$(*)$-graph

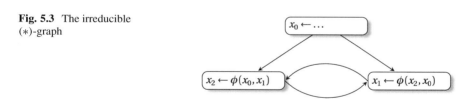

unnecessary because other ϕ-functions are optimized away. Consider the example in Fig. 5.2. We look for a definition of x from block E. If Algorithm FindDef-FromTop considers the blocks in a unfavourable order, e.g., E, D, C, B, A, D, C, some unnecessary ϕ-functions cannot be removed by Phi-Necessary, as shown in Fig. 5.2b. While the ϕ-function in block C can be eliminated by the local criterion applied by Phi-Necessary, the ϕ-function in block B remains. This is because the depth-first search carried out by FindDefFromTop will not visit block B a second time. To remove the remaining ϕ-functions, the local criterion can be iteratively applied to all placed ϕ-functions until a fixpoint is reached. For reducible control flow, this then produces the minimal number of placed ϕ-functions. The classical $(*)$-graph in Fig. 5.3 illustrates that this does not hold for the irreducible case. This is similar to the rules discussed in Sect. 3.3.1.

5.4 Further Reading

The algorithms presented in this chapter are independent of the transformation that violated SSA and can be used as a black box: For every variable for which SSA was violated, a routine is called that restores SSA. Both algorithms rely on computed def-use chains because they traverse all uses from a SSA variable to find suitable definitions; however, they differ in their prerequisites and their runtime behaviour:

The first algorithm (Choi et al. [69]) is based on the iterated dominance frontiers like the classical SSA construction algorithm by Cytron et al. [90]. Hence, it is less suited for optimizations that also change the flow of control since that would require recomputing the iterated dominance frontiers. On the other hand, by using the iterated dominance frontiers, the algorithm can find the reaching definitions quickly by scanning the dominance tree upwards. Furthermore, one could also envision applying incremental algorithms to construct the dominance tree [238, 264] and the dominance frontier [265] to account for changes in the control flow. This has not yet been done and no practical experiences have been reported so far.

The second algorithm is based on the algorithm by Braun et al. [47] which is an extension of the construction algorithm that Click describes in his thesis [75] to construct SSA from an abstract syntax tree. It does not depend on additional analysis information such as iterated dominance frontiers or the dominance tree. Thus, it is well suited for transformations that change the CFG because no information needs to be recomputed. On the other hand, it might be slower to find the reaching definitions because they are searched by a depth-first search in the CFG.

Both approaches construct *pruned* SSA, i.e., they do not add dead ϕ-functions. The first approach produces minimal SSA by the very same arguments Cytron et al. [90] give. The second only guarantees minimal SSA for reducible CFGs. This follows from the iterative application of the function Phi-Necessary which implements the two simplification rules presented by Aycock and Horspool [14], who showed that their iterative application yields minimal SSA on reducible graphs. These two local rules can be extended to a non-local one which has to find strongly connected ϕ-components that all refer to the same exterior variable. Such a non-local check also eliminates unnecessary ϕ-functions in the irreducible case [47].

Chapter 6
Functional Representations of SSA

Lennart Beringer

This chapter discusses alternative representations of SSA using the terminology and structuring mechanisms of functional programming languages. The reading of SSA as a discipline of functional programming arises from a correspondence between dominance and syntactic scope that subsequently extends to numerous aspects of control and data-flow structure.

The development of functional representations of SSA is motivated by the following considerations:

1. Relating the core ideas of SSA to concepts from other areas of compiler and programming language research provides conceptual insights into the SSA discipline and thus contributes to a better understanding of the practical appeal of SSA to compiler writers.
2. Reformulating SSA as a functional program makes explicit some of the syntactic conditions and semantic invariants that are implicit in the definition and use of SSA. Indeed, the introduction of SSA itself was motivated by a similar goal: to represent aspects of program structure—namely the def-use relationships—explicitly in syntax, by enforcing a particular naming discipline. In a similar way, functional representations directly enforce invariants such as "all ϕ-functions in a block must be of the same arity," "the variables assigned to by these ϕ-functions must be distinct," "ϕ-functions are only allowed to occur at the beginning of a basic block," or "each use of a variable should be dominated by its (unique) definition." Constraints such as these would typically have to be validated or (re-)established after each optimization phase of an SSA-based compiler, but are typically enforced *by construction* if a functional representation is chosen.

L. Beringer (✉)
Princeton University, Princeton, NJ, USA
e-mail: eberinge@cs.princeton.edu

© The Author(s), under exclusive license to Springer Nature Switzerland AG 2022
F. Rastello, F. Bouchez Tichadou (eds.), *SSA-based Compiler Design*,
https://doi.org/10.1007/978-3-030-80515-9_6

Consequently, less code is required, improving the robustness, maintainability, and code readability of the compiler.

3. The intuitive meaning of "unimplementable" ϕ-instructions is complemented by a concrete execution model, facilitating the rapid implementation of interpreters. This enables the compiler developers to experimentally validate SSA-based analyses and transformations at their genuine language level, without requiring SSA destruction. Indeed, functional intermediate code can often be directly emitted as a program in a high-level mainstream functional language, giving the compiler writer access to existing interpreters and compilation frameworks. Thus, rapid prototyping is supported and high-level evaluation of design decisions is enabled.

4. Formal frameworks of program analysis that exist for functional languages become applicable. Type systems provide a particularly attractive formalism due to their declarativeness and compositional nature. As type systems for functional languages typically support higher-order functions, they can be expected to generalize more easily to interprocedural analyses than other static analysis formalisms.

5. We obtain a formal basis for comparing variants of SSA—such as the variants discussed elsewhere in this book—for translating between these variants, and for constructing and destructing SSA. Correctness criteria for program analyses and associated transformations can be stated in a uniform manner and can be proved to be satisfied using well-established reasoning principles.

Rather than discussing all these considerations in detail, the purpose of the present chapter is to informally highlight particular aspects of the correspondence and then point the reader to some more advanced material. Our exposition is example-driven but leads to the identification of concrete correspondence pairs between the imperative/SSA world and the functional world.

Like the remainder of the book, our discussion is restricted to code occurring in a single procedure.

6.1 Low-Level Functional Program Representations

Functional languages represent code using declarations of the form

$$\textbf{function } f(x_0, \ldots, x_n) = e \tag{6.1}$$

where the syntactic category of expression e conflates the notions of expressions and commands of imperative languages. Typically, e may contain further nested or (mutually) recursive function declarations. A declaration of the form (6.1) binds the formal parameters x_i and the function name f within e.

6.1.1   *Variable Assignment Versus Binding*

A language construct provided by almost all functional languages is the *let-binding*:

$$\textbf{let } x = e_1 \textbf{ in } e_2 \textbf{ end}$$

The effect of this expression is to evaluate e_1 and bind the resulting value to variable x for the duration of the evaluation of e_2. The code affected by this binding, e_2, is called the *static scope* of x and is easily syntactically identifiable. In the following, we occasionally indicate scopes by code-enclosing boxes and list the variables that are in scope using subscripts.

For example, the scope associated with the top-most binding of v to 3 in code

$$
\begin{array}{l}
\textbf{let } v = 3 \textbf{ in} \\
\quad \boxed{
\begin{array}{l}
\textbf{let } y = (\textbf{let } v = 2 \times v \textbf{ in } \boxed{4 \times v}_{v} \textbf{ end}) \\
\textbf{in } \boxed{y \times v}_{v,y} \quad \textbf{end}
\end{array}
}_{v} \\
\textbf{end}
\end{array}
\qquad (6.2)
$$

spans both inner let-bindings, the scopes of which are themselves not nested inside one other as the inner binding of v occurs in the e_1 position of the let-binding for y.

In contrast to an assignment in an imperative language, a let-binding for variable x hides any previous value bound to x for the duration of evaluating e_2 but does not permanently overwrite it. Bindings are treated in a stack-like fashion, resulting in a tree-shaped nesting structure of boxes in our code excerpts. For example, in the above code, the inner binding of v to value $2 \times 3 = 6$ shadows the outer binding of v to value 3 precisely for the duration of the evaluation of the expression $4 \times v$. Once this evaluation has terminated (resulting in the binding of y to 24), the binding of v to 3 becomes visible again, yielding the overall result of 72.

The concepts of binding and static scope ensure that functional programs enjoy the characteristic feature of SSA, namely the fact that each use of a variable is uniquely associated with a point of definition. Indeed, the point of definition for a use of x is given by the *nearest enclosing binding of x*. Occurrences of variables in an expression that are not enclosed by a binding are called *free*. A well-formed procedure declaration contains all free variables of its body among its formal parameters. Thus, the notion of scope makes explicit the invariant that each use of a variable should be dominated by its (unique) definition.

In contrast to SSA, functional languages achieve the association of definitions to uses without imposing the global uniqueness of variables, as witnessed by the duplicate binding occurrences for v in the above code. As a consequence of this decoupling, functional languages enjoy a strong notion of *referential transparency*: The choice of x as the variable holding the result of e_1 depends only on the free variables of e_2. For example, we may rename the inner v in code (6.2) to z without altering the meaning of the code:

$$\textbf{let } v = 3 \textbf{ in}$$

$$\boxed{\begin{array}{l} \textbf{let } y = (\textbf{let } z = 2 \times v \textbf{ in } \boxed{4 \times z}_{v,z} \textbf{ end}) \\ \textbf{in } \boxed{y \times v}_{v,y} \textbf{ end} \end{array}}_{v} \tag{6.3}$$

$$\textbf{end}$$

Note that this conversion formally makes the outer v visible for the expression $4 \times z$, as indicated by the index v, z decorating its surrounding box.

In order to avoid altering the meaning of the program, the choice of the newly introduced variable has to be such that confusion with other variables is avoided. Formally, this means that a renaming

$$\textbf{let } x = e_1 \textbf{ in } e_2 \textbf{ end} \qquad \text{to} \qquad \textbf{let } y = e_1 \textbf{ in } e_2[y \leftrightarrow x] \textbf{ end}$$

can only be carried out if y is not a *free* variable of e_2. Moreover, in the event that e_2 already contains some preexisting bindings to y, the *substitution* of x by y in e_2 (denoted by $e_2[y \leftrightarrow x]$ above) first renames these preexisting bindings in a suitable manner. Also note that the renaming only affects e_2—any occurrences of x or y in e_1 refer to conceptually *different* but identically named variables, but the static scoping discipline ensures these will never be confused with the variables involved in the renaming. In general, the semantics-preserving renaming of bound variables is called α-renaming. Typically, program analyses for functional languages are compatible with α-renaming in that they behave equivalently for fragments that differ only in their choice of bound variables, and program transformations α-rename bound variables whenever necessary.

A consequence of referential transparency, and thus a property typically enjoyed by functional languages, is *compositional equational reasoning*: the meaning of a piece of code e is only dependent on its free variables and can be calculated from the meaning of its subexpressions. For example, the meaning of a phrase **let** $x = e_1$ **in** e_2 **end** only depends on the free variables of e_1 and on the free variables of e_2 *other than* x. Hence, languages with referential transparency allow one to replace a subexpression by some semantically equivalent phrase without altering the meaning of the surrounding code. Since semantic preservation is a core requirement of program transformations, the suitability of SSA for formulating and implementing such transformations can be explained by the proximity of SSA to functional languages.

6.1.2 *Control Flow: Continuations*

The correspondence between let-bindings and points of variable definition in assignments extends to other aspects of program structure, in particular to code in *continuation-passing-style* (CPS), a program representation routinely used in compilers for functional languages [9].

Satisfying a roughly similar purpose as return addresses or function pointers in imperative languages, a continuation specifies how the execution should proceed once the evaluation of the current code fragment has terminated. Syntactically, continuations are expressions that may occur in functional position (i.e., are typically applied to argument expressions), as is the case for the variable k in the following modification from code (6.2):

$$
\begin{aligned}
&\textbf{let } v = 3 \textbf{ in} \\
&\quad \textbf{let } y = (\textbf{let } v = 2 \times v \textbf{ in } 4 \times v \textbf{ end}) \\
&\quad \textbf{in } k(y \times v) \textbf{ end} \\
&\textbf{end}
\end{aligned}
\qquad (6.4)
$$

In effect, k represents any function that may be applied to the result of expression (6.2).

Surrounding code may specify the concrete continuation by binding k to a suitable expression. It is common practice to write these continuation-defining expressions in λ-notation, i.e., in the form $\lambda x.e$ where x typically occurs free in e. The effect of the expression is to act as the (unnamed) function that sends x to $e(x)$, i.e., formal parameter x represents the place-holder for the argument to which the continuation is applied. Note that x is α-renameable, as λ acts as a binder. For example, a client of the above code fragment wishing to multiply the result by 2 may insert code (6.4) in the e_2 position of a let-binding for k that contains $\lambda x.\, 2 \times x$ in its e_1-position, as in the following code:

$$
\begin{aligned}
&\textbf{let } k = \lambda x.\, 2 \times x \\
&\textbf{in let } v = 3 \textbf{ in} \\
&\quad\quad \textbf{let } y = (\textbf{let } z = 2 \times v \textbf{ in } 4 \times z \textbf{ end}) \\
&\quad\quad \textbf{in } k(y \times v) \textbf{ end} \\
&\quad\textbf{end} \\
&\textbf{end}
\end{aligned}
\qquad (6.5)
$$

When the continuation k is applied to the argument $y \times v$, the (dynamic) value $y \times v$ (i.e., 72) is substituted for x in the expression $2 \times x$, just like in an ordinary function application.

Alternatively, the client may wrap fragment (6.4) in a function definition with formal argument k and construct the continuation in the calling code, where he would be free to choose a different name for the continuation-representing variable:

$$
\begin{aligned}
&\textbf{function } f(k) = \\
&\quad \textbf{let } v = 3 \textbf{ in} \\
&\qquad \textbf{let } y = (\textbf{let } z = 2 \times v \textbf{ in } 4 \times z \textbf{ end}) \\
&\qquad \textbf{in } k(y \times v) \textbf{ end} \\
&\quad \textbf{end} \\
&\quad \textbf{in let } k = \lambda x.\, 2 \times x \textbf{ in } f(k) \textbf{ end} \\
&\textbf{end}
\end{aligned}
\tag{6.6}
$$

This makes CPS form a discipline of programming using higher-order functions, as continuations are constructed "on the fly" and communicated as arguments of other function calls. Typically, the caller of f is itself parametric in *its* continuation, as in

$$
\begin{aligned}
&\textbf{function } g(k) = \\
&\quad \textbf{let } k' = \lambda x.\, k(x + 7) \textbf{ in } f(k') \textbf{ end}
\end{aligned}
\tag{6.7}
$$

where f is invoked with a newly constructed continuation k' that applies the addition of 7 to its formal argument x (which at runtime will hold the result of f) before passing the resulting value on as an argument to the outer continuation k. In a similar way, the function

$$
\begin{aligned}
&\textbf{function } h(y, k) = \\
&\quad \textbf{let } x = 4 \textbf{ in} \\
&\qquad \textbf{let } k' = \lambda z.\, k(z \times x) \\
&\qquad \textbf{in if } y > 0 \\
&\qquad\quad \textbf{then let } z = y \times 2 \textbf{ in } k'(z) \textbf{ end} \\
&\qquad\quad \textbf{else let } z = 3 \textbf{ in } k'(z) \textbf{ end} \\
&\qquad \textbf{end} \\
&\quad \textbf{end}
\end{aligned}
\tag{6.8}
$$

constructs from k a continuation k' that is invoked (with different arguments) in each branch of the conditional. In effect, the sharing of k' amounts to the definition of a control-flow merge point, as indicated by the CFG corresponding to h in Fig. 6.1a. Contrary to the functional representation, the top-level continuation parameter k is not explicitly visible in the CFG—it roughly corresponds to the frame slot that holds the return address in an imperative procedure call.

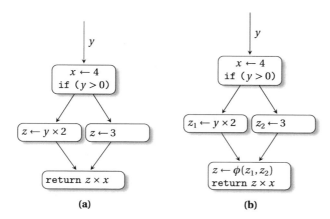

Fig. 6.1 Control-flow graph for code (6.8) (**a**), and SSA representation (**b**)

The SSA form of this CFG is shown in Fig. 6.1b. If we apply similar renamings of z to z_1 and z_2 in the two branches of (6.8), we obtain the following fragment:

$$
\begin{aligned}
&\textbf{function } h(y, k) = \\
&\quad \textbf{let } x = 4 \textbf{ in} \\
&\qquad \textbf{let } k' = \lambda z.\, k(z \times x) \\
&\qquad \textbf{in if } y > 0 \\
&\qquad\quad \textbf{then let } z_1 = y \times 2 \textbf{ in } k'(z_1) \textbf{ end} \\
&\qquad\quad \textbf{else let } z_2 = 3 \textbf{ in } k'(z_2) \textbf{ end} \\
&\qquad \textbf{end} \\
&\quad \textbf{end}
\end{aligned}
\tag{6.9}
$$

We observe that the role of the formal parameter z of continuation k' is exactly that of a ϕ-function: to unify the arguments stemming from various calls sites by binding them to a common name for the duration of the ensuing code fragment—in this case just the return expression. As expected from the above understanding of scope and dominance, the scopes of the bindings for z_1 and z_2 coincide with the dominance regions of the identically named imperative variables: both terminate at the point of function invocation/jump to the control-flow merge point.

The fact that transforming (6.8) into (6.9) only involves the referentially transparent process of α-renaming indicates that program (6.8) already contains the essential structural properties that SSA distills from an imperative program.

Programs in CPS equip *all* function declarations with continuation arguments. By interspersing ordinary code with continuation-forming expressions as shown above, they model the flow of control exclusively by communicating, constructing, and invoking continuations.

6.1.3   Control Flow: Direct Style

An alternative to the explicit passing of continuation terms via additional function arguments is the *direct style*, in which we represent code as a set of locally named tail-recursive functions, for which the last operation is a call to a function, eliminating the need to save the return address.

In direct style, no continuation terms are constructed dynamically and then passed as function arguments, and we hence exclusively employ λ-free function definitions in our representation. For example, code (6.8) may be represented as

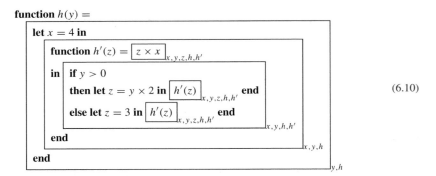

$$(6.10)$$

where the local function h' plays a similar role to the continuation k' and is jointly called from both branches. In contrast to the CPS representation, however, the body of h' returns its result directly rather than by passing it on as an argument to some continuation. Also note that neither the declaration of h nor that of h' contains additional continuation parameters. Thus, rather than handing its result directly over to some caller-specified receiver (as communicated by the continuation argument k), h simply returns control back to the caller, who is then responsible for any further execution. Roughly speaking, the effect is similar to the imperative compilation discipline of always setting the return address of a procedure call to the instruction pointer immediately following the call instruction.

A stricter format is obtained if the granularity of local functions is required to be that of basic blocks:

$$
\begin{aligned}
&\textbf{function } h(y) = \\
&\quad \textbf{let } x = 4 \textbf{ in} \\
&\qquad \textbf{function } h'(z) = z \times x \\
&\qquad \textbf{in if } y > 0 \\
&\qquad\quad \textbf{then function } h_1() = \textbf{let } z = y \times 2 \textbf{ in } h'(z) \textbf{ end} \\
&\qquad\qquad\quad \textbf{in } h_1() \textbf{ end} \\
&\qquad\quad \textbf{else function } h_2() = \textbf{let } z = 3 \textbf{ in } h'(z) \textbf{ end} \\
&\qquad\qquad\quad \textbf{in } h_2() \textbf{ end} \\
&\quad \textbf{end} \\
&\textbf{end}
\end{aligned}
\tag{6.11}
$$

Now, function invocations correspond precisely to jumps, reflecting more directly the CFG from Fig. 6.1.

The choice between CPS and direct style is orthogonal to the granularity level of functions: both CPS and direct style are compatible with the strict notion of basic blocks but also with more relaxed formats such as extended basic blocks. In the extreme case, *all* control-flow points are explicitly named, i.e., one local function or continuation is introduced per instruction. The exploration of the resulting design space is ongoing and has so far not yet led to a clear consensus in the literature. In our discussion below, we employ the arguably easier-to-read direct style, but the gist of the discussion applies equally well to CPS.

Independent of the granularity level of local functions, the process of moving from the CFG to the SSA form is again captured by suitably α-renaming the bindings of z in h_1 and h_2:

$$
\begin{aligned}
&\textbf{function } h(y) = \\
&\quad \textbf{let } x = 4 \textbf{ in} \\
&\qquad \textbf{function } h'(z) = z \times x \\
&\qquad \textbf{in if } y > 0 \\
&\qquad\quad \textbf{then function } h_1() = \textbf{let } z_1 = y \times 2 \textbf{ in } h'(z_1) \textbf{ end} \\
&\qquad\qquad\quad \textbf{in } h_1() \textbf{ end} \\
&\qquad\quad \textbf{else function } h_2() = \textbf{let } z_2 = 3 \textbf{ in } h'(z_2) \textbf{ end} \\
&\qquad\qquad\quad \textbf{in } h_2() \textbf{ end} \\
&\quad \textbf{end} \\
&\textbf{end}
\end{aligned}
\tag{6.12}
$$

Again, the role of the formal parameter z of the control-flow merge point function h' is identical to that of a ϕ-function. In accordance with the fact that the basic blocks representing the arms of the conditional do not contain ϕ-functions, the local functions h_1 and h_2 have empty parameter lists—the free occurrence of y in the body of h_1 is bound at the top level by the formal argument of h.

6.1.4 Let-Normal Form

For both direct style and CPS the correspondence to SSA is most pronounced for code in *let-normal form*: Each intermediate result must be explicitly named by a variable, and function arguments must be names or constants. Syntactically, let-normal form isolates basic instructions in a separate category of primitive terms a and then requires let-bindings to be of the form **let** $x = a$ **in** e **end**. In particular, neither jumps (conditional or unconditional) nor let-bindings are primitive. Let-normalized form is obtained by repeatedly rewriting code as follows:

$$
\begin{array}{l}
\mathbf{let}\ x = \mathbf{let}\ y = e\ \mathbf{in}\ \boxed{e'}_{y}\ \mathbf{end} \\[2pt]
\mathbf{in}\ \boxed{e''}_{x} \\[2pt]
\mathbf{end}
\end{array}
\quad\text{into}\quad
\begin{array}{l}
\mathbf{let}\ y = e \\[2pt]
\mathbf{in}\ \boxed{\mathbf{let}\ x = e'\ \mathbf{in}\ \boxed{e''}_{x,y}\ \mathbf{end}}_{y} \\[2pt]
\mathbf{end,}
\end{array}
$$

subject to the side condition that y is not free in e''. For example, let-normalizing code (6.3) pulls the let-binding for z to the outside of the binding for y, yielding

$$
\begin{array}{l}
\mathbf{let}\ v = 3\ \mathbf{in} \\[2pt]
\quad \boxed{\begin{array}{l}
\mathbf{let}\ z = 2 \times v\ \mathbf{in} \\[2pt]
\quad \boxed{\mathbf{let}\ y = 4 \times z\ \mathbf{in}\ \boxed{y \times v}_{v,y,z}\ \mathbf{end}}_{v,z} \\[2pt]
\mathbf{end}
\end{array}}_{v} \\[2pt]
\mathbf{end}
\end{array}
\tag{6.13}
$$

Programs in let-normal form thus do not contain let-bindings in the e_1-position of outer let-expressions. The stack discipline in which let-bindings are managed is simplified as scopes are nested inside each other. While still enjoying referential transparency, let-normal code is in closer correspondence to imperative code than non-normalized code as the chain of nested let-bindings directly reflects the sequence of statements in a basic block, interspersed occasionally by the definition of continuations or local functions. Exploiting this correspondence, we apply a simplified notation in the rest of this chapter where the scope-identifying boxes (including their indices) are omitted and chains of let-normal bindings are abbreviated by single comma-separated (and ordered) let-blocks. Using this convention, code (6.13) becomes

$$
\begin{array}{l}
\mathbf{let}\ v = 3, \\[2pt]
\quad z = 2 \times v, \\[2pt]
\quad y = 4 \times z \\[2pt]
\mathbf{in}\ y \times v\ \mathbf{end}
\end{array}
\tag{6.14}
$$

Summarizing our discussion up to this point, Table 6.1 collects some correspondences between functional and imperative/SSA concepts.

Table 6.1 Correspondence pairs between functional form and SSA (part I)

Functional concept	Imperative/SSA concept
Variable binding in let $\cdot$ · · ·	Assignment (point of definition)
α-renaming $\cdot$ · · ·	Variable renaming
Unique association of binding occurrences to uses $\cdot$ · · ·	Unique association of defs to uses
Formal parameter of continuation/local function $\cdot$ · · ·	ϕ-function (point of definition)
Lexical scope of bound variable $\cdot$ · · ·	Dominance region

Table 6.2 Correspondence pairs between functional form and SSA: program structure

Functional concept	Imperative/SSA concept
Immediate subterm relationship $\cdot$ · · ·	Direct control-flow successor relationship
Arity of function f_i $\cdot$ · · ·	Number of ϕ-functions at beginning of b_i
Distinctness of formal param. of f_i $\cdot$ · · ·	Distinctness of LHS-variables in the ϕ-block of b_i
Number of call sites of function f_i $\cdot$ · · ·	Arity of ϕ-functions in block b_i
Parameter lifting/dropping $\cdot$ · · ·	Addition/removal of ϕ-function
Block floating/sinking $\cdot$ · · ·	Reordering according to dominator tree structure
Potential nesting structure $\cdot$ · · ·	Dominator tree

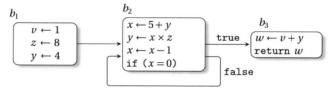

Fig. 6.2 Functional construction of SSA: running example

6.2   Functional Construction and Destruction of SSA

The relationship between SSA and functional languages is extended by the correspondences shown in Table 6.2. We discuss some of these aspects by considering the translation into SSA, using the program in Fig. 6.2 as a running example.

6.2.1   Initial Construction Using Liveness Analysis

A simple way to represent this program in let-normalized direct style is to introduce one function f_i for each basic block b_i. The body of each f_i arises by introducing one let-binding for each assignment and converting jumps into function calls. In order to determine the formal parameters of these functions we perform a liveness analysis. For each basic block b_i, we choose an arbitrary enumeration of its live-in variables. We then use this enumeration as the list of formal parameters in the declaration of the function f_i, and also as the list of actual arguments in calls to f_i. We collect all function definitions in a block of mutually tail-recursive functions at the top level:

$$
\begin{aligned}
\textbf{function } f_1() &= \textbf{let } v = 1, \; z = 8, \; y = 4 \\
&\quad \textbf{in } f_2(v, z, y) \textbf{ end} \\
\textbf{and } f_2(v, z, y) &= \textbf{let } x = 5 + y, \; y = x \times z, \; x = x - 1 \\
&\quad \textbf{in if } x = 0 \textbf{ then } f_3(y, v) \textbf{ else } f_2(v, z, y) \textbf{ end} \\
\textbf{and } f_3(y, v) &= \textbf{let } w = y + v \textbf{ in } w \textbf{ end} \\
\textbf{in } f_1() \textbf{ end}
\end{aligned}
\tag{6.15}
$$

The resulting program has the following properties:

- All function declarations are *closed*: The free variables of their bodies are contained in their formal parameter lists;[1]
- Variable names are not unique, but the unique association of definitions to uses is satisfied;
- Each subterm e_2 of a let-binding **let** $x = e_1$ **in** e_2 **end** corresponds to the direct control-flow successor of the assignment to x.

If desired, we may α-rename to make names globally unique. As the function declarations in code (6.15) are closed, all variable renamings are independent from each other. The resulting code (6.16) corresponds precisely to the SSA program shown in Fig. 6.3 (see also Table 6.2): Each formal parameter of a function f_i is the target of one ϕ-function for the corresponding block b_i. The arguments of these ϕ-functions are the arguments in the corresponding positions in the calls to f_i. As the number of arguments in each call to f_i coincides with the number of formal parameters of f_i, the ϕ-functions in b_i are all of the same arity, namely the number of call sites to f_i. In order to coordinate the relative positioning of the arguments of the ϕ-functions, we choose an arbitrary enumeration of these call sites.

$$
\begin{aligned}
\textbf{function } f_1() &= \textbf{let } v_1 = 1, \; z_1 = 8, \; y_1 = 4 \\
&\quad \textbf{in } f_2(v_1, z_1, y_1) \textbf{ end} \\
\textbf{and } f_2(v_2, z_2, y_2) &= \textbf{let } x_1 = 5 + y_2, \; y_3 = x_1 \times z_2, \; x_2 = x_1 - 1 \\
&\quad \textbf{in if } x_2 = 0 \textbf{ then } f_3(y_3, v_2) \textbf{ else } f_2(v_2, z_2, y_3) \textbf{ end} \\
\textbf{and } f_3(y_4, v_3) &= \textbf{let } w_1 = y_4 + v_3 \textbf{ in } w_1 \textbf{ end} \\
\textbf{in } f_1() \textbf{ end}
\end{aligned}
\tag{6.16}
$$

---

[1] Apart from the function identifiers f_i, which can always be chosen distinct from the variables.

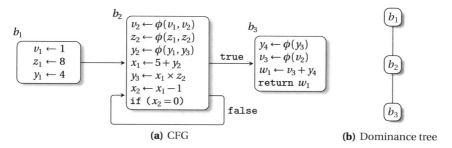

Fig. 6.3 Pruned (non-minimal) SSA form

Under this perspective, the above construction of parameter lists amounts to equipping each b_i with ϕ-functions for all its live-in variables, with subsequent renaming of the variables. Thus, the above method corresponds to the construction of *pruned* (but not minimal) SSA—see Chap. 2.

While resulting in a legal SSA program, the construction clearly introduces more ϕ-functions than necessary. Each superfluous ϕ-function corresponds to the situation where all call sites to some function f_i pass identical arguments. The technique for eliminating such arguments is called λ-*dropping* and is the inverse of the more widely known transformation λ-*lifting*.

6.2.2 λ-Dropping

λ-dropping may be performed before or after variable names are made distinct, but for our purpose, the former option is more instructive. The transformation consists of two phases, *block sinking* and *parameter dropping*.

6.2.2.1 Block Sinking

Block sinking analyses the static call structure to identify which function definitions may be moved inside each other. For example, whenever our set of function declarations contains definitions $f(x_1, \ldots, x_n) = e_f$ and $g(y_1, \ldots, y_m) = e_g$ where $f \neq g$ and such that all calls to f occur in e_f or e_g, we can move the declaration for f into that of g—note the similarity to the notion of dominance. If applied aggressively, block sinking indeed amounts to making the entire dominance tree structure explicit in the program representation. In particular, algorithms for computing the dominator tree from a CFG discussed elsewhere in this book can be applied to identify block sinking opportunities, where the CFG is given by the call graph of functions.

In our example (6.15), f_3 is only invoked from within f_2, and f_2 is only called in the bodies of f_2 and f_1 (see the dominator tree in Fig. 6.3 (right)). We may thus move the definition of f_3 into that of f_2, and the latter one into f_1.

Several options exist as to where f should be placed in its host function. The first option is to place f at the beginning of g, by rewriting to

$$\textbf{function } g(y_1, \ldots, y_m) = \textbf{function } f(x_1, \ldots, x_n) = e_f$$
$$\textbf{in } e_g \textbf{ end}$$

This transformation does not alter the semantic of the code, as the declaration of f is closed: moving f into the scope of the formal parameters $y_1, \ldots, y_m$ (and also into the scope of g itself) does not alter the bindings to which variable uses inside e_f refer.

Applying this transformation to example (6.15) yields the following code:

$$
\begin{aligned}
&\textbf{function } f_1() = \\
&\quad \textbf{function } f_2(v, z, y) = \\
&\qquad \textbf{function } f_3(y, v) = \textbf{let } w = y + v \textbf{ in } w \textbf{ end} \\
&\qquad \textbf{in let } x = 5 + y, \; y = x \times z, \; x = x - 1 \\
&\qquad\quad \textbf{in if } x = 0 \textbf{ then } f_3(y, v) \textbf{ else } f_2(v, z, y) \textbf{ end} \\
&\qquad \textbf{end} \\
&\quad \textbf{in let } v = 1, \; z = 8, \; y = 4 \textbf{ in } f_2(v, z, y) \textbf{ end} \\
&\quad \textbf{end} \\
&\textbf{in } f_1() \textbf{ end}
\end{aligned}
\qquad (6.17)
$$

An alternative strategy is to insert f near the end of its host function g, in the vicinity of the calls to f. This brings the declaration of f additionally into the scope of all let-bindings in e_g. Again, referential transparency and preservation of semantics are respected as the declaration on f is closed. In our case, the alternative strategy yields the following code:

$$
\begin{aligned}
&\textbf{function } f_1() = \\
&\quad \textbf{let } v = 1, \; z = 8, \; y = 4 \\
&\quad \textbf{in function } f_2(v, z, y) = \\
&\qquad \textbf{let } x = 5 + y, \; y = x \times z, \; x = x - 1 \\
&\qquad \textbf{in if } x = 0 \\
&\qquad\quad \textbf{then function } f_3(y, v) = \textbf{let } w = y + v \textbf{ in } w \textbf{ end} \\
&\qquad\qquad\quad \textbf{in } f_3(y, v) \textbf{ end} \\
&\qquad\quad \textbf{else } f_2(v, z, y) \\
&\qquad \textbf{end} \\
&\quad \textbf{in } f_2(v, z, y) \textbf{ end} \\
&\quad \textbf{end} \\
&\textbf{in } f_1() \textbf{ end}
\end{aligned}
\qquad (6.18)
$$

In general, one would insert f directly prior to its call if g contains only a single call site for f. In the event that g contains multiple call sites for f, these are (due to their tail-recursive positioning) in different arms of a conditional, and we would insert f directly prior to this conditional.

Both outlined placement strategies result in code whose nesting structure reflects the dominance relationship of the imperative code. In our example, code (6.17) and (6.18) both nest f_3 inside f_2 inside f_1, in accordance with the dominator tree of the imperative program show on Fig. 6.3.

6.2.2.2 Parameter Dropping

The second phase of λ-dropping, *parameter dropping*, removes superfluous parameters based on the syntactic scope structure. Removing a parameter x from the declaration of some function f has the effect that any use of x inside the body of f will not be bound by the declaration of f any longer, but by the (innermost) binding for x that *contains* the declaration of f. In order to ensure that removing the parameter does not alter the meaning of the program, we thus have to ensure that this f-containing binding for x also contains any call to f, since the binding applicable at the call site determines the value that is passed as the actual argument (before x is deleted from the parameter list).

For example, the two parameters of f_3 in (6.18) can be removed without altering the meaning of the code, as we can statically predict the values they will be instantiated with: Parameter y will be instantiated with the result of $x \times z$ (which is bound to y in the first line in the body of f_2), and v will always be bound to the value passed via the parameter v of f_2. In particular, the bindings for y and v at the *declaration* site of f_3 are identical to those applicable at the *call* to f_3.

In general, we may drop a parameter x from the declaration of a possibly recursive function f if the following two conditions are met:

1. The tightest scope for x enclosing the declaration of f coincides with the tightest scope for x surrounding each call site to f outside of its declaration.
2. The tightest scope for x enclosing any recursive call to f (i.e., a call to f in the body of f) is the one associated with the formal parameter x in the declaration of f.

The rationale for these clauses is that removing x from f's parameter list means that any free occurrence of x in f's body is now bound outside of f's declaration. Therefore, for each call to f to be correct, one needs to ensure that this outside binding coincides with the one containing the call.

Similarly, we may simultaneously drop a parameter x occurring in *all* declarations of a block of mutually recursive functions $f_1, \ldots, f_n$, if the scope for x at the point of declaration of the block coincides with the tightest scope in force at any call site to some f_i outside the block, and if in each call to some f_i inside some f_j, the tightest scope for x is the one associated with the formal parameter x of f_j. In both cases, dropping a parameter means removing it from the list of formal parameter lists of the function declarations concerned, and also from the argument lists of the corresponding function calls.

In code (6.18), these conditions sanction the removal of both parameters from the non-recursive function f_3. The scope applicable for v at the site of declaration of f_3 and also at its call site is the one rooted at the formal parameter v of f_2. In case of y, the common scope is the one rooted at the let-binding for y in the body of f_2. We thus obtain the following code:

$$
\begin{aligned}
&\textbf{function } f_1() \;=\\
&\quad \textbf{let } v = 1,\; z = 8,\; y = 4\\
&\quad \textbf{in function } f_2(v, z, y) =\\
&\qquad\quad \textbf{let } x = 5 + y,\; y = x \times z,\; x = x - 1\\
&\qquad\quad \textbf{in if } x = 0\\
&\qquad\qquad\quad \textbf{then function } f_3() = \textbf{let } w = y + v \textbf{ in } w \textbf{ end}\\
&\qquad\qquad\qquad\quad \textbf{in } f_3() \textbf{ end}\\
&\qquad\qquad\quad \textbf{else } f_2(v, z, y)\\
&\qquad\quad \textbf{end}\\
&\qquad \textbf{in } f_2(v, z, y) \textbf{ end}\\
&\quad \textbf{end}\\
&\quad \textbf{in } f_1() \textbf{ end}
\end{aligned}
\tag{6.19}
$$

Considering the recursive function f_2 next we observe that the recursive call is in the scope of the let-binding for y in the body of f_2, preventing us from removing y. In contrast, neither v nor z has binding occurrences in the body of f_2. The scopes applicable at the external call site to f_2 coincide with those applicable at its site of declaration and are given by the scopes rooted in the let-bindings for v and z. Thus, parameters v and z may be removed from f_2:

$$
\begin{aligned}
&\textbf{function } f_1() \;=\\
&\quad \textbf{let } v = 1,\; z = 8,\; y = 4\\
&\quad \textbf{in function } f_2(y) =\\
&\qquad\quad \textbf{let } x = 5 + y,\; y = x \times z,\; x = x - 1\\
&\qquad\quad \textbf{in if } x = 0\\
&\qquad\qquad\quad \textbf{then function } f_3() = \textbf{let } w = y + v \textbf{ in } w \textbf{ end}\\
&\qquad\qquad\qquad\quad \textbf{in } f_3() \textbf{ end}\\
&\qquad\qquad\quad \textbf{else } f_2(y)\\
&\qquad\quad \textbf{end}\\
&\qquad \textbf{in } f_2(y) \textbf{ end}\\
&\quad \textbf{end}\\
&\quad \textbf{in } f_1() \textbf{ end}
\end{aligned}
\tag{6.20}
$$

Interpreting the uniquely renamed variant of (6.20) back in SSA yields the desired code with a single ϕ-function, for variable y at the beginning of block b_2, see Fig. 6.4. The reason that this ϕ-function cannot be eliminated (the redefinition of y in the loop) is precisely the reason why y survives parameter dropping.

Given this understanding of parameter dropping we can also see why inserting functions near the end of their hosts during block sinking (as in code (6.18)) is in general preferable to inserting them at the beginning of their hosts (as in code (6.17)): The placement of function declarations in the vicinity of their calls potentially enables the dropping of more parameters, namely those that are let-bound in the body of the host function.

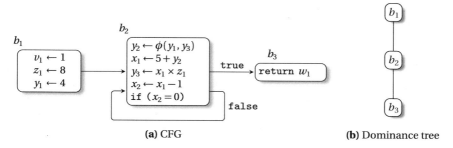

Fig. 6.4 SSA code after λ-dropping

An immediate consequence of this strategy is that blocks with a single direct predecessor indeed do not contain ϕ-functions. Such a block b_f is necessarily dominated by its direct predecessor b_g, hence we can always nest f inside g. Inserting f in e_g directly prior to its call site implies that condition (1) is necessarily satisfied for all parameters x of f. Thus, all parameters of f can be dropped and no ϕ-function is generated.

6.2.3 Nesting, Dominance, Loop-Closure

As we observed above, analysing whether function definitions may be nested inside one another is tantamount to analysing the imperative dominance structure: function f_i may be moved inside f_j exactly if all non-recursive calls to f_i come from within f_j, i.e., exactly if all paths from the initial program point to block b_i traverse b_j, i.e., exactly if b_j dominates b_i. This observation is merely the extension to function identifiers of our earlier remark that lexical scope coincides with the dominance region, and that points of definition/binding occurrences should dominate the uses. Indeed, functional languages do not distinguish between code and data when aspects of binding and use of variables are concerned, as witnessed by our use of the let-binding construct for binding code-representing expressions to the variables k in our syntax for CPS.

Thus, the dominator tree immediately suggests a function nesting scheme, where all children of a node are represented as a single block of mutually recursive function declarations.[2]

---

[2] Refinements of this representation will be sketched in Sect. 6.3.

The choice as to where functions are placed corresponds to variants of SSA. For example, in loop-closed SSA form (see Chaps. 14 and 10), SSA names that are defined in a loop must not be used outside the loop. To this end, special-purpose unary ϕ-nodes are inserted for these variables at the loop exit points. As the loop is unrolled, the arity of these trivial ϕ-nodes grows with the number of unrollings, and the program continuation is always supplied with the value the variable obtained in the final iteration of the loop. In our example, the only loop-defined variable used in f_3 is y—and we already observed in code (6.17) how we can prevent the dropping of y from the parameter list of f_3: we insert f_3 at the *beginning* of f_2, preceding the let-binding for y. Of course, we would still like to drop as many parameters from f_2 as possible, hence we apply the following placement policy during block sinking: Functions that are targets of loop-exiting function calls and have live-in variables that are defined in the loop are placed at the beginning of the loop headers. Other functions are placed at the end of their hosts. Applying this policy to our original program (6.15) yields (6.21).

$$
\begin{aligned}
&\textbf{function } f_1() = \\
&\quad \textbf{let } v = 1,\ z = 8,\ y = 4 \\
&\quad \textbf{in function } f_2(v, z, y) = \\
&\qquad\quad \textbf{function } f_3(y, v) = \textbf{let } w = y + v \textbf{ in } w \textbf{ end} \\
&\qquad\quad \textbf{in let } x = 5 + y,\ y = x \times z,\ x = x - 1 \\
&\qquad\qquad\quad \textbf{in if } x = 0 \textbf{ then } f_3(y, v) \textbf{ else } f_2(v, z, y) \textbf{ end} \\
&\qquad\quad \textbf{end} \\
&\qquad \textbf{in } f_2(v, z, y) \textbf{ end} \\
&\quad \textbf{end} \\
&\textbf{in } f_1() \textbf{ end}
\end{aligned}
\tag{6.21}
$$

We may now drop v (but not y) from the parameter list of f_3, and v and z from f_2, to obtain code (6.22).

$$
\begin{aligned}
&\textbf{function } f_1() = \\
&\quad \textbf{let } v = 1,\ z = 8,\ y = 4 \\
&\quad \textbf{in function } f_2(y) = \\
&\qquad\quad \textbf{function } f_3(y) = \textbf{let } w = y + v \textbf{ in } w \textbf{ end} \\
&\qquad\quad \textbf{in let } x = 5 + y,\ y = x \times z,\ x = x - 1 \\
&\qquad\qquad\quad \textbf{in if } x = 0 \textbf{ then } f_3(y) \textbf{ else } f_2(y) \textbf{ end} \\
&\qquad\quad \textbf{end} \\
&\qquad \textbf{in } f_2(y) \textbf{ end} \\
&\quad \textbf{end} \\
&\textbf{in } f_1() \textbf{ end}
\end{aligned}
\tag{6.22}
$$

The SSA form corresponding to (6.22) contains the desired loop-closing ϕ-node for y at the beginning of b_3, as shown in Fig. 6.5a. The nesting structure of both (6.21) and (6.22) coincides with the dominance structure of the original imperative code and its loop-closed SSA form.

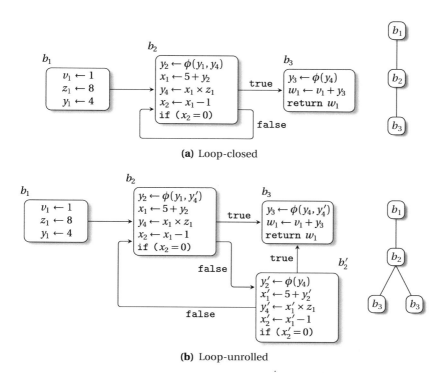

(a) Loop-closed

(b) Loop-unrolled

Fig. 6.5 Loop-closed (**a**) and loop-unrolled (**b**) forms of running example program, corresponding to codes (6.22) and (6.23), respectively

We unroll the loop by duplicating the body of f_2, *without duplicating the declaration of* f_3:

$$
\begin{aligned}
&\textbf{function } f_1() = \\
&\quad \textbf{let } v = 1, \ z = 8, \ y = 4 \\
&\quad \textbf{in function } f_2(y) = \\
&\qquad \textbf{function } f_3(y) = \textbf{let } w = y + v \textbf{ in } w \textbf{ end} \\
&\qquad \textbf{in let } x = 5 + y, \ y = x \times z, \ x = x - 1 \\
&\qquad\quad \textbf{in if } x = 0 \textbf{ then } f_3(y) \\
&\qquad\quad \textbf{else function } f_{2'}(y) = \\
&\qquad\qquad\quad \textbf{let } x = 5 + y, \ y = x \times z, \ x = x - 1 \\
&\qquad\qquad\quad \textbf{in if } x = 0 \textbf{ then } f_3(y) \textbf{ else } f_2(y) \textbf{ end} \\
&\qquad\qquad \textbf{in } f_{2'}(y) \textbf{ end} \\
&\qquad \textbf{end} \\
&\quad \textbf{in } f_2(y) \textbf{ end} \\
&\quad \textbf{end} \\
&\textbf{in } f_1() \textbf{ end}
\end{aligned}
\qquad (6.23)
$$

Both calls to f_3 are in the scope of the declaration of f_3 and contain the appropriate loop-closing arguments. In the SSA reading of this code—shown in Fig. 6.5b— the first instruction in b_3 has turned into a non-trivial ϕ-node. As expected, the parameters of this ϕ-node correspond to the two control-flow arcs leading into b_3, one for each call site to f_3 in code (6.23). Moreover, the call and nesting structure of (6.23) is indeed in agreement with the control flow and dominance structure of the loop-unrolled SSA representation.

6.2.4 Destruction of SSA

The above example code excerpts where variables are not made distinct exhibit a further pattern: The argument list of any call coincides with the list of formal parameters of the invoked function. This discipline is not enjoyed by functional programs in general, and is often destroyed by optimizing program transformations. However, programs that do obey this discipline can be immediately converted to imperative non-SSA form. Thus, the task of SSA destruction amounts to converting a functional program with arbitrary argument lists into one where argument lists and formal parameter lists coincide for each function. This can be achieved by introducing additional let-bindings of the form **let** $x = y$ **in** e **end**. For example, a call $f(v, z, y)$ where f is declared as **function** $f(x, y, z) = e$ may be converted to

$$\textbf{let } x = v, \ a = z, \ z = y, \ y = a \textbf{ in } f(x, y, z)$$

in correspondence to the move instructions introduced in imperative formulations of SSA destruction (see Chaps. 3 and 21). Appropriate transformations can be formulated as manipulations of the functional representation, although the target format is not immune to α-renaming and thus only syntactically a functional language. For example, we can give a local algorithm that considers each call site individually and avoids the "lost-copy" and "swap" problems (cf. Chap. 21): Instead of introducing let-bindings for all parameter positions of a call, the algorithm scales with the number and size of cycles that span identically named arguments and parameters (like the cycle between y and z above), and employs a single additional variable (called a in the above code) to break all these cycles one by one.

6.3 Refined Block Sinking and Loop Nesting Forests

As discussed when outlining λ-dropping, block sinking is governed by the dominance relation between basic blocks. Thus, a typical dominance tree with root b and subtrees rooted at $b_1, \ldots, b_n$ is most naturally represented as a block of function declarations for the f_i, nested inside the declaration of f:

$$
\begin{aligned}
&\textbf{function } f(\ldots) = \textbf{let} \ldots < \textit{body of } b > \ldots \textbf{in} \\
&\quad \textbf{function } f_1(\ldots) = e_1 \qquad \triangleright \textit{body of } b_1, \textit{with calls to } b, b_i \\
&\qquad\qquad \vdots \\
&\quad \textbf{and } f_n(\ldots) = e_n \qquad \triangleright \textit{body of } b_n, \textit{with calls to } b, b_i \\
&\textbf{in} \ldots < \textit{calls to } b, b_i \textit{ from } b > \ldots \textbf{end}
\end{aligned}
\tag{6.24}
$$

By exploiting additional control-flow structure between the b_i, it is possible to obtain refined placements, namely placements that correspond to notions of *loop nesting forests* that have been identified in the SSA literature.

These refinements arise if we enrich the above dominance tree by adding arrows $b_i \rightarrow b_j$ whenever the CFG contains a directed edge from one of the dominance successors of b_i (i.e., the descendants of b_i in the dominance tree) to b_j.

In the case of a *reducible* CFG, the resulting graph contains only trivial loops. Ignoring these self-loops, we perform a post-order DFS (or more generally a reverse topological ordering) among the b_i and stagger the function declarations according to the resulting order. As an example, consider the CFG in Fig. 6.6a and its enriched dominance tree shown in Fig. 6.6b. A possible (but not unique) ordering of the children of b is $[b_5, b_1, b_3, b_2, b_4]$, resulting in the nesting shown in code (6.25).

$$
\begin{aligned}
&\textbf{function } f(\ldots) = \textbf{let} \ldots < \textit{body of } f > \ldots \textbf{in} \\
&\quad \textbf{function } f_5(\ldots) = e_5 \\
&\quad \textbf{in function } f_1(\ldots) = e_1 \qquad \triangleright \textit{contains calls to } f \textit{ and } f_5 \\
&\qquad \textbf{in function } f_3(\ldots) = e_3 \\
&\qquad\quad \textbf{in function } f_2(\ldots) = \\
&\qquad\qquad \textbf{let} \ldots < \textit{body of } f_2 > \ldots \\
&\qquad\qquad \textbf{in function } g(\ldots) = e_g \qquad \triangleright \textit{contains call to } f_3 \\
&\qquad\qquad \textbf{in} \ldots < \textit{calls to } f_5 \textit{ and } g > \ldots \textbf{end} \\
&\qquad\quad \textbf{in function } f_4(\ldots) = e_4 \quad \triangleright \textit{contains calls to } f_3 \textit{ and } f_4 \\
&\qquad\quad \textbf{in} \ldots < \textit{calls to } f_1, f_2, f_4, f_5 > \ldots \textbf{end}
\end{aligned}
\tag{6.25}
$$

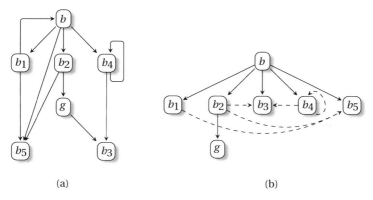

(a) (b)

Fig. 6.6 Example placement for reducible flow graphs. (**a**) Control-flow graph; (**b**) Overlay of CFG (dashed edges) arcs between dominated blocks onto dominance graph (solid edges)

The code respects the dominance relationship in much the same way as the naive placement, but additionally makes f_1 inaccessible from within e_5, and makes f_3 inaccessible from within f_1 or f_5. As the reordering does not move function declarations *inside* each other (in particular: no function declaration is brought into or moved out of the scope of the formal parameters of any other function) the reordering does not affect the potential to subsequently perform parameter dropping.

Declaring functions using λ-abstraction brings further improvements. This enables us not only to syntactically distinguish between loops and non-recursive control-flow structures using the distinction between **let** and **letrec** present in many functional languages, but also to further restrict the visibility of function names. Indeed, while b_3 is immediately dominated by b in the above example, its only control-flow predecessors are b_2/g and b_4. We would hence like to make the declaration of f_3 local to the *tuple* (f_2, f_4), i.e., invisible to f. This can be achieved by combining **let/letrec** bindings with pattern matching, if we insert the shared declaration of f_3 *between* the declaration of the names f_2 and f_4 and the λ-bindings of their formal parameters p_i:

$$
\begin{aligned}
&\textbf{letrec } f = \lambda\, p.\, \textbf{let} \ldots < \textit{body of } f > \ldots \textbf{in} \\
&\quad \textbf{let } f_5 = \lambda\, p_5.\, e_5 \\
&\quad \textbf{in let } f_1 = \lambda\, p_1.\, e_1 \qquad\qquad \triangleright \textit{contains calls to } f \textit{ and } f_5 \\
&\qquad \textbf{in letrec } (f_2, f_4) = \\
&\qquad\quad \textbf{let } f_3 = \lambda\, p_3.\, e_3 \\
&\qquad\quad \textbf{in } (\lambda\, p_2.\, \textbf{let} \ldots < \textit{body of } f_2 > \ldots \textbf{in} \qquad\qquad\qquad (6.26) \\
&\qquad\qquad\qquad \textbf{let } g = \lambda\, p_g.\, e_g \qquad \triangleright \textit{contains call to } f_3 \\
&\qquad\qquad\qquad \textbf{in} \ldots < \textit{calls to } f_5 \textit{ and } g > \ldots \textbf{end} \\
&\qquad\qquad , \lambda\, p_4.\, e_4) \qquad\qquad \triangleright \textit{contains calls to } f_3 \textit{ and } f_4 \\
&\qquad \textbf{end} \qquad \triangleright \textit{declaration of tuple } (f_2, f_4) \textit{ ends here} \\
&\quad \textbf{in} \ldots < \textit{calls to } f_1, f_2, f_4, f_5 > \ldots \textbf{end}
\end{aligned}
$$

The recursiveness of f_4 is inherited by the function pair (f_2, f_4) but f_3 remains non-recursive. In general, the role of f_3 is played by any merge point b_i that is not directly called from the dominator node b.

In the case of *irreducible* CFGs, the enriched dominance tree is no longer acyclic (even when ignoring self-loops). In this case, the functional representation not only depends on the chosen DFS order but additionally on the partitioning of the enriched graph into loops. As each loop forms a *strongly connected component* (SCC), different partitionings are possible, corresponding to different notions of *loop nesting forests*. Of the various loop nesting forest strategies proposed in the SSA literature [236], the scheme introduced by Steensgaard [271] is particularly appealing from a functional perspective.

In Steensgaard's notion of loops, the headers H of a loop $L = (B, H)$ are precisely the entry nodes of its body B, i.e., those nodes in B that have a direct predecessor outside of B. For example, G_0 shown in Fig. 6.7 contains the outer loop $L_0 = (\{u, v, w, x\}, \{u, v\})$, whose constituents B_0 are determined as the maximal

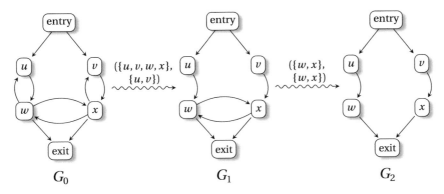

Fig. 6.7 Illustration of Steensgaard's construction of loop nesting forests

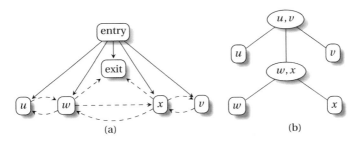

Fig. 6.8 Illustration of Steensgaard's construction of loop nesting forests: (**a**) CFG-enriched dominance tree; (**b**) resulting loop nesting forest

SCC of G_0. Removing the back edges of L_0 from G_0 (i.e., edges from B_0 to H_0) yields G_1, whose (only) SCC determines a further inner loop $L_1 = (\{w, x\}, \{w, x\})$. Removing the back edges of L_1 from G_1 results in the acyclic G_2, terminating the process.

Figure 6.8a shows the CFG-enriched dominance tree of G_0. The body of loop L_0 is easily identified as the maximal SCC, and likewise the body of L_1 once the cycles (u, w) and (x, v) are broken by the removal of the back edges $w \to u$ and $x \to v$.

The loop nesting forest resulting from Steensgaard's construction is shown in Fig. 6.8b. Loops are drawn as ellipses decorated with the appropriate header nodes and nested in accordance with the containment relation $B_1 \subset B_0$ between the bodies.

In the functional representation, a loop $L = (B, \{h_1, \ldots, h_n\})$ yields a function declaration block for functions $h_1, \ldots, h_n$, with private declarations for the non-headers from $B \setminus H$. In our example, loop L_0 provides entry points for the headers u and v but not for its non-headers w and x. Instead, the loop comprised of the latter nodes, L_1, is nested inside the definition of L_0, in accordance with the loop nesting forest.

function entry$(\ldots) =$
  let $\ldots < $ *body of* **entry** $ > \ldots$
  in letrec $(u, v) = $ $\triangleright$ *define outer loop* L_0, *with headers* u, v
     letrec $(w, x) = $ $\triangleright$ *define inner loop* L_1, *with headers* w, x
        let exit $= \lambda p_{exit}. \ldots < $ *body of* **exit** $ >$
        in $(\lambda p_w. \ldots < $ *body of* w, *with calls to* u, x, *and* **exit** $ > \ldots$
          , $\lambda p_x. \ldots < $ *body of* x, *with calls to* w, v, *and* **exit** $ > \ldots)$ (6.27)
      end $\triangleright$ *end of inner loop*
    in $(\lambda p_u. \ldots < $ *body of* u, *with call to* w $ > \ldots$
     , $\lambda p_v. \ldots < $ *body of* v, *with call to* x $ > \ldots)$
    end $\triangleright$ *end of outer loop*
  in $\ldots < $ *calls from* **entry** *to* u *and* v $ > \ldots$

By placing L_1 inside L_0 according to the scheme from code (6.26) and making **exit** private to L_1, we obtain the representation (6.27), which captures all the essential information of Steensgaard's construction. Effectively, the functional reading of the loop nesting forest extends the earlier correspondence between the nesting of individual functions and the dominance relationship to groups of functions and basic blocks: loop L_0 *dominates* L_1 in the sense that any path from **entry** to a node in L_1 passes through L_0; more specifically, any path from **entry** to a *header* of L_1 passes through a *header* of L_0.

In general, each step of Steensgaard's construction may identify several loops, as a CFG may contain several maximal SCCs. As the bodies of these SCCs are necessarily non-overlapping, the construction yields a forest comprised of trees shaped like the loop nesting forest in Fig. 6.8b. As the relationship between the trees is necessarily acyclic, the declarations of the function declaration tuples corresponding to the trees can be placed according to the loop-extended notion of dominance.

6.4 Further Reading and Concluding Remarks

Further Reading
Shortly after the introduction of SSA, O'Donnell [213] and Kelsey [160] noted the correspondence between let-bindings and points of variable declaration and its extension to other aspects of program structure using continuation-passing style. Appel [10, 11] popularized the correspondence using a direct-style representation, building on his earlier experience with continuation-based compilation [9].

Continuations and low-level functional languages have been an object of intensive study since their inception about four decades ago [174, 294]. For retrospective accounts of the historical development, see Reynolds [246] and Wadsworth [299]. Early studies of CPS and direct style include work by Reynolds and Plotkin [229, 244, 245]. Two prominent examples of CPS-based compilers are those by Sussman et al. [279] and Appel [9]. An active area of research concerns the relative merit of the various functional representations, algorithms for formal conversion between these formats, and their efficient compilation to machine code, in particular with respect to their integration with program analyses and optimizing transformations [92, 161, 243]. A particularly appealing variant is that of Kennedy [161], where the approach to explicitly name control-flow points such as merge points is taken to its logical conclusion. By mandating *all* control-flow points to be explicitly named, a uniform representation is obtained that allows optimizing transformations to be implemented efficiently, avoiding the administrative overhead to which some of the alternative approaches are susceptible.

Occasionally, the term *direct style* refers to the combination of tail-recursive functions and let-normal form, and the conditions on the latter notion are strengthened so that only variables may appear as branch conditions. Variations of this discipline include *administrative normal form* (A-normal form, ANF) [121], B-form [280], and SIL [285].

Closely related to continuations and direct-style representation are *monadic* intermediate languages as used by Benton et al. [24] and Peyton-Jones et al. [223]. These partition expressions into categories of *values* and *computations*, similar to the isolation of primitive terms in let-normal form [229, 245]. This allows one to treat side-effects (memory access, IO, exceptions, etc.) in a uniform way, following Moggi [200], and thus simplifies reasoning about program analyses and the associated transformations in the presence of impure language features.

Lambda-lifting and dropping are well-known transformations in the functional programming community, and are studied in-depth by Johnsson [155] and Danvy et al. [91].

Rideau et al. [247] present an in-depth study of SSA destruction, including a verified implementation in the proof assistant Coq for the "windmills" problem, i.e., the task of correctly introducing ϕ-compensating assignments. The local algorithm to avoid the lost-copy problem and swap problem identified by Briggs et al. [50] was given by Beringer [25]. In this solution, the algorithm to break the cycles is in line with the results of May [196].

We are not aware of previous work that transfers the analysis of loop nesting forests to the functional setting, or of loop analyses in the functional world that correspond to loop nesting forests. Our discussion of Steensgaard's construction was based on a classification of loop nesting forests by Ramalingam [236], which also served as the source of the example in Fig. 6.7. Two alternative constructions discussed by Ramalingam are those by Sreedhar, Gao and Lee [266], and Havlak [142]. To us, it appears that a functional reading of Sreedhar, Gao, and Lee's construction would essentially yield the nesting mentioned at the beginning of Sect. 6.3. Regarding Havlak's construction, the fact that entry points of loops are not necessarily classified as headers appears to make an elegant representation in functional form at least challenging.

Extending the syntactic correspondences between SSA and functional languages, similarities may be identified between their characteristic program analysis frameworks, *data-flow analyses* and *type systems*. Chakravarty et al. [62] prove the correctness of a functional representation of Wegmann and Zadeck's SSA-based sparse conditional constant propagation algorithm [303]. Beringer et al. [26] consider data-flow equations for liveness and read-once variables, and formally translate their solutions to properties of corresponding typing derivations. Laud et al. [177] present a formal correspondence between data-flow analyses and type systems but consider a simple imperative language rather than SSA.

At present, intermediate languages in functional compilers do not provide syntactic support for expressing nesting forest directly. Indeed, most functional compilers do not perform advanced analyses of nested loops. As an exception to this rule, the MLton compiler (http://mlton.org) implements Steensgaard's algorithm for detecting loop nesting forests, leading to a subsequent analysis of the loop unrolling and loop switching transformations [278].

Concluding Remarks

In addition to low-level functional languages, alternative representations for SSA have been proposed, but their discussion is beyond the scope of this chapter. Glesner [129] employs an encoding in terms of abstract state machines [134] that disregards the sequential control flow inside basic blocks but retains control flow between basic blocks to prove the correctness of a code generation transformation. Later work by the same author uses a more direct representation of SSA in the theorem prover Isabelle/HOL for studying further SSA-based analyses.

Matsuno and Ohori [195] present a formalism that captures core aspects of SSA in a notion of (non-standard) types while leaving the program text in unstructured non-SSA form. Instead of classifying values or computations, types represent the definition points of variables. Contexts associate program variables at each program point with types in the standard fashion, but the non-standard notion of types means that this association models the reaching-definitions relationship rather than characterizing the values held in the variables at runtime. Noting a correspondence between the types associated with a variable and the sets of def-use paths, the authors admit types to be formulated over type variables whose introduction and use corresponds to the introduction of ϕ-nodes in SSA.

Finally, Pop et al.'s model [233] dispenses with control flow entirely and instead views programs as sets of equations that model the assignment of values to variables in a style reminiscent of partial recursive functions. This model is discussed in more detail in Chap. 10.

Part II
Analysis

Chapter 7
Introduction

Markus Schordan and Fabrice Rastello

The objective of Part II is to describe several program analyses and optimizations that benefit from SSA form.

In particular, we illustrate how SSA form makes an analysis more convenient because of its similarity to *functional programs*. Technically, the def-use chains explicitly expressed through the SSA graph, but also the static-single information property , are what make SSA so convenient.

We also illustrate how SSA form can be used to design *sparse* analyses that are faster and more efficient than their dense counterparts. This is especially important for just-in-time compilation but also for complex inter-procedural ahead-of-time analysis.

Finally, as already mentioned in Chap. 2, SSA form can come in different flavours. The vanilla flavour is strict SSA, or in equivalent terms, SSA form with the dominance property. The most common SSA construction algorithm exploits this dominance property by two means. First, it serves to compute join sets for ϕ-placement in a very efficient way using the dominance frontier. Second, it allows variable renaming using a folding scheme along the dominance tree. The notion of dominance and dominance frontier are two *structural properties* that make SSA form singular for compiler analysis and transformations.

The following paragraphs provide a short overview of the chapters that constitute this part.

M. Schordan
Lawrence Livermore National Laboratory, Livermore, CA, USA
e-mail: schordan1@llnl.gov

F. Rastello (✉)
Inria, Grenoble, France
e-mail: fabrice.rastello@inria.fr

© The Author(s), under exclusive license to Springer Nature Switzerland AG 2022   91
F. Rastello, F. Bouchez Tichadou (eds.), *SSA-based Compiler Design*,
https://doi.org/10.1007/978-3-030-80515-9_7

Propagating Information Using SSA

There are several analyses that propagate information through the SSA graph. Chapter 8 gives a general description of the mechanism. It shows how SSA form facilitates the design and implementation of analyses equivalent to traditional data-flow analyses and how the SSA property serves to reduce analysis time and memory consumption. The presented *Propagation Engine* is an extension of the well-known approach by Wegman and Zadeck [303] for sparse conditional constant propagation. The basic algorithm is not limited to constant propagation and can be used to solve a broad class of data-flow problems more efficiently than the iterative work list algorithm for solving data-flow equations. The basic idea is to directly propagate information computed at the unique definition of a variable to all its uses.

Liveness

A data-flow analysis potentially iterates up to as many times as the maximum loop depth of a given program until it stabilizes and terminates. In contrast, the properties of SSA form mean liveness analysis can be accelerated without the requirement of any iteration to reach a fixed point: it only requires at most two passes over the CFG. Also, an extremely simple liveness check is possible by providing a query system to answer questions such as "is variable v live at location q?". Chapter 9 shows how the loop nesting forest and dominance property can be exploited to devise very efficient liveness analyses.

Loop Tree and Induction Variables

Chapter 10 illustrates how capturing properties of the SSA graph itself (circuits) can be used to determine a very specific subset of program variables: induction variables. The induction variable analysis is based on the detection of self-references in the SSA representation and the extraction of the loop tree, which can be performed on the SSA graph as well. The presented algorithm translates the SSA representation into a representation of polynomial functions, describing the sequence of values that SSA variables hold during the execution of loops. The number of iterations is computed as the minimum solution of a polynomial inequality with integer solutions, also called Diophantine inequality.

Redundancy Elimination

The elimination of redundant computations is an important compiler optimization. A computation is *fully redundant* if it has also occurred earlier regardless of control flow, and a computation is *partially redundant* if it has occurred earlier only on certain paths. Following the program flow, once we are past the dominance frontiers, any further occurrence of a redundant expression is partially redundant, whereas any occurrence before the dominance frontier is fully redundant. The difference for the optimization is that fully redundant expressions can be deleted, whereas for (strictly) partially redundant expressions insertions are required to eliminate the redundancy. Thus, since partial redundancies start at dominance frontiers they are related to SSA's ϕ statements. Chapter 11 shows how the dominance frontier that is used to insert a minimal number of ϕ-function for SSA construction can also be used to minimize redundant computations. It presents the SSAPRE algorithm and its variant, the speculative PRE, which (possibly speculatively) perform partial redundancy elimination (PRE). It also discusses a direct and very useful application of the presented technique, register promotion via PRE. Unfortunately, the SSAPRE algorithm is not capable of recognizing redundant computations among lexically different expressions that yield the same value. Therefore, redundancy elimination based on value analysis (GVN) is also discussed, although a description of the value-based partial redundancy elimination (GVN-PRE) algorithm of VanDrunen [295], which subsumes both PRE and GVN, is missing.

Alias Analysis

The book is lacking an extremely important compiler analysis, namely alias analysis. Disambiguating pointers improves the precision and performance of many other analyses and optimizations. To be effective, flow-sensitivity and inter-procedurality are required but, with standard iterative data-flow analysis, lead to serious scalability issues. The main difficulty with making alias analysis sparse is that it is a chicken and egg problem. For each may-alias between two variables, the associated information interferes. Assuming the extreme scenario where any pair of variables is a pair of aliases, one necessarily needs to assume that the modification of one potentially modifies the other. In other words, the def-use chains are completely dense and there is no benefit in using any extended SSA form such as HSSA form (see Chap. 16). The idea is thus to decompose the analysis into phases where sophistication increases with sparsity. This is precisely what the *staged flow-sensitive analysis* of Hardekopf and Lin [138] achieves with two "stages." The first stage, performing a not-so-precise auxiliary pointer analysis, creates def-use chains used to enable sparsity in the second stage. The second stage, the primary analysis, is then a flow-sensitive pointer analysis. An important aspect is that as long as the auxiliary pointer analysis in the first stage is sound, the primary flow-sensitive analysis in the second stage will also be sound and will be at least as precise as a traditional "non-sparse" flow-sensitive analysis. The sparsity improves the runtime of the analysis, but it does not reduce precision.

Another way to deal with pointers in SSA form is to use a variant of SSA, called partial SSA, which requires variables to be divided into two classes: one class that contains only variables that are never referenced by pointers, and another class containing all those variables that can be referenced by pointers (address-taken variables). To avoid complications involving pointers in SSA form, only variables in the first class are put into SSA form. This technique is used in modern compilers such as GCC and LLVM.

Chapter 8
Propagating Information Using SSA

Florian Brandner and Diego Novillo

A central task of compilers is to *optimize* a given input program such that the resulting code is more efficient in terms of execution time, code size, or some other metric of interest. However, in order to perform these optimizations, typically some form of *program analysis* is required to determine if a given program transformation is applicable, to estimate its profitability, and to guarantee its correctness.

Data-flow analysis is a simple yet powerful approach to program analysis that is utilized by many compiler frameworks and program analysis tools today. We will introduce the basic concepts of traditional data-flow analysis in this chapter and will show how the *Static Single Assignment* form (SSA) facilitates the design and implementation of equivalent analyses. We will also show how the SSA property allows us to reduce the compilation time and memory consumption of the data-flow analyses that this program representation supports.

Traditionally, data-flow analysis is performed on a *control-flow graph* representation (CFG) of the input program. Nodes in the graph represent operations, and edges represent the potential flow of program execution. Information on certain *program properties* is propagated among the nodes along the control-flow edges until the computed information stabilizes, i.e., no *new* information can be inferred from the program.

The *propagation engine* presented in the following sections is an extension of the well-known approach by Wegman and Zadeck for *sparse conditional constant propagation* (also known as SSA-CCP). Instead of using the CFG, they represent the input program as an *SSA graph* as defined in Chap. 14: operations are again

F. Brandner (✉)
Télécom Paris, Institut Polytechnique de Paris, Palaiseau, France
e-mail: florian.brandner@telecom-paris.fr

D. Novillo
Google, Toronto, ON, Canada
e-mail: dnovillo@google.com

© The Author(s), under exclusive license to Springer Nature Switzerland AG 2022
F. Rastello, F. Bouchez Tichadou (eds.), *SSA-based Compiler Design*,
https://doi.org/10.1007/978-3-030-80515-9_8

represented as nodes in this graph; however, the edges represent *data dependencies* instead of control flow. This representation allows selective propagation of program properties among data-dependent graph nodes only. As before, the processing stops when the information associated with the graph nodes stabilizes. The basic algorithm is not limited to constant propagation and can also be applied to solve a large class of other data-flow problems efficiently. However, not all data-flow analyses can be modelled. Chapter 13 addresses some of the limitations of the SSA-based approach.

The remainder of this chapter is organized as follows. First, the basic concepts of (traditional) data-flow analysis are presented in Sect. 8.1. This will provide the theoretical foundation and background for the discussion of the SSA-based propagation engine in Sect. 8.2. We then provide an example of a data-flow analysis that can be performed efficiently by the aforementioned engine, namely copy propagation in Sect. 8.3.

8.1 Preliminaries

Data-flow analysis is at the heart of many compiler transformations and optimizations but also finds application in a broad spectrum of analysis and verification tasks in program analysis tools such as program checkers, profiling tools, and timing analysis tools. This section gives a brief introduction to the basics of data-flow analysis.

As noted before, data-flow analysis derives information from certain interesting program properties that may help to optimize the program. Typical examples of interesting properties are: the set of *live* variables at a given program point, the particular constant value a variable may take, or the set of program points that are reachable at runtime. Liveness information, for example, is critical during register allocation, while the two latter properties help to simplify computations and to identify dead code.

The analysis results are gathered from the input program by propagating information among its operations considering all potential execution paths. The propagation is typically performed iteratively until the computed results stabilize. Formally, a data-flow problem can be specified using a *monotone framework* that consists of:

- A *complete lattice* representing the property space
- A *flow graph* resembling the control flow of the input program
- A set of *transfer functions* modelling the effect of individual operations on the property space

Property Space A key concept for data-flow analysis is the representation of the property space via *partially ordered sets* $(L, \sqsubseteq)$, where L represents some interesting program property and $\sqsubseteq$ represents a reflexive, transitive, and antisymmetric relation. Using the $\sqsubseteq$ relation, *upper* and *lower bounds*, as well as *least upper* and *greatest lower bounds*, can be defined for subsets of L.

A particularly interesting class of partially ordered sets are *complete lattices*, where all subsets have a least upper bound as well as a greatest lower bound. Those bounds are unique and are denoted by $\bigsqcup$ and $\bigsqcap$, respectively. In the context of program analysis, the former is often referred to as the *join operator*, while the latter is termed the *meet operator*. Complete lattices have two distinguished elements, the *least element* and the *greatest element*, often denoted by $\bot$ and $\top$, respectively.

An *ascending chain* is a totally ordered subset $\{l_1, \ldots, l_n\}$ of a complete lattice. A chain is said to *stabilize* if there exists an index m, where $\forall i > m : l_i = l_m$. An analogous definition can be given for *descending chains*.

Program Representation The functions of the input program are represented as control-flow graphs, where the nodes represent operations, or instructions, and edges denote the potential flow of execution at runtime. Data-flow information is then propagated from one node to another adjacent node along the respective graph edge using *in* and *out* sets associated with every node. If there exists only one edge connecting two nodes, data can be simply copied from one set to the other. However, if a node has multiple incoming edges, the information from those edges has to be combined using the meet or join operator.

Sometimes, it is helpful to reverse the flow graph to propagate information, i.e., reverse the direction of the edges in the control-flow graph. Such analyses are termed *backward analyses*, while those using the regular flow graph are *forward analyses*.

Transfer Functions   Aside from the control flow, the operations of the program need to be accounted for during analysis. Usually, these operations change the way data is propagated from one control-flow node to the other. Every operation is thus mapped to a *transfer function*, which transforms the information available from the *in* set of the flow graph node of the operation and stores the result in the corresponding *out* set.

8.1.1   Solving Data-Flow Problems

Putting all those elements together—a complete lattice, a flow graph, and a set of transfer functions—yields an instance of a monotone framework. This framework describes a set of *data-flow equations* whose solution will ultimately converge to the solution of the data-flow analysis. A very popular and intuitive way to solve these equations is to compute the *maximal (minimal) fixed point* (MFP) using an iterative work list algorithm. The work list contains edges of the flow graph that have to be revisited. Visiting an edge consists of first combining the information from the *out* set of the source node with the *in* set of the target node, using the meet or join operator, and then applying the transfer function of the target node. The obtained information is then propagated to all direct successors of the target node by appending the corresponding edges to the work list. The algorithm terminates when the data-flow information stabilizes, as the work list then becomes empty.

A single flow edge can be appended several times to the work list in the course of the analysis. It may even happen that an infinite feedback loop prevents the algorithm from terminating. We are thus interested in bounding the number of times a flow edge is processed. Recalling the definition of chains from before (see Sect. 8.1), the *height* of a lattice is defined by the length of its longest chain. We can ensure termination for lattices fulfiling the *ascending chain condition*, which ensures that the lattice has finite height. Given a lattice with finite height h and a flow graph $G = (V, E)$, it is easy to see that the MFP solution can be computed in $O(|E| \cdot h)$ time, where $|E|$ represents the number of edges. Since the number of edges is bounded by the number of graph nodes $|V|$, more precisely, $|E| \leq |V|^2$, this gives a $O(|V|^2 \cdot h)$ general algorithm to solve data-flow analyses. Note that the height of the lattice often depends on properties of the input program, which might ultimately yield bounds worse than cubic in the number of graph nodes. For instance, the lattice for copy propagation consists of the cross product of many smaller lattices, each representing the potential values of a variable occurring in the program. The total height of the lattice thus directly depends on the number of variables in the program.

In terms of memory consumption, we have to propagate data-flow to all relevant program points. Nodes are required to hold information even when it is not directly related to the node; hence, each node must store complete *in* and *out* sets.

8.2 Data-Flow Propagation Under SSA Form

SSA form allows us to solve a large class of data-flow problems more efficiently than the iterative work list algorithm presented previously. The basic idea is to directly propagate information computed at the unique definition of a variable to all its uses. In this way, intermediate program points that neither define nor use the variable of interest do not have to be taken into consideration, thus reducing memory consumption and compilation time.

8.2.1 Program Representation

Data-flow analyses under SSA form rely on a specialized program representation based on *SSA graphs* (see Chap. 14). Besides the data dependencies, the SSA graph captures the *relevant* join nodes of the CFG of the program. A join node is relevant for the analysis whenever the value of two or more definitions may reach a use

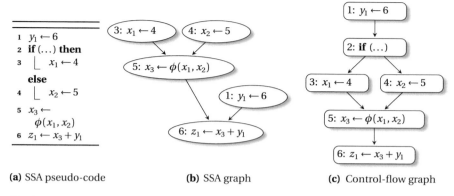

(a) SSA pseudo-code   **(b)** SSA graph   **(c)** Control-flow graph

Fig. 8.1 Example program and its SSA graph

by passing through that join. The SSA form properties ensure that a ϕ-function is placed at the join node and that any use of the variable that the ϕ-function defines has been properly updated to refer to the correct name.

Consider, for example, the code excerpt shown in Fig. 8.1, along with its corresponding SSA graph and CFG. Assume we are interested in propagating information from the assignment of variable y_1, at the beginning of the code, down to its unique use at the end. The traditional CFG representation causes the propagation to pass through several intermediate program points. These program points are concerned only with computations of the variables x_1, x_2, and x_3 and are thus irrelevant for y_1. The SSA graph representation, on the other hand, propagates the desired information directly from definition to use sites, without any intermediate step. At the same time, we also find that the control-flow join following the conditional is properly represented by the ϕ-function defining the variable x_3 in the SSA graph.

Even though the SSA graph captures data dependencies and the relevant join nodes in the CFG, it lacks information on other *control dependencies*. However, analysis results can often be improved significantly by considering the additional information that is available from the control dependencies in the CFG. As an example, consider once more the code in Fig. 8.1, and assume that the condition associated with the if-statement is known to be false for all possible program executions. Consequently, the ϕ-function will select the value of x_2 in all cases, which is known to be of constant value 5. However, due to the shortcomings of the SSA graph, this information cannot be derived. It is thus important to use both the control-flow graph and the SSA graph during data-flow analysis in order to obtain the best possible results.

8.2.2 Sparse Data-Flow Propagation

Similar to monotone frameworks for traditional data-flow analysis, frameworks for *sparse data-flow propagation* under SSA form can be defined using:

- A *complete lattice* representing the property space
- A set of *transfer functions* for the operations of the program
- A *control-flow graph* capturing the execution flow of the program
- An *SSA graph* representing data dependencies

We again seek a maximal (minimal) fixed-point solution (MFP) using an iterative work list algorithm. However, in contrast to the algorithm described previously, data-flow information is not propagated along the edges of the control-flow graph but along the edges of the SSA graph. For regular uses, the propagation is straightforward due to the fact that every use receives its value from a unique definition. Special care has to be taken only for ϕ-functions, which select a value among their operands depending on the incoming control-flow edges. The data-flow information of the incoming operands has to be combined using the meet or join operator of the lattice. As data-flow information is propagated along SSA edges that have a single source, it is sufficient to store the data-flow information with the SSA graph node. The *in* and *out* sets used by the traditional approach—see Sect. 8.1— are obsolete, since ϕ-functions already provide the required buffering. In addition, the control-flow graph is used to track which operations are not reachable under any program execution and thus can be safely ignored during the computation of the fixed-point solution.

The algorithm is shown in Algorithm 8.1 and processes two work lists, the *CFGWorkList*, which contains edges of the control-flow graph, and the *SSAWork-List*, which contains edges from the SSA graph. It proceeds by removing the top element of either of those lists and processing the respective edge. Throughout the main algorithm, operations of the program are visited to update the work lists and propagate information using Algorithm 8.2.

The *CFGWorkList* is used to track edges of the CFG that were encountered to be *executable*, i.e., where the data-flow analysis cannot rule out that a program execution traversing the edge exists. Once the algorithm has determined that a CFG edge is executable, it will be processed by Step 3 of the main algorithm. First, all ϕ-functions of its target node need to be reevaluated due to the fact that Algorithm 8.2 discarded the respective operands of the ϕ-functions so far—because the control-flow edge was not yet marked executable. Similarly, the operation of the target node has to be evaluated when the target node is encountered to be executable for the first time, i.e., the currently processed control-flow edge is the first of its incoming edges that is marked executable. Note that this is only required the *first* time the node is encountered to be executable, due to the processing of operations in Step 4b, which thereafter triggers the reevaluation automatically when necessary through the SSA graph.

Algorithm 8.1: Sparse data-flow propagation

---

1. Initialization

    • Every edge in the CFG is marked not executable;
    • The *CFGWorkList* is seeded with the outgoing edges of the CFG's *start* node;
    • The *SSAWorkList* is empty.

2. Remove the top element of one of the two work lists.
3. If the element is a control-flow edge, proceed as follows:

    • Mark the edge as executable;
    • Visit every ϕ-function associated with the target node (Algorithm 8.2);
    • If the target node was reached the first time via the *CFGWorkList*, visit all its operations (Algorithm 8.2);
    • If the target node has a single, non-executable outgoing edge, append that edge to the *CFGWorkList*.

4. If the element is an edge from the SSA graph, process the target operation as follows:

    a. When the target operation is a ϕ-function visit that ϕ-function;
    b. For other operations, examine the executable flag of the incoming edges of the respective CFG node; Visit the operation if any of the edges is executable.

5. Continue with step 2 until both work lists become empty.

Algorithm 8.2: Visiting an operation

---

1. Propagate data-flow information depending on the operation's kind:

    a. ϕ-functions:
       Combine the data-flow information from the node's operands where the corresponding control-flow edge is executable.
    b. Conditional branches:
       Examine the branch's condition(s) using the data-flow information of its operands; Determine all outgoing edges of the branch's CFG node whose condition is potentially satisfied; Append the CFG edges that were non-executable to the *CFGWorkList*.
    c. Other operations:
       Update the operation's data-flow information by applying its transfer function.

2. Whenever the data-flow information of an operation changes, append all outgoing SSA graph edges of the operation to the *SSAWorkList*.

Regular operations as well as ϕ-functions are visited by Algorithm 8.2 when the corresponding control-flow graph node has become executable, or whenever the data-flow information of one of their direct predecessors in the SSA graph has changed. At ϕ-functions, the information from multiple control-flow paths is combined using the usual meet or join operator. However, only those operands where the associated control-flow edge is marked executable are considered. Conditional branches are handled by examining their conditions based on the data-flow information computed so far. Depending on whether those conditions are satisfiable or not, control-flow edges are appended to the *CFGWorkList* to ensure that all reachable operations are considered during the analysis. Finally, all regular

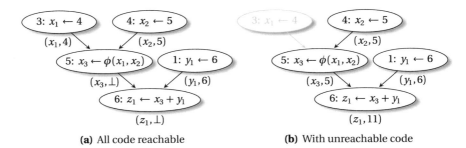

Fig. 8.2 Sparse conditional data-flow propagation using SSA graphs

operations are processed by applying the relevant transfer function and possibly propagating the updated information to all uses by appending the respective SSA graph edges to the *SSAWorkList*.

As an example, consider the program shown in Fig. 8.1 and the constant propagation problem. First, assume that the condition of the if-statement cannot be statically evaluated, we thus have to assume that all its successors in the CFG are reachable. Consequently, all control-flow edges in the program will eventually be marked executable. This will trigger the evaluation of the constant assignments to the variables x_1, x_2, and y_1. The transfer functions immediately yield that the variables are all constant, holding the values 4, 5, and 6, respectively. This new information will trigger the reevaluation of the ϕ-function of variable x_3. As both of its incoming control-flow edges are marked executable, the combined information yields $4 \sqcap 5 = \bot$, i.e., the value is known not to be a particular constant value. Finally, also the assignment to variable z_1 is reevaluated, but the analysis shows that its value is not a constant as depicted in Fig. 8.2a. If, however, the if-condition is known to be false for all possible program executions, a more precise result can be computed, as shown in Fig. 8.2b. Neither the control-flow edge leading to the assignment of variable x_1 nor its outgoing edge leading to the ϕ-function of variable x_3 is marked executable. Consequently, the reevaluation of the ϕ-function considers the data-flow information of its second operand x_2 only, which is known to be constant. This enables the analysis to show that the assignment to variable z_1 is, in fact, constant as well.

8.2.3 Discussion

During the course of the propagation algorithm, every edge of the SSA graph is processed at least once, whenever the operation corresponding to its definition is found to be executable. Afterwards, an edge can be revisited several times depending on the height h of the lattice representing the property space of the analysis. On the other hand, edges of the control-flow graph are processed at most once. This leads to an upper bound in execution time of $O(|E_{SSA}| \cdot h + |E_{CFG}|)$, where E_{SSA} and E_{CFG} represent the edges of the SSA graph and the control-flow graph, respectively. The size of the SSA graph increases with respect to the original non-SSA program. Measurements indicate that this growth is linear, yielding a bound that is comparable to the bound of traditional data-flow analysis. However, in practice, the SSA-based propagation engine outperforms the traditional approach. This is due to the direct propagation from the definition of a variable to its uses, without the costly intermediate steps that have to be performed on the CFG. The overhead is also reduced in terms of memory consumption: instead of storing the *in* and *out* sets capturing the complete property space on every program point, it is sufficient to associate every node in the SSA graph with the data-flow information of the corresponding variable only, leading to considerable savings in practice.

8.3   Example—Copy Propagation

Even though data-flow analysis based on SSA graphs has its limitations, it is still a useful and effective solution for interesting problems, as we will show in the following example. Copy propagation under SSA form is, in principle, very simple. Given the assignment $x \leftarrow y$, all we need to do is to traverse the immediate uses of x and replace them with y, thereby effectively eliminating the original copy operation. However, such an approach will not be able to propagate copies past ϕ-functions, particularly those in loops. A more powerful approach is to split copy propagation into two phases: First, a data-flow analysis is performed to find copy-related variables throughout the program. Second, a rewrite phase eliminates spurious copies and renames variables.

The analysis for copy propagation can be described as the problem of propagating the *copy of value* of variables. Given a sequence of copies as shown in Fig. 8.3a, we say that y_1 is a *copy of* x_1 and z_1 is a *copy of* y_1. The problem with this representation is that there is no apparent link from z_1 to x_1. In order to handle transitive copy relations, all transfer functions operate on copy of values instead of the direct source of the copy. If a variable is not found to be a copy of anything

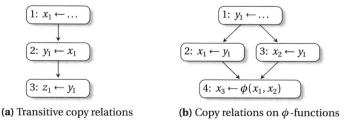

(a) Transitive copy relations   **(b)** Copy relations on ϕ-functions

Fig. 8.3 Analysis of copy-related variables

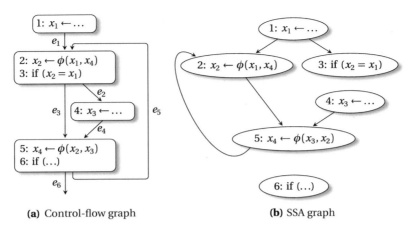

(a) Control-flow graph   **(b)** SSA graph

Fig. 8.4 ϕ-functions in loops often obfuscate copy relations

else, its copy of value is the variable itself. For the above example, this yields that both y_1 and z_1 are copies of x_1, which in turn is a copy of itself. The lattice of this data-flow problem is thus similar to the lattice used previously for constant propagation. The lattice elements correspond to variables of the program instead of integer numbers. The least element of the lattice represents the fact that a variable is a copy of itself.

Similarly, we would like to obtain the result that x_3 is a copy of y_1 for the example of Fig. 8.3b. This is accomplished by choosing the join operator such that a copy relation is propagated whenever the copy of values of all the operands of the ϕ-function matches. When visiting the ϕ-function for x_3, the analysis finds that x_1 and x_2 are both copies of y_1, and consequently propagates that x_3 is also a copy of y_1.

The next example shows a more complex situation where copy relations are obfuscated by loops—see Fig. 8.4. Note that the actual visiting order depends on the shape of the CFG and immediate uses; in other words, the ordering used here is meant for illustration only. Processing starts at the operation labeled 1, with both work lists empty and the data-flow information $\top$ associated with all variables:

1. Assuming that the value assigned to variable x_1 is not a copy, the data- flow information for this variable is lowered to $\bot$, the SSA edges leading to operations 2 and 3 are appended to the *SSAWorkList*, and the control-flow graph edge e_1 is appended to the *CFGWorkList*.
2. Processing the control-flow edge e_1 from the work list causes the edge to be marked executable and the operations labeled 2 and 3 to be visited. Since edge e_5 is not yet known to be executable, the processing of the ϕ-function yields a copy relation between x_2 and x_1. This information is utilized in order to determine which outgoing control-flow graph edges are executable for the conditional branch. Examining the condition shows that only edge e_3 is executable and thus needs to be added to the work list.

3. Control-flow edge e_3 is processed next and marked executable for the first time. Furthermore, the ϕ-function labeled 5 is visited. Due to the fact that edge e_4 is not known to be executable, this yields a copy relation between x_4 and x_1 (via x_2). The condition of the branch labeled 6 cannot be analysed and thus causes its outgoing control-flow edges e_5 and e_6 to be added to the work list.
4. Now, control-flow edge e_5 is processed and marked executable. Since the target operations are already known to be executable, only the ϕ-function is revisited. However, variables x_1 and x_4 have the same copy of value x_1, which is identical to the previous result computed in Step 2. Thus, neither of the two work lists is modified.
5. Assuming that the control-flow edge e_6 leads to the exit node of the control-flow graph, the algorithm stops after processing the edge without modifications to the data-flow information computed so far.

The straightforward implementation of copy propagation would have needed multiple passes to discover that x_4 is a copy of x_1. The iterative nature of the propagation, along with the ability to discover non-executable code, allows one to handle even obfuscated copy relations. Moreover, this kind of propagation will only reevaluate the subset of operations affected by newly computed data-flow information instead of the complete control-flow graph once the set of executable operations has been discovered.

8.4 Further Reading

Traditional data-flow analysis is well established and well described in numerous papers. The book by Nielsen, Nielsen, and Hankin [208] gives an excellent introduction to the theoretical foundations and practical applications. For reducible flow graphs, the order in which operations are processed by the work list algorithm can be optimized [144, 157, 208], allowing one to derive tighter complexity bounds. However, relying on reducibility is problematic because the flow graphs are often *not* reducible even for proper structured languages. For instance, reversed control-flow graphs for backward problems can be—and in fact almost always are—irreducible even for programs with reducible control-flow graphs, for instance because of loops with multiple exits. Furthermore, experiments have shown that the tighter bounds do not necessarily lead to improved compilation times [85].

Apart from computing a fixed-point (MFP) solution, traditional data-flow equations can also be solved using a more powerful approach called the *meet over all paths* (MOP) solution, which computes the *in* data-flow information for a basic block by examining *all* possible paths from the start node of the control-flow graph. Even though more powerful, computing the MOP solution is often harder or even undecidable [208]. Consequently, the MFP solution is preferred in practice.

The sparse propagation engine [210, 303], as presented in the chapter, is based on the underlying properties of the SSA form. Other intermediate representations offer similar properties. *Static Single Information* form (SSI) first introduced by Ananian [8] and then later fixed by Singer [261] has been designed with the objective of allowing both backward and forward problems. It uses σ-functions, which are placed at program points where data-flow information for backward problems needs to be merged [260]. But as illustrated by dead-code elimination (backward sparse data-flow analysis on SSA), the fact that σ-functions are necessary for backward propagation problems (or ϕ-functions for forward) is a misconception. This subtle point is further explained in Sect. 13.2. Bodík also uses an extended SSA form, *e*-SSA, his goal being to eliminate array bounds checks [32]. *e*-SSA is a simple extension that allows one to also propagate information from conditionals. Chapter 13 revisits the concepts of static single information and proposes a generalization that subsumes all those extensions to SSA form. Ruf [250] introduces the *value dependence graph*, which captures both control and data dependencies. He derives a sparse representation of the input program, which is suited for data-flow analysis, using a set of transformations and simplifications.

The *sparse evaluation graph* by Choi et al. [67] is based on the same basic idea as the approach presented in this chapter: intermediate steps are eliminated by bypassing irrelevant CFG nodes and merging the data-flow information only when necessary. Their approach is closely related to the placement of ϕ-functions and similarly relies on the dominance frontier during construction. A similar approach, presented by Johnson and Pingali [153], is based on single-entry/single-exit regions. The resulting graph is usually less sparse but is also less complex to compute. Ramalingam [237] further extends those ideas and introduces the *compact evaluation graph*, which is constructed from the initial CFG using two basic transformations. The approach is superior to the sparse representations by Choi et al. as well as the approach presented by Johnson and Pingali.

The previous approaches derive a sparse graph suited for data-flow analysis using graph transformations applied to the CFG. Duesterwald et al. instead examine the data-flow equations, eliminate redundancies, and apply simplifications to them [107].

Chapter 9
Liveness

Benoît Boissinot and Fabrice Rastello

This chapter illustrates the use of strict SSA properties to simplify and accelerate *liveness analysis*, which determines, for all variables, the set of program points where these variables are *live*, i.e., their values are potentially used by subsequent operations. Liveness information is essential to solve storage assignment problems, eliminate redundancies, and perform code motion. For instance, optimizations like software pipelining, trace scheduling, register-sensitive redundancy elimination (see Chap. 11), if-conversion (see Chap. 20), as well as register allocation (see Chap. 22) heavily rely on liveness information.

Traditionally, liveness information is obtained by data-flow analysis: liveness sets are computed for all basic blocks and variables simultaneously by solving a set of data-flow equations. These equations are usually solved by an iterative algorithm, propagating information backwards through the control-flow graph (CFG) until a fixed point is reached and the liveness sets stabilize. The number of iterations depends on the control-flow structure of the considered program, and more precisely on the structure of its loops.

In this chapter, we show that, for strict SSA form programs, the live range of a variable has valuable properties that can be expressed in terms of the loop nesting forest of the CFG and its corresponding directed acyclic graph, the *forward-CFG*.

B. Boissinot
ENS Lyon, Lyon, France
e-mail: benoit.boissinot@ens-lyon.org

F. Rastello (✉)
Inria, Grenoble, France
e-mail: fabrice.rastello@inria.fr

© The Author(s), under exclusive license to Springer Nature Switzerland AG 2022   107
F. Rastello, F. Bouchez Tichadou (eds.), *SSA-based Compiler Design*,
https://doi.org/10.1007/978-3-030-80515-9_9

Informally speaking, and restricted to reducible CFGs, those properties for a variable v are:

- v is live at a program point q if and only if v is live at the entry h of the largest loop/basic block (highest node in the loop nesting forest) that contains q but not the definition of v.
- v is live at h if and only if there is a path in the forward-CFG from h to a use of v that does not contain the definition.

A direct consequence of this property is the possible design of a data-flow algorithm that computes liveness sets *without the requirement of any iteration* to reach a fixed point: at most, two passes over the CFG are necessary. The first pass, very similar to traditional data-flow analysis, computes partial liveness sets by traversing the forward-CFG backwards. The second pass refines the partial liveness sets and computes the final solution by propagating forwards along the loop nesting forest. For the sake of clarity, we first present the algorithm for reducible CFGs. Irreducible CFGs can be handled with a slight variation of the algorithm, with no need to modify the CFG itself.

Another approach to liveness analysis more closely follows the classical definition of liveness: a variable is live at a program point q if q belongs to a path of the CFG leading from a definition of that variable to one of its uses without passing through another definition of the same variable. Therefore, the set of program points where a variable is live, in other words its live range, can be computed using a backward traversal starting on its uses and stopping when its definition is reached (unique under SSA).

One application of the properties of live ranges under strict SSA form is the design of a simple liveness check algorithm. In contrast to classical data-flow analyses, a liveness check does not provide the set of variables live at a block, but its characteristic function. A liveness check provides a query system to answer questions such as "Is variable v live at location q?" Its main features are:

1. The algorithm itself consists of two parts, a *pre-computation* part, and an *online* part executed at each liveness query. It is not based on setting up and subsequently solving data-flow equations.
2. The pre-computation is *independent of variables,* it only depends on the structure of the control-flow graph; Hence, pre-computed information *remains valid* when variables or their uses are added or removed.
3. An actual query uses the def-use chain of the variable in question and determines the answer essentially by testing membership in pre-computed sets of basic blocks.

We will first need to repeat basic definitions relevant to our context and provide the theoretical foundations in the next section, before presenting multiple algorithms to compute liveness sets: The two-pass data-flow algorithm in Sect. 9.2 and the algorithms based on path exploration in Sect. 9.4. We present the liveness check algorithm last, in Sect. 9.3.

9.1 Definitions

Liveness is a property that relates program points to sets of variables which are considered to be *live* at these program points. Intuitively, a variable is considered live at a given program point when its value will be used in the future of any dynamic execution. Statically, liveness can be approximated by following paths backwards on the control-flow graph, connecting the uses of a given variable to its definitions—or, in the case of SSA forms, to its unique definition. The variable is said to be *live* at all program points along these paths. For a CFG node q, representing an instruction or a basic block, a variable v is *live-in* at q if there is a path, not containing the definition of v, from q to a node where v is used (including q itself). It is *live-out* at q if it is live-in at some direct successor of q.

The computation of live-in and live-out sets at the entry and the exit of basic blocks is usually termed *liveness analysis*. It is indeed sufficient to consider only these sets at basic block boundaries, since liveness within a basic block is trivial to recompute from its live-out set with a backward traversal of the block (whenever the definition of a variable is encountered, it is pruned from the live-out set). *Live-ranges* are closely related to liveness. Instead of associating program points with sets of live variables, the live range of a variable specifies the set of program points where that variable is live. Live ranges of programs under strict SSA form exhibit certain useful properties (see Chap. 2), some of which can be exploited for register allocation (see Chap. 22).

The special behaviour of ϕ-functions often causes confusion about where exactly its operands are actually used and defined. For a regular operation, variables are used and defined where the operation takes place. However, the semantics of ϕ-functions (and, in particular, the actual place of ϕ-uses) should be defined carefully, especially when dealing with SSA destruction. In algorithms for SSA destruction (see Chap. 21), a use in a ϕ-function is considered live somewhere inside the corresponding direct predecessor block, but, depending on the algorithm and, in particular, the way copies are inserted, it may or may not be considered as live-out for that predecessor block. Similarly, the definition of a ϕ-function is always considered to be at the beginning of the block, but, depending on the algorithm, it may or may not be marked as live-in for the block. To make the description of algorithms easier, we follow the same definition as the one used in Chap. 21, Sect. 21.2:

Definition 9.1 (Liveness for ϕ-Function Operands—Multiplexing Mode) For a ϕ-function $a_0 = \phi(a_1, \ldots, a_n)$ in block B_0, where a_i comes from block B_i:

- Its definition-operand is considered to be at the entry of B_0, in other words variable a_0 is live-in of B_0.
- Its use operands are at the exit of the corresponding direct predecessor basic blocks, in other words, variable a_i is live-out of basic block B_i.

This corresponds to placing a copy $a_0 \leftarrow a_i$ on each edge from B_i to B_0. The data-flow equations given below and the algorithms presented follow the same semantics. They require minor modifications when other ϕ-semantics are desired.

9.2 Data-Flow Approaches

A well-known and frequently used approach to compute the live-in and live-out sets of basic blocks is backward data-flow analysis (see Chap. 8, Sect. 8.1). The liveness sets are given by a set of equations that relate *upward-exposed* uses and definitions to live-in and live-out sets. We say a use is *upward-exposed* in a block when there is no local definition preceding it, i.e., the live range "escapes" the block at the top. The sets of upward-exposed uses and definitions do not change during liveness analysis and can thus, for each block B, be pre-computed:

$\text{Defs}(B)$:	variables defined (ϕexcluded) in B
$\text{Uses}(B)$:	variables used in B (ϕ excluded)
$\text{UpwardExposed}(B)$:	variables used in B without any preceding definition in B
$\text{PhiDefs}(B)$:	variables defined by a ϕ at the entry of B
$\text{PhiUses}(B)$:	variables used in a ϕ at the entry of a direct successor of B

With these pre-computed sets, the data-flow equations can be written as:

$$\text{LiveIn}(B) = \text{PhiDefs}(B) \cup \text{UpwardExposed}(B) \cup (\text{LiveOut}(B) \setminus \text{Defs}(B))$$

$$\text{LiveOut}(B) = \bigcup\nolimits_{S \in \text{directsuccs}(B)} (\text{LiveIn}(S) \setminus \text{PhiDefs}(S)) \cup \text{PhiUses}(B)$$

Informally, the live-in of block B are the variables defined in the ϕ-functions of B, those used in B (and not defined in B), and those which are just "passing through." On the other hand, the live-out are those that must be live for a direct successor S, i.e., either live-in of S (but not defined in a ϕ-function of S) or used in a ϕ-function of S.

9.2.1 Liveness Sets on Reducible Graphs

Instead of computing a fixed point, we show that liveness information can be derived in two passes over the control-flow graph. The first version of the algorithm requires the CFG to be reducible. We then show that arbitrary control-flow graphs can be handled elegantly and with no additional cost, except for a cheap pre-processing step on the loop nesting forest.

Fig. 9.1 An example of a
reducible CFG. Forward-CFG
is represented using full
edges; back edges are
thickened. Backward pass on
forward-CFG sets v as live-in
of node 5, but not of node 2.
Forward pass on loop nesting
forest then sets v as live at
node 6 but not at node 7

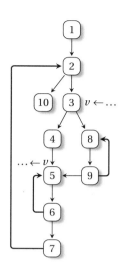

The key properties of live ranges under strict SSA form on a reducible CFG that
we exploit for this purpose can be outlined as follow:

1. Let q be a CFG node that does not contain the definition d of a variable, and h be
   the header of the maximal loop containing q but not d. If such a maximal loop
   does not exist, then let $h = q$.
      The variable is live-in at q if and only if there exists a forward path from h to
   a use of the variable without going through the definition d.
2. If a variable is live-in at the header of a loop then it is live at all nodes inside the
   loop.

As an example, consider the code of Fig. 9.1. For $q = 6$, the header of the largest
loop containing 6 but not the definition d in 3 is $h = 5$. As a forward path (down
edges) exists from 3 to 5, variable v is live-in at 5. It is thus also live in all nodes
inside the loop, in particular, in node 6. On the other hand, for $q = 7$, the largest
"loop" containing 7 but not 3 is 7 itself. As there is no forward path from 7 to any
use (node 5), v is not live-in of 7 (note that v is not live-in of 2 either).

Those two properties pave the way for describing the two steps that make up our
liveness set algorithm:

1. A backward pass propagates partial liveness information upwards using a post-
   order traversal of the forward-CFG;
2. The partial liveness sets are then refined by traversing the loop nesting forest,
   propagating liveness from loop-headers down to all basic blocks within loops.

Algorithm 9.1 shows the necessary initialization and the high-level structure to
compute liveness in two passes.

Algorithm 9.1: Two-pass liveness analysis: reducible CFG

---
1 **foreach** basic block B **do**
2 | mark B as unprocessed
3 DAG_DFS(r)   ▷ r is the CFG root node
4 **foreach** root node L of the loop nesting forest **do**
5 | LoopTree_DFS(L)

The post-order traversal is shown in Algorithm 9.2, which performs a simple depth-first search and gives partial liveness sets to every basic block of the CFG. The algorithm roughly corresponds to the pre-computation step of the traditional iterative data-flow analysis; however, back edges are not considered during the traversal. Recalling the definition of liveness for ϕ-functions, PhiUses(B) denotes the set of live-out variables from basic block B due to uses by ϕ-functions in direct successors of B. Similarly, PhiDefs(B) denotes the set of variables defined by a ϕ-function in B.

Algorithm 9.2: Partial liveness, with post-order traversal

---
1 **Function** *DAG_DFS(block B)*
2 | **foreach** $S \in$ directsuccs(B) such that (B, S) is not a back edge **do**
3 | | **if** S is unprocessed **then** DAG_DFS(S)
4 | *Live* $\leftarrow$ PhiUses(B)
5 | **foreach** $S \in$ directsuccs(B) such that (B, S) is not a back edge **do**
6 | | *Live* $\leftarrow$ *Live* $\cup$ (LiveIn(S) \ PhiDefs(S))
7 | LiveOut(B) $\leftarrow$ *Live*
8 | **foreach** program point p in B, backwards, **do**
9 | | remove variables defined at p from *Live*
10 | | add uses at p to *Live*
11 | LiveIn(B) $\leftarrow$ *Live* $\cup$ PhiDefs(B)
12 | mark B as processed

The next phase, which traverses the loop nesting forest, is shown in Algorithm 9.3. The live-in and live-out sets of all basic blocks within a loop are unified with the liveness sets of its loop-header.

Algorithm 9.3: Propagate live variables within loop bodies

---
1 **Function** *LoopTree_DFS(node N of the loop nesting forest)*
2 | **if** N is a loop node **then**
3 | | $B_N \leftarrow$ Block(N)   ▷ *The loop-header of N*
4 | | *LiveLoop* $\leftarrow$ LiveIn(B_N) \ PhiDefs(B_N)
5 | | **foreach** $M \in$ Children(N) **do**
6 | | | $B_M \leftarrow$ Block(M)   ▷ *Loop-header or block*
7 | | | LiveIn(B_M) $\leftarrow$ LiveIn(B_M) $\cup$ *LiveLoop*
8 | | | LiveOut(B_M) $\leftarrow$ LiveOut(B_M) $\cup$ *LiveLoop*
9 | | | LoopTree_DFS(M)

Fig. 9.2 Bad case for
iterative data-flow analysis

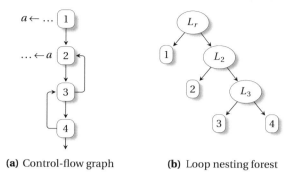

(**a**) Control-flow graph   (**b**) Loop nesting forest

Example 9.1 The CFG of Fig. 9.2a is a pathological case for iterative data-flow analysis. The pre-computation phase does not mark variable a as live throughout the two loops. An iteration is required for every loop nesting level until the final solution is computed. In our algorithm, after the CFG traversal, the traversal of the loop nesting forest (Fig. 9.2b) propagates the missing liveness information from the loop-header of loop L_2 down to all blocks within the loop's body and all inner loops, i.e., blocks 3 and 4 of L_3.

9.2.1.1   Correctness

The first pass propagates the liveness sets using a post-order traversal of the forward-CFG, G_{fwd}, obtained by removing all back edges from the CFG G. The first two lemmas show that this pass correctly propagates liveness information to the loop-headers of the original CFG.

Lemma 9.1 *Let G be a reducible CFG, v an SSA variable, and d its definition. If L is a maximal loop not containing d, then v is live-in at the loop-header h of L iff there is a path in G_{fwd} (i.e., back edge free) from h to a use of v that does not go through d.*

Lemma 9.2 *Let G be a reducible CFG, v an SSA variable, and d its definition. Let p be a node of G such that all loops containing p also contain d. Then v is live-in at p iff there is a path in G_{fwd}, from p to a use of v that does not go through d.*

Pointers to formal proofs are provided in the last section of this chapter. The important property used in the proof is the dominance property that requires the full live range of a variable to be dominated by its definition d. As a consequence, any back edge part of the live range is dominated by d, and the associated loop cannot contain d.

Algorithm 9.2, which propagates liveness information along the DAG G_{fwd}, can only mark variables as live-in if they are indeed live-in. Furthermore, if, after this propagation, a variable v is missing in the live-in set of a CFG node p, Lemma 9.2

shows that p belongs to a loop that does not contain the definition of v. Let L be such a maximal loop. According to Lemma 9.1, v is correctly marked as live-in at the header of L. The next lemma shows that the second pass of the algorithm (Algorithm 9.3) correctly adds variables to the live-in and live-out sets where they are missing.

Lemma 9.3 *Let G be a reducible CFG, L a loop, and v an SSA variable. If v is live-in at the loop-header of L, it is live-in and live-out at every CFG node in L.*

The intuition is straightforward: a loop is a strongly connected component, and because d is live-in of L, d cannot be part of L.

9.2.2   Liveness Sets on Irreducible Flow Graphs

The algorithms based on loops described above are only valid for reducible graphs. We can also derive an algorithm that works for irreducible graphs, as follows: transform the irreducible graph into a reducible graph, such that the liveness in both graphs is *equivalent*. First of all we would like to stress two points:

1. We do not require the transformed graph to be *semantically equivalent* to the original one, only isomorphism of liveness is required.
2. We do not actually modify the graph in practice, but Algorithm 9.2 can be changed to simulate the modification of some edges on the fly.

There are loop nesting forest representations with possibly multiple headers per irreducible loop. For the sake of clarity (and simplicity of implementation), we consider a representation where each loop has a unique entry node as header. In this case, the transformation simply relies on redirecting any edge $s \rightarrow t$ arriving in the middle of a loop to the header of the outermost loop (if it exists) that contains t but not s. The example of Fig. 9.3 illustrates this transformation, with the modified edge highlighted. Considering the associated loop nesting forest (with nodes 2, 5, and 8 as loop-headers), edge $9 \rightarrow 6$ is redirected to node 5.

Obviously the transformed code does not have the same semantics as the original one. But, because a loop is a strongly connected component, the dominance relationship is unchanged. As an example, the immediate dominator of node 5 is 3, in both the original and the transformed CFG. For this reason, any variable live-in of loop L_5—thus live everywhere in the loop—will be live on any path from 3 to the loop. Redirecting an incoming edge to another node of the loop—in particular, the header—does not change this behaviour.

To avoid building this transformed graph explicitly, an elegant alternative is to modify the CFG traversal in Algorithm 9.2. Whenever an entry-edge $s \rightarrow t$ is encountered during the traversal, instead of visiting t, we visit the header of the largest loop containing t and not s. This header node is nothing else than the highest ancestor of t in the loop nesting forest that is not an ancestor of s. We represent

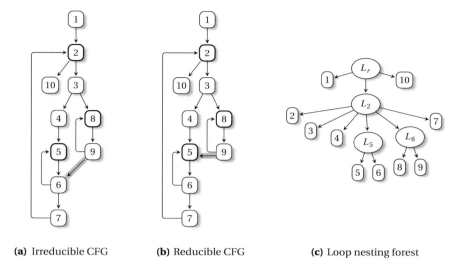

(a) Irreducible CFG   **(b)** Reducible CFG   **(c)** Loop nesting forest

Fig. 9.3 A reducible CFG derived from an irreducible CFG, using the loop nesting forest. The transformation redirects edges arriving inside a loop to the loop-header (here $9 \rightarrow 6$ into $9 \rightarrow 5$)

such as node as $t.OLE(s)$ (for "Outermost Loop Excluding"). As an example in Fig. 9.3, considering edge $9 \rightarrow 6$, the highest ancestor of 6 not an ancestor of 9 is $9.OLE(6) = L_5$. Overall, the transformation amounts to replacing all occurrences of S by $S.OLE(B)$ at Lines 3 and 6 of Algorithm 9.2, which allows it to handle irreducible control-flow graphs.

9.2.3   Computing the Outermost Excluding Loop (OLE)

Our approach potentially involves many outermost excluding loop queries, especially for the liveness check algorithm, as developed further. An efficient implementation of OLE is required. The technique proposed here and shown in Algorithm 9.5 is to pre-compute the set of ancestors from the loop-tree for every node. A simple set operation can then find the node we are looking for: the ancestors of the definition node are removed from the ancestors of the query point. From the remaining ancestors, we pick the shallowest. Using bitsets for encoding the set of ancestors of a given node, indexed with a topological order of the loop-tree, these operations are easily implemented. The removal is a bit inversion followed by a bitwise "and" operation, and the shallowest node is found by searching for the first set bit in the bitset. Since the number of loops (and thus the number loop-headers) is rather small, the bitsets are themselves small as well and this optimization does not result in much wasted space.

Consider a topological indexing of loop-headers: n.LTindex (n being a loop-header) or reciprocally i.node (i being an index). For each node, we associate a bitset (indexed by loop-headers) of all its ancestors in the loop-tree: n.ancestors. This can be computed using any topological traversal of the loop-tree by a call of DFS_COMPUTE_ANCESTORS(L_r). Note that some compiler intermediate representations sometimes consider L_r as a loop-header. Considering so in DFS_COMPUTE_ANCESTORS will not spoil the behaviour of *OLE*.

Algorithm 9.4: Compute the loop nesting forest ancestors

---

1 **Function** *DFS_compute_ancestors(node n)*
2   **if** $n \neq L_r$ **then**
3 | n.ancestors ← n.LTparent.ancestors
4   **else**
5 | n.ancestors ← ∅   ▷ *empty bitset*
6   **if** n.isLoopHeader **then**
7 | n.ancestors.add(n.LTindex)
8   **foreach** s in n.LTchildren **do**
9 | DFS_COMPUTE_ANCESTORS(s)

Using this information, finding the outermost excluding loop can be done by simple bitset operations as in Algorithm 9.5.

Algorithm 9.5: Outermost excluding loop

---

1 **Function** *OLE(node self, node b)*
2   nCA ← bitset_and(*self*.ancestors, bitset_not(b.ancestors))
3   **if** nCA.isempty **then**
4 | **return** *self*
5   **else**
6 | **return** nCA.bitset_leading_set.node

Example 9.2 Consider the example of Fig. 9.3c again and suppose the loops L_2, L_8, and L_5 are, respectively, indexed 0, 1, and 2. Using big-endian notations for bitsets, Algorithm 9.4 would give binary labels 110 to node 9 and 101 to node 6. The outermost loop containing 6 but not 9 is given by the leading bit of $101 \wedge \neg 110 = 001$, i.e., L_5.

9.3 Liveness Check Using Loop Nesting Forest and Forward Reachability

In contrast to liveness sets, the liveness check does not provide the set of variables live at a block, but provides a query system to answer questions such as "is variable v live at location q?" Such a framework is well suited to tree scan based register allocation (see Chap. 22), SSA destruction (see Chap. 21), or Hyperblock scheduling (see Chap. 18). Most register-pressure aware algorithms such as code-motion are not designed to take advantage of a liveness check query system and still require sets. This query system can obviously be built on top of pre-computed liveness sets. Queries in $O(1)$ are possible, at least for basic block boundaries, providing the use of sparsesets or bitsets to allow for efficient element-wise queries. If sets are only stored at basic block boundaries, to allow a query system at instruction granularity, it is possible to use the list of uses of variables or backward scans. Constant time, worst-case complexity is lost in this scenario and liveness sets that have to be incrementally updated at each (even minor) code transformation can be avoided and replaced by less memory-consuming data structures that only depend on the CFG.

In the following, we consider the live-in query of variable a at node q. To avoid notational overhead, let a be defined in the CFG node $d = def(a)$ and let $u \in uses(a)$ be a node where a is used. Suppose that q is strictly dominated by d (otherwise v cannot be live at q). Lemmas 9.1–9.3 given in Sect. 9.2.1.1 can be rephrased as follows:

1. Let h be the header of the maximal loop containing q but not d. Let h be q if such maximal loop does not exist. Then v is live-in at h if and only if there exists a forward path that goes from h to u.
2. If v is live-in at the header of a loop then it is live at any node inside the loop.

In other words, v is live-in at q if and only if there exists a forward path from h to u where h, if it exists, is the header of the maximal loop containing q but not d, and q itself otherwise. Given the forward control-flow graph and the loop nesting forest, finding out if a variable is live at some program point can be done in two steps. First, if there exists a loop containing the program point q and not the definition, pick the header of the biggest such loop instead as the query point. Then check for reachability from q to any use of the variable in the forward CFG. As explained in Sect. 9.2.2, for irreducible CFG, the *modified-forward CFG* that redirects any edge $s \rightarrow t$ to the loop-header of the outermost loop containing t but excluding s ($t.OLE(s)$), has to be used instead. Correctness is proved from the theorems used for liveness sets.

Algorithm 9.6 puts a little bit more effort into providing a query system at instruction granularity. If q is in the same basic block as d (lines 8–13), then v is live at q if and only if there is a use outside the basic block, or inside but after q. If h is a loop-header then v is live at q if and only if a use is forward reachable from h (lines 19–20). Otherwise, if the use is in the same basic block as q it must be after q to bring the variable live at q (lines 17–18). In this pseudo-code, upper case is used for basic blocks while lower case is used for program points at instruction granularity. "def(a)" is an operand. "uses(a)" is a set of operands. "basicBlock(u)" returns the basic block containing the operand u. Given the semantics of the ϕ-function instruction, the basic block returned by this function for a ϕ-function operand can be different from the block where the instruction textually occurs. Also, "u.order" provides the corresponding (increasing) ordering in the basic block. For a ϕ-function operand, the ordering number might be greater than the maximum ordering of the basic block if the semantics of the ϕ-function places the uses on outgoing edges of the direct predecessor block. $Q.OLE(D)$ corresponds to Algorithm 9.5 given in Sect. 9.2.3. forwardReachable(H, U), which tells whether U is reachable in the modified-forward CFG, will be described later.

Algorithm 9.6: Live-in check

---

1 **Function** *IsLiveIn(programPoint q, var a)*
2     $d \leftarrow def(a)$
3     $D \leftarrow \text{basicBlock}(d)$
4     $Q \leftarrow \text{basicBlock}(q)$
5   **if** not $\big(D$ sdom Q or $(D = Q$ and order$(d) < $ order$(q))\big)$ **then**
6   **return** *false*
7   **if** $Q = D$ **then**
8   **foreach** u **in** $uses(a)$ **do**
9           $U \leftarrow \text{basicBlock}(u)$
10   **if** $U \neq D$ or order$(q) \leq $ order(u) **then**
11   **return** true
12   **return** *false*
13     $H \leftarrow Q.OLE(D)$
14   **foreach** u **in** $uses(a)$ **do**
15        $U \leftarrow \text{basicBlock}(u)$
16   **if** (not isLoopHeader(H)) and $U = Q$ and order$(u) < $ order(q) **then**
17   **continue**
18   **if** forward Reachable(H, U) **then**
19   **return** true
20   **return** *false*

The live-out check algorithm, given by Algorithm 9.7, only differs from live-in check in lines 5, 11, and 17, which involve ordering comparisons. In line 5, if q is equal to d it cannot be live-in while it might be live-out; in lines 11 and 17 if q is at a use point it makes it live-in but not necessarily live-out.

Algorithm 9.7: Live-out check

---

1 **Function** *IsLiveOut(programPoint q, var a)*
2      $d \leftarrow def(a)$
3      $D \leftarrow \text{basicBlock}(d)$
4      $Q \leftarrow \text{basicBlock}(q)$
5   **if** not $\big(D$ sdom Q or $(D = Q$ and order$(d) \leq$ order$(q))\big)$ **then**
6   **return** *false*   ▷ *q must be dominated by the definition*
7   **if** $Q = D$ **then**
8   **foreach** u **in** *uses(a)* **do**
9              $U \leftarrow \text{basicBlock}(u)$
10   **if** $U \neq D$ or order$(q) <$ order(u) **then**
11   **return** true
12   **return** *false*
13      $H \leftarrow Q.OLE\,(D)$
14   **foreach** u **in** *uses(a)* **do**
15          $U \leftarrow \text{basicBlock}(u)$
16   **if** (not isLoopHeader(H)) and $U = Q$ and order$(u) \leq$ order(q) **then**
17   **continue**
18   **if** forwardReachable(H, U) **then**
19   **return** true
20   **return** *false*

9.3.1 Computing Modified-Forward Reachability

The liveness check query system relies on pre-computations for efficient *OLE* and forwardReachable queries. The outermost excluding loop is identical to the one used for liveness sets. We explain how to compute modified-forward reachability here (i.e., forward reachability on transformed CFG to handle irreducibility). In practice we do not explicitly build the modified-forward graph. To efficiently compute modified-forward reachability we simply need to traverse the modified-forward graph in reverse topological order. A post-order initiated by a call to the recursive function DFS_Compute_forwardReachable(r) (Algorithm 9.8) will do the job. Bitsets can be used to efficiently implement sets of basic blocks. Once forward reachability has been pre-computed this way, forwardReachable(H, U) returns true if and only if $U \in H$.forwardReachable.

9.4 Liveness Sets Using Path Exploration

Another, maybe more intuitive, way of calculating liveness sets is closely related to the definition of the live range of a given variable. As recalled earlier, a variable is live at a program point p, if p belongs to a path of the CFG leading from a definition

---

Algorithm 9.8: Computation of modified-forward reachability using a traversal along a reverse topological order

---

1 **Function** *DFS_Compute_forwardReachable(block N)*
2   N.forwardReachable $\leftarrow \varnothing$
3   N.forwardReachable.add(N)
4   **foreach** $S \in$ directsuccs(N) **if** (N, S) is not a back edge **do**
5          $H \leftarrow S.OLE(N)$
6   **if** H.forwardReachable $= \perp$ **then**
7   DFS_Compute_forwardReachable(H)
8   N.forwardReachable $\leftarrow N$.forwardReachable $\cup$ H.forwardReachable

---

of that variable to one of its uses without passing through the definition. Therefore, the live range of a variable can be computed using a backward traversal starting at its uses and stopping when its (unique) definition is reached.

Actual implementation of this idea could be done in several ways. In particular, the order along which use operands are processed, in addition to the way liveness sets are represented, can substantially impact the performance. The one we choose to develop here allows the use of a simple stack-like set representation which avoids any expensive set-insertion operations and set-membership tests. The idea is to process use operands variable by variable. In other words, the processing of different variables is not intermixed, i.e., the processing of one variable is completed before the processing of another variable begins.

Depending on the particular compiler framework, a pre-processing step that performs a full traversal of the program (i.e., the instructions) might be required in order to derive the def-use chains for all variables, i.e., a list of all uses for each SSA variable. The traversal of the variable list and processing of its uses thanks to def-use chains is depicted in Algorithm 9.9.

Note that, in strict SSA form, in a given block, no use can appear before a definition. Thus, if v is live-out or used in a block B, it is live-in iff it is not defined in B. This leads to the code of Algorithm 9.10 for path exploration. Here, the liveness sets are implemented using a stack-like data structure.

Algorithm 9.9: Compute liveness sets by variable using def-use chains

---

1 **Function** *Compute_LiveSets_SSA_ByVar(CFG)*
2   **foreach** variable v **do**
3   **foreach** block B such that $v \in$ Uses(B) $\cup$ PhiUses(B) **do**
4   **if** $v \in$ PhiUses(B) **then**   ▷ *Used in the ϕ of a direct successor block*
5   LiveOut(B) = LiveOut(B) $\cup \{v\}$
6   Up_and_Mark_Stack(B, v)

Algorithm 9.10: Optimized path exploration using a stack-like data structure

---

1 **Function** *Up_and_Mark_Stack(B, v)*
2   **if** $v \in$ Defs(B) **then return**   ▷ *Killed in the block, stop*
3   **if** top(LiveIn(B)) = v **then return**   ▷ *Propagation already done, stop*
4   push(LiveIn(B), v)
5   **if** $v \in$ PhiDefs(B) **then return**   ▷ *Do not propagate φ definitions*
6   **foreach** $P \in$ CFG_directpreds(B) **do**   ▷ *Propagate backwards*
7   **if** top(LiveOut(P)) ≠ v **then**
8   push(LiveOut(P), v)
9   Up_and_Mark_Stack(P, v)

9.5 Further Reading

Liveness information is usually computed with iterative data-flow analysis, which goes back to Kildall [166]. The algorithms are, however, not specialized to the computation of liveness sets and overhead may incur. Several strategies are possible, leading to different worst-case complexities and performances in practice. Round-robin algorithms propagate information according to a fixed block ordering derived from a depth-first spanning tree, and iterate until it stabilizes. The complexity of this scheme was analysed by Kam et al. [157]. Node listing algorithms specify, a priori, the overall sequence of nodes, where repetitions are allowed, along which data-flow equations are applied. Kennedy [162] devises for structured flow graphs, node listings of size $2|V|$, with $|V|$ the number of control-flow nodes, and mentions the existence of node listings of size $O(|V| \log(|V|))$ for reducible flow graphs. Worklist algorithms focus on blocks that may need to be updated because the liveness sets of their successors (for backward problems) changed. Empirical results by Cooper et al. [85] indicate that the order in which basic blocks are processed is critical and directly impacts the number of iterations. They showed that, in practice, a mixed solution called "single stack worklist" based on a worklist initialized with a round-robin order is the most efficient for liveness analysis.

Alternative ways to solve data-flow problems belong to the family of elimination-based algorithms [251]. Through recursive reductions of the CFG, variables of the data-flow system are successively eliminated and equations are reduced until the CFG reduces to a single node. The best, but unpractical, worst-case complexity elimination algorithm has an almost-linear complexity $O(|E|\alpha(|E|))$. It requires the CFG (resp. the reverse CFG) to be reducible for a forward (resp. backward) analysis. For non-reducible flow graphs, none of the existing approaches can guarantee a worst-case complexity better than $O(|E|^3)$. In practice, irreducible CFGs are rare, but liveness analysis is a backward data-flow problem, which frequently leads to irreducible reverse CFGs.

Gerlek et al. [126] use the so-called λ-operators to collect upward-exposed uses at control-flow split points. More specifically, the λ-operators are placed at the iterated dominance frontiers, computed on the reverse CFG, of the set of uses of a variable. These λ-operators and the other uses of variables are chained together and liveness is efficiently computed on this graph representation. The technique of Gerlek et al. can be considered as a precursor of the live variable analysis based on the Static Single Information (SSI) form conjectured by Singer [261] and revisited by Boissinot et al. [37]. In both cases, the insertion of pseudo-instructions guarantees that any definition is post-dominated by a use.

Another approach to computing liveness was proposed by Appel [10]. Instead of computing the liveness information for all variables at the same time, variables are handled individually by exploring paths in the CFG starting from variable uses. Using logic programming, McAllester [197] presented an equivalent approach to show that liveness analysis can be performed in time proportional to the number of instructions and variables. However, his theoretical analysis is limited to a restricted input language with simple conditional branches and instructions. A more generalized analysis is given in Chapter 2 of the habilitation thesis of Rastello [239], in terms of both theoretical complexity and practical evaluation (Sect. 9.4 describes a path-exploration technique restricted to SSA programs).

The loop nesting forest considered in this chapter corresponds to the one obtained using Havlak's algorithm [142]. A more generalized definition exists and corresponds to the *minimal* loop nesting forest as defined by Ramalingam [236]. The handling of any minimal loop nesting forest is also detailed in Chapter 2 of [239].

Handling of irreducible CFG can be done through CFG transformations such as node splitting [2, 149]. Such a transformation can lead to an exponential growth in the number of nodes. Ramalingam [236] proposed a transformation (different from the one presented here but also without any exponential growth) that only maintains the dominance property (not the full semantic).

Finding the maximal loop not containing a node s but containing a node t (*OLE*) is a problem similar to finding the least common ancestor (LCA) of the two nodes s and t in the rooted loop-nested forest: the loop in question is the only direct child of LCA(s, t), ancestor of t. As described in [23], an LCA query can be reduced to a Range Minimum Query (RMQ) problem that can itself be answered in $O(1)$, with a pre-computation of $O(n)$. The adaptation of LCA to provide an efficient algorithm for *OLE* queries is detailed in Chapter 2 of [239].

This chapter is a short version of Chapter 2 of [239] which, among other details, contains formal proofs and handling of different ϕ-function semantics. Sparsesets are described by Cooper and Torczon [83].

Chapter 10
Loop Tree and Induction Variables

Sebastian Pop and Albert Cohen

This chapter presents an extension of SSA whereby the extraction of the reducible loop tree can be done only on the SSA graph itself. This extension also captures reducible loops in the CFG. This chapter first illustrates this property and then shows its usefulness through the problem of induction variable recognition.

10.1 Part of the CFG and Loop Tree Can Be Exposed Through the SSA

During the construction of the SSA representation based on a CFG representation, a large part of the CFG information is translated into the SSA representation. As the construction of the SSA has precise rules to place the ϕ-nodes in special points of the CFG (i.e., at the merge of control-flow branches), by identifying patterns of uses and definitions, it is possible to expose a part of the CFG structure through the SSA representation.

Furthermore, it is possible to identify higher-level constructs inherent to the CFG representation, such as strongly connected components of basic blocks (or reducible loops), based only on the patterns of the SSA definitions and uses. The induction variable analysis presented in this chapter is based on the detection of self-references in the SSA representation and on its characterization.

S. Pop
Amazon Web Services, Austin, Texas, USA
e-mail: spop@amazon.com

A. Cohen (✉)
Google, Paris, France
e-mail: albertcohen@google.com

© The Author(s), under exclusive license to Springer Nature Switzerland AG 2022
F. Rastello, F. Bouchez Tichadou (eds.), *SSA-based Compiler Design*,
https://doi.org/10.1007/978-3-030-80515-9_10

This first section shows that the classical SSA representation is not sufficient to represent the semantics of the original program. We will see the minimal amount of information that has to be added to the classical SSA representation in order to represent the loop information: similar to the ϕ_{exit}-function used in the gated SSA presented in Chap. 14, the loop-closed SSA form adds an extra variable at the end of a loop for each variable defined in a loop and used after the loop.

10.1.1 An SSA Representation Without the CFG

In the classical definition of SSA, the CFG provides the skeleton of the program: basic blocks contain assignment statements defining SSA variable names, and the basic blocks with multiple direct predecessors contain ϕ-nodes. Let us look at what happens when, starting from a classical SSA representation, we remove the CFG.

In order to remove the CFG, imagine a pretty printer function that dumps only the arithmetic instructions of each basic block and skips the control instructions of an imperative program by traversing the CFG structure in any order. Does the representation obtained from this pretty printer contain enough information to enable us to compute the same thing as the original program?[1] Let us see what happens with an example in its CFG-based SSA representation:

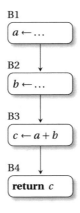

After removing the CFG structure, listing the definitions in an arbitrary order, we could obtain this:

---

[1] To simplify the discussion, we consider the original program to be free of side effect instructions.

---

return c
$b \leftarrow \ldots$ $\triangleright$ *some computation independent of a*
$c \leftarrow a + b$
$a \leftarrow \ldots$ $\triangleright$ *some computation independent of b*

And this SSA code is sufficient, in the absence of side effects, to recover an order of computation that leads to the same result as in the original program. For example, the evaluation of this sequence of statements would produce the same result:

$b \leftarrow \ldots$ $\triangleright$ *some computation independent of a*
$a \leftarrow \ldots$ $\triangleright$ *some computation independent of b*
$c \leftarrow a + b$
return c

10.1.2 *Natural Loop Structures on the SSA*

We will now see how to represent the natural loops in the SSA form by systematically adding extra ϕ-nodes at the end of loops, together with extra information about the loop-exit predicate. Supposing that the original program contains a loop:

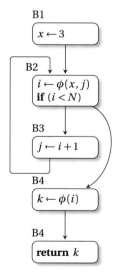

By pretty printing with a random order traversal, we could obtain this SSA code:

$$x \leftarrow 3$$
return k
$$i \leftarrow \phi(x, j)$$
$$k \leftarrow \phi(i)$$
$$j \leftarrow i + 1$$

Note that some information is lost in this pretty printing: the exit condition of the loop has been lost. We will have to record this information in the extension of the SSA representation. However, the loop structure still appears through the cyclic definition of the induction variable i. To expose it, we can rewrite this SSA code using simple substitutions, such as:

$$i \leftarrow \phi(3, i + 1)$$
$$k \leftarrow \phi(i)$$
return k

Thus, we have the definition of the SSA name i defined as a function of itself. This pattern is characteristic of the existence of a loop. We remark that there are two kinds of ϕ-nodes used in this example:

- **Loop-ϕ nodes** "$i = \phi(x, j)$" (also denoted $i = \phi_{entry}(x, j)$ as in Chap. 14) have an argument that contains a self-reference j and an invariant argument x: here the defining expression "$j = i + 1$" contains a reference to the same loop-ϕ definition i, while x (here 3) is not part of the circuit of dependencies that involves i and j. Note that it is possible to define a canonical SSA form by limiting the number of arguments of loop-ϕ nodes to two.
- **Close-ϕ nodes** "$k = \phi_{exit}(i)$" (also denoted $k = \phi_{exit}(i)$ as in Chap. 14) capture the last value of a name defined in a loop. Names defined in a loop can only be used within that loop or in the arguments of a close-ϕ node (which is "closing" the set of uses of the names defined in that loop). In a canonical SSA form, it is possible to limit the number of arguments of close-ϕ nodes to one.

10.1.3 Improving the SSA Pretty Printer for Loops

As we have seen in the example above, the exit condition of the loop disappeared during the basic pretty printing of the SSA. To capture the semantics of the computation of the loop, we have to specify this condition in the close-ϕ-node when we exit the loop, so as to be able to derive which value will be available at the end of the loop. With our extension, which adds the loop-exit condition to the syntax of the close-ϕ, the SSA pretty printing of the above example would be:

$x \leftarrow 3$
$i \leftarrow \phi_{entry}(x, j)$
$j \leftarrow i + 1$
$k \leftarrow \phi_{exit}(i \geq N, i)$
return k

So k is defined as the "first value" of i satisfying the loop- exit condition, "$i \geq N$." In the case of finite loops, this is well defined as being the first element satisfying the loop-exit condition of the sequence defined by the corresponding loop-ϕ node.

Below we will look at an algorithm that translates the SSA representation into a representation of polynomial functions, describing the sequence of values that SSA names take during the execution of a loop. The algorithm is restricted to the loops that are reducible. All such loops are labeled. Note that irreducible control flow is not forbidden: only the loops carrying self-definitions must be reducible.

10.2   Analysis of Induction Variables

The purpose of the induction variable analysis is to provide a characterization of the sequences of values taken by a variable during the execution of a loop. This characterization can be an exact function of the canonical induction variable of the loop (i.e., a loop counter that starts at zero with a step of one for each iteration of the loop) or an approximation of the values taken during the execution of the loop represented by values in an abstract domain. In this section, we will see a possible characterization of induction variables in terms of sequences. The domain of sequences will be represented by *chains of recurrences*: as an example, a canonical induction variable with an initial value 0 and a stride 1 that would occur in the loop with label x will be represented by the chain of recurrences $\{0, +, 1\}_x$.

10.2.1   Stride Detection

The first phase of the induction variable analysis is the detection of the strongly connected components of the SSA. This can be performed by traversing the use-def SSA chains and detecting that some definitions are visited twice. For a self-referring use-def chain, it is possible to derive the step of the corresponding induction variable as the overall effect of one iteration of the loop on the value of the loop-ϕ node. When the step of an induction variable depends on another cyclic definition, one has to further analyse the inner cycle. The analysis of the induction variable ends when all the inner cyclic definitions used for the computation of the step have been analysed. Note that it is possible to construct SSA graphs with strongly connected

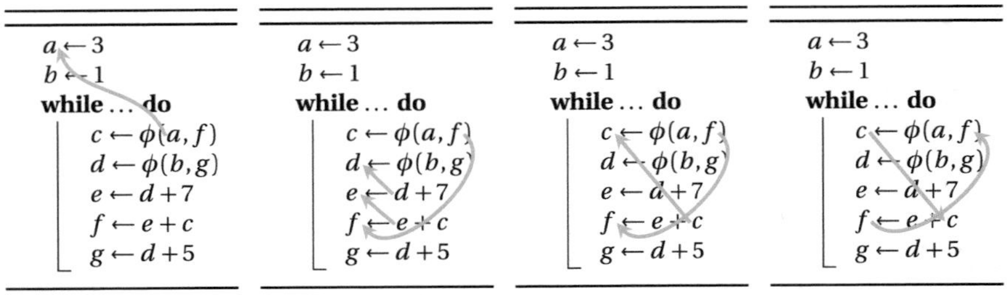

| (a) Init. condition edge | (b) Searching for c | (c) Backtrack, found c | (d) Cyclic chain |

Fig. 10.1 Detection of the cyclic definition using a depth-first search traversal of the use-def chains

components that are impossible to characterize with the chains of recurrences. This is precisely the case of the following example which shows two inter-dependent circuits, the first involving a and b with step $c + 2$, and the second involving c and d with step $a + 2$. This leads to an endless loop, which must be detected.

$$a \leftarrow \phi_{entry}(0, b)$$
$$c \leftarrow \phi_{entry}(1, d)$$
$$b \leftarrow c + 2$$
$$d \leftarrow a + 3$$

Let us now look at an example, presented in Fig. 10.1, to see how the stride detection works. The arguments of a ϕ-node are analysed to determine whether they contain self-references or are pointing towards the initial value of the induction variable. In this example, (a) represents the use-def edge that points towards the invariant definition. When the argument to be analysed points towards a longer use-def chain, the full chain is traversed, as shown in (b), until a ϕ-node is reached. In this example, the ϕ-node that is reached in (b) is different to the ϕ-node from which the analysis started, and so in (c) a search starts on the uses that have not yet been analysed. When the original ϕ-node is found, as in (c), the cyclic def-use chain provides the step of the induction variable: the step is "$+e$" in this example. Knowing the symbolic expression for the step of the induction variable may not be enough, as we will see next; one has to instantiate all the symbols ("e" in the current example) defined in the varying loop to precisely characterize the induction variable.

10.2.2 Translation to Chains of Recurrences

Once the def-use circuit and its corresponding overall loop update expression have been identified, it is possible to translate the sequence of values of the induction variable to a chain of recurrences. The syntax of a polynomial chain of recurrence is $\{base, +, step\}_x$, where $base$ and $step$ may be arbitrary expressions or constants, and x is the loop with which the sequence is associated. As a chain of recurrences represents the sequence of values taken by a variable during the execution of a loop, the associated expression of a chain of recurrences is given by $\{base, +, step\}_x(\ell_x) = base + step \times \ell_x$, which is a function of ℓ_x, the number of times the body of loop x has been executed.

When $base$ or $step$ translates to sequences varying in outer loops, the resulting sequence is represented by a multivariate chain of recurrences. For example, $\{\{0, +, 1\}_x, +, 2\}_y$ defines a multivariate chain of recurrences with a step of 1 in loop x and a step of 2 in loop y, where loop y is enclosed in loop x. When $step$ translates into a sequence varying in the same loop, the chain of recurrences represents a polynomial of a higher degree. For example, $\{3, +, \{8, +, 5\}_x\}_x$ represents a polynomial evolution of degree 2 in loop x. In this case, the chain of recurrences is also written omitting the extra braces: $\{3, +, 8, +, 5\}_x$. The semantics of a chain of recurrences is defined using the binomial coefficient $\binom{n}{p} = \frac{n!}{p!(n-p)!}$, by the equation:

$$\{c_0, +, c_1, +, c_2, +, \ldots, +, c_n\}_x(\vec{\ell_x}) = \sum_{p=0}^{n} c_p \binom{\ell_x}{p},$$

with $\vec{\ell}$ the iteration domain vector (the iteration loop counters of all the loops in which the chain of recurrences variates), and ℓ_x the iteration counter of loop x. This semantics is very useful in the analysis of induction variables, as it makes it possible to split the analysis into two phases, with a symbolic representation as a partial intermediate result:

1. First, the analysis leads to an expression where the step part "s" is left in a symbolic form, i.e., $\{c_0, +, s\}_x$.
2. Then, by instantiating the step, i.e., $s = \{c_1, +, c_2\}_x$, the chain of recurrences is that of a higher-degree polynomial, i.e., $\{c_0, +, \{c_1, +, c_2\}_x\}_x = \{c_0, +, c_1, +, c_2\}_x$.

10.2.3 Instantiation of Symbols and Region Parameters

The last phase of the induction variable analysis consists in the instantiation (or further analysis) of symbolic expressions left from the previous phase. This includes the analysis of induction variables in outer loops, computing the last value of the counter of a preceding loop, and the propagation of closed-form expressions for loop invariants defined earlier. In some cases, it becomes necessary to leave in a symbolic form every definition outside a given region, and these symbols are then called parameters of the region.

Let us look again at the example of Fig. 10.1 to see how the sequence of values of the induction variable c is characterized with the notation of the chains of recurrences. The first step, after the cyclic definition is detected, is the translation of this information into a chain of recurrences: in this example, the initial value (or base of the induction variable) is a and the step is e, and so c is represented by a chain of recurrences $\{a, +, e\}_1$ that is varying in loop number 1. The symbols are then instantiated: a is trivially replaced by its definition, leading to $\{3, +, e\}_1$. The analysis of e leads to the chain of recurrences $\{8, +, 5\}_1$, which is then used in the chain of recurrences of c, $\{3, +, \{8, +, 5\}_1\}_1$, and is equivalent to $\{3, +, 8, +, 5\}_1$, a polynomial of degree two:

$$F(\ell) = 3\binom{\ell}{0} + 8\binom{\ell}{1} + 5\binom{\ell}{2}$$
$$= \frac{5}{2}\ell^2 + \frac{11}{2}\ell + 3.$$

10.2.4 Number of Iterations and Computation of the End-of-Loop Value

One of the important static analyses for loops is to evaluate their trip count, i.e., the number of times the loop body is executed before the exit condition becomes true. In common cases, the loop-exit condition is a comparison of an induction variable against some constant, parameter, or other induction variables. The number of iterations is then computed as the minimum solution of a polynomial inequality with integer solutions, also called a Diophantine inequality. When one or more coefficients of the Diophantine inequality are parameters, the solution is left in parametric form. The number of iterations can also be an expression varying in an outer loop, in which case it can be characterized using a chain of recurrences.

Consider a scalar variable varying in an outer loop with strides dependent on the value computed in an inner loop. The expression representing the number of iterations in the inner loop can then be used to express the evolution function of the scalar variable varying in the outer loop.

For example, the following code:

```
x ← 0
for i = 0; i < N; i + + do                        ▷ loop₁
    for j = 0; j < M; j + + do                    ▷ loop₂
        x ← x + 1
```

would be written in loop-closed SSA form as

$$x_0 \leftarrow 0$$
$$i \leftarrow \phi^1_{entry}(0, i + 1)$$
$$x_1 \leftarrow \phi^1_{entry}(x_0, x_2)$$
$$x_4 \leftarrow \phi^1_{exit}(i < N, x_1)$$
$$j \leftarrow \phi^2_{entry}(0, j + 1)$$
$$x_3 \leftarrow \phi^2_{entry}(x_1, x_3 + 1)$$
$$x_2 \leftarrow \phi^2_{exit}(j < M, x_3)$$

---

x_3 represents the value of variable x at the end of the original imperative program. The analysis of scalar evolutions for variable x_4 would trigger the analysis of scalar evolutions for all the other variables defined in the loop-closed SSA form, as follows:

- First, the analysis of variable x_4 would trigger the analysis of i, N, and x_1.

  - The analysis of i leads to $i = \{0, +, 1\}_1$, i.e., the canonical loop counter l_1 of $loop_1$.
  - N is a parameter and is left in its symbolic form.
  - The analysis of x_1 triggers the analysis of x_0 and x_2:

    The analysis of x_0 leads to $x_0 = 0$.
    Analysing x_2 triggers the analysis of j, M, and x_3:

    - $j = \{0, +, 1\}_2$, i.e., the canonical loop counter l_2 of $loop_2$.
    - M is a parameter.
    - $x_3 = \phi^2_{entry}(x_1, x_3 + 1) = \{x_1, +, 1\}_2$.

    $x_2 = \phi^2_{entry}(j < M, x_3)$ is then computed as the last value of x_3 after $loop_2$, i.e., it is the chain of recurrences of x_3 applied to the first iteration of $loop_2$ that does not satisfy $j < M$ or equivalently $l_2 < M$. The corresponding Diophantine inequality $l_2 \geq M$ has a minimum solution $l_2 = M$. So, to finish the computation of the scalar evolution of x_2, we apply M to the scalar evolution of x_3, leading to $x_2 = \{x_1, +, 1\}_2(M) = x_1 + M$.

  - The scalar evolution analysis of x_1 then leads to $x_1 = \phi^1_{entry}(x_0, x_2) = \phi^1_{entry}(x_0, x_1 + M) = \{x_0, +, M\}_1 = \{0, +, M\}_1$.

- Finally, the analysis of x_4 ends with $x_4 = \phi^1_{exit}(i < N, x_1) = \{0, +, M\}_1(N) = M \times N$.

10.3   Further Reading

Induction variable detection has been studied extensively in the past because of its central role in loop optimizations. Wolfe [306] designed the first SSA-based induction variable recognition technique. It abstracts the SSA graph and classifies inductions according to a wide spectrum of patterns.

When operating on a low-level intermediate representation with arbitrary gotos, detecting the natural loops is the first step in the analysis of induction variables. In general, and when operating on low-level code in particular, it is preferable to use analyses that are more robust to complex control flows that do not resort to an early classification into predefined patterns. Chains of recurrences [15, 167, 318] have been proposed to characterize the sequence of values taken by a variable during the execution of a loop [293], and it has proven to be more robust to the presence of complex, unstructured control flows, to the characterization of induction variables over modulo-arithmetic such as unsigned wrap-around types in C, and to implementation in a production compiler [232].

The formalism and presentation of this chapter are derived from the thesis work of Sebastian Pop. The manuscript [231] contains pseudo-code and links to the implementation of scalar evolutions in GCC since version 4.0. The same approach has also influenced the design of LLVM's scalar evolution, but the implementation is different.

Induction variable analysis is used in dependence tests for scheduling and parallelization [308] and more recently, the extraction of short vector to SIMD instructions [212].[2] The Omega test [235] and parametric integer linear programming [115] have typically been used to reason about system parametric affine Diophantine inequalities. But in many cases, simplications and approximations can lead to polynomial decision procedures [16]. Modern parallelizing compilers tend to implement both kinds, depending on the context and aggressiveness of the optimization.

Substituting an induction variable with a closed-form expression is also useful for the removal of the cyclic dependencies associated with the computation of the induction variable itself [127]. Other applications include enhancements to strength reduction and loop-invariant code motion [127], and induction variable canonicalization (reducing induction variables to a single one in a given loop) [185].

The number of iterations of loops can also be computed based on the characterization of induction variables. This information is essential to advanced loop analyses, such as value-range propagation [210], and enhanced dependence tests for scheduling and parallelization [16, 235]. It also enables more opportunities for scalar optimization when the induction variable is used after its defining loop. Loop

---

[2] Note, however, that the computation of closed-form expressions is not required for dependence testing itself [309].

transformations also benefit from the replacement of the end-of-loop value as this removes scalar dependencies between consecutive loops. Another interesting use of the end-of-loop value is the estimation of the worst-case execution time (WCET) where an attempt is made to obtain an upper bound approximation of the time necessary for a program to terminate.

Chapter 11
Redundancy Elimination

Fred Chow

Redundancy elimination is an important category of optimizations performed by modern optimizing compilers. In the course of program execution, certain computations may be repeated multiple times and yield the same results. Such redundant computations can be eliminated by saving and reusing the results of the earlier computations instead of recomputing them later.

There are two types of redundancies: *full* redundancy and *partial* redundancy. A computation is fully redundant if the computation has occurred earlier regardless of the flow of control. The elimination of full redundancy is also called common subexpression elimination. A computation is partially redundant if the computation has occurred only along certain paths. Full redundancy can be regarded as a special case of partial redundancy where the redundant computation occurs regardless of the path taken.

There are two different views of a computation related to redundancy: how it is computed and the computed value. The former relates to the operator and the operands it operates on, which translates to how it is represented in the program representation. The latter refers to the value generated by the computation in the static sense.[1] As a result, algorithms for finding and eliminating redundancies can be classified into those that are *syntax-driven* and those that are *value-driven*. In syntax-driven analyses, two computations are the same if they are the same operation applied to the same operands that are program variables or constants. In this case, redundancy can arise only if the variables' values have not changed

---

[1] All values referred to in this chapter are static values viewed with respect to the program code. A static value can map to different dynamic values during program execution.

F. Chow (✉)
Huawei, Fremont, CA, USA
e-mail: fchow99@comcast.net

© The Author(s), under exclusive license to Springer Nature Switzerland AG 2022
F. Rastello, F. Bouchez Tichadou (eds.), *SSA-based Compiler Design*,
https://doi.org/10.1007/978-3-030-80515-9_11

between the occurrences of the computation. In value-based analyses, redundancy arises whenever two computations yield the same value. For example, $a + b$ and $a + c$ compute the same result if b and c can be determined to hold the same value. In this chapter, we deal mostly with syntax-driven redundancy elimination. The last section will extend our discussion to value-based redundancy elimination.

In our discussion on syntax-driven redundancy elimination, our algorithm will focus on the optimization of a lexically identical expression, like $a + b$, that appears in the program. During compilation, the compiler will repeat the redundancy elimination algorithm on all the other lexically identified expressions in the program.

The style of the program representation can impact the effectiveness of the algorithm applied. We distinguish between *statements* and *expressions*. Expressions compute to values without generating any side effect. Statements have side effects as they potentially alter memory contents or control flow, and are not candidates for redundancy elimination. In dealing with lexically identified expressions, we advocate a maximal expression tree form of program representation. In this style, a large expression tree such as $a + b * c - d$ is represented as is without having to specify any assignments to temporaries for storing the intermediate values of the computation.[2] We also assume the Conventional SSA Form of program representation, in which each ϕ-web (see Chap. 2) is interference-free and the live ranges of the SSA versions of each variable do not overlap. We further assume the HSSA (see Chap. 16) form that completely models the aliasing in the program.

11.1 Why Partial Redundancy Elimination and SSA Are Related

Figure 11.1 shows the two most basic forms of partial redundancy. In Fig. 11.1a, $a + b$ is redundant when the right path is taken. In Fig. 11.1b, $a + b$ is redundant whenever the back edge (see Sect. 4.4.1) of the loop is taken. Both are examples of *strictly* partial redundancies , in which insertions are required to eliminate the redundancies. In contrast, a full redundancy can be deleted without requiring any insertion. *Partial redundancy elimination (PRE)* is powerful because it subsumes global common subexpressions and loop-invariant code motion.

We can visualize the impact on redundancies of a single computation, as shown in Fig. 11.2. In the region of the control-flow graph dominated by the occurrence of $a + b$, any further occurrence of $a + b$ is fully redundant, assuming a and b are not modified. Following the program flow, once we are past the dominance frontiers, any further occurrence of $a + b$ is partially redundant. In constructing SSA form, dominance frontiers are where ϕs are inserted. Since partial redundancies start at dominance frontiers, partial redundancy elimination should borrow techniques

---

[2] The opposite of maximal expression tree form is the triplet form in which each arithmetic operation always defines a temporary.

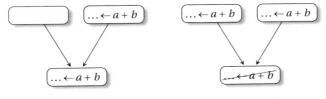

(a) First example program, before and after PRE

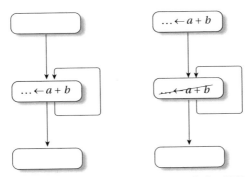

(b) Second example program, before and after PRE

Fig. 11.1 Two basic examples of partial redundancy elimination

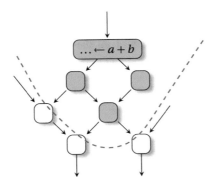

Fig. 11.2 Dominance frontiers (dashed) are boundaries between fully (highlighted basic blocks) and partially (normal basic blocks) redundant regions

from SSA ϕs insertion. In fact, the same sparse approach to modelling the use-def relationships among the occurrences of a program variable can be used to model the redundancy relationships among the different occurrences of $a + b$.

The algorithm that we present, named SSAPRE, performs PRE efficiently by taking advantage of the use-def information inherent in its input Conventional SSA Form. If an occurrence $a_j + b_j$ is redundant with respect to $a_i + b_i$, SSAPRE

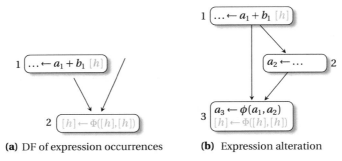

(a) DF of expression occurrences   (b) Expression alteration

Fig. 11.3 Examples of Φ-insertion

builds a redundancy edge that connects $a_i + b_i$ to $a_j + b_j$. To expose potential partial redundancies, we introduce the operator Φ at the dominance frontiers of the occurrences, which has the effect of factoring the redundancy edges at merge points in the control-flow graph.[3] The resulting *factored redundancy graph* (FRG) can be regarded as the SSA form for expressions.

To make the *expression SSA form* more intuitive, we introduce the hypothetical temporary h, which can be thought of as the temporary that will be used to store the value of the expression. The FRG can be viewed as the SSA graph for h. Observe that we have not yet determined where h should be defined or used. In referring to the FRG, a *use* node will refer to a node in the FRG that is not a definition.

The SSA form for h is constructed in two steps similar to ordinary SSA form: the Φ-insertion step followed by the Renaming step. In the Φ-insertion step, we insert Φs at the dominance frontiers of all the expression occurrences, to ensure that we do not miss any possible placement positions for the purpose of PRE, as in Fig. 11.3a. We also insert Φs caused by expression alteration. Such Φs are triggered by the occurrence of ϕs for any of the operands in the expression. In Fig. 11.3b, the Φ at block 3 is caused by the ϕ for a in the same block, which in turns reflects the assignment to a in block 2.

The Renaming step assigns SSA versions to h such that occurrences renamed to identical h-versions will compute to the same values. We conduct a pre-order traversal of the dominator tree similar to the Renaming step in SSA construction for variables, but with the following modifications: (1) In addition to a renaming stack for each variable, we maintain a renaming stack for the expression; (2) Entries on the expression stack are popped as our dominator tree traversal backtracks past the blocks where the expression originally received the version. Maintaining the variable and expression stacks together allows us to decide efficiently whether two occurrences of an expression should be given the same h-version.

---

[3] Adhering to the SSAPRE convention, we use lower case ϕs in the SSA form of variables and upper case Φs in the SSA form for expressions.

Fig. 11.4 Examples of
expression renaming

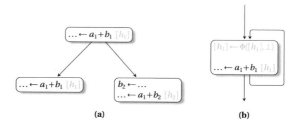

(a)   (b)

There are three kinds of occurrences of the expression in the program: (real) the occurrences in the original program, which we call *real* occurrences; (Φ-def) the inserted Φs; and (Φ-use) the use operands of the Φs, which are regarded as occurring at the ends of the direct predecessor blocks of their corresponding edges. During the visitation in Renaming, a Φ is always given a new version. For a non-Φ, i.e., cases (real) and (Φ-use), we check the current version of every variable in the expression (the version on the top of each variable's renaming stack) against the version of the corresponding variable in the occurrence on the top of the expression's renaming stack. If all the variable versions match, we assign it the same version as the top of the expression's renaming stack. If one of the variable versions does not match, for case (real), we assign it a new version, as in the example of Fig. 11.4a; for case (Φ-use), we assign the special class ⊥ to the Φ-use to denote that the value of the expression is unavailable at that point, as in the example of Fig. 11.4b. If a new version is assigned, we push the version on the expression stack.

The FRG captures all the redundancies of $a + b$ in the program. In fact, it contains just the right amount of information for determining the optimal code placement. Because strictly partial redundancies can only occur at the Φ-nodes, insertions for PRE only need to be considered at the Φs.

11.2   How SSAPRE Works

Referring to the expression being optimized as X, we use the term *placement* to denote the set of points in the *optimized* program where X's computation occurs. In contrast, the *original computation points* refer to the points in the *original* program where X's computation took place. The *original* program will be transformed to the *optimized* program by performing a set of *insertions* and *deletions*.

The objective of SSAPRE is to find a placement that satisfies the following four criteria, in this order:

- **Correctness** : X is fully available at all the original computation points.
- **Safety**: There is no insertion of X on any path that did not originally contain X.
- **Computational optimality** : No other safe and correct placement can result in fewer computations of X on any path from entry to exit in the program.
- **Lifetime optimality** : Subject to computational optimality, the life range of the temporary introduced to store X is minimized.

Each occurrence of X at its original computation point can be qualified with exactly one of the following attributes: (1) *fully redundant*; (2) *strictly partially redundant*; (3) *non-redundant*.

As a code placement problem, SSAPRE follows the same two-step process used in all PRE algorithms. The first step determines the best set of insertion points that render fully redundant as many strictly partially redundant occurrences as possible. The second step deletes fully redundant computations, taking into account the effects of the inserted computations. As we consider this second step to be well understood, the challenge lies in the first step for coming up with the best set of insertion points. The first step will tackle the safety, computational optimality, and lifetime optimality criteria, while the correctness criterion is delegated to the second step. For the rest of this section, we only focus on the first step for finding the best insertion points, which is driven by the strictly partially redundant occurrences.

We assume that all critical edges in the control-flow graph have been removed by inserting empty basic blocks at such edges (see Algorithm 3.5). In the SSAPRE approach, insertions are only performed at Φ-uses. When we say a Φ is a candidate for insertion, it means we will consider insertions at its use operands to render X available at the entry to the basic block containing that Φ. An insertion at a Φ-use means inserting X at the incoming edge corresponding to that Φ operand. In reality, the actual insertion is done at the end of the direct predecessor block.

11.2.1 The Safety Criterion

As we have pointed out at the end of Sect. 11.1, insertions only need to be considered at the Φs. The safety criterion implies that we should only insert at Φs where X is *downsafe* (fully anticipated). Thus, we perform data-flow analysis on the FRG to determine the *downsafe* attribute for Φs. Data-flow analysis can be performed with linear complexity on SSA graphs, which we illustrate with the Downsafety computation.

A Φ is not *downsafe* if there is a control-flow path from that Φ along which the expression is not computed before program exit or before being altered by the redefinition of one of its variables. Except for loops with no exit, this can only happen in one of the following cases: (dead) there is a path to exit or an alteration of the expression along which the Φ result version is not used; or (transitive) the

Φ result version appears as the operand of another Φ that is not *downsafe*. Case (dead) represents the initialization for our backward propagation of ¬*downsafe*; all other Φs are initially marked *downsafe*. The Downsafety propagation is based on case (transitive). Since a real occurrence of the expression blocks the case (transitive) propagation, we define a *has_real_use* flag attached to each Φ operand and set this flag to true when the Φ operand is defined by another Φ and the path from its defining Φ to its appearance as a Φ operand crosses a real occurrence. The propagation of ¬*downsafe* is blocked whenever the *has_real_use* flag is true. Figure 11.1 gives the Downsafety propagation algorithm. The initialization of the *has_real_use* flags is performed in the earlier Renaming phase.

Algorithm 11.1: Downsafety propagation

---

1 **foreach** $f \in \{$Φs in the program$\}$ **do**
2   **if** ∃ path P to program exit or alteration of expression along which f is not used **then**
        $downsafe(f) \leftarrow$ false
3

4 **foreach** $f \in \{$Φs in the program$\}$ **do**
5   **if not** $downsafe(f)$ **then**
6   **foreach** operand ω of f **do**
7   **if not** $has\_real\_use\ (\omega)$ **then** Reset_downsafe(ω)

8 **Function** $Reset\_downsafe(X)$
9   **if** def(X) is not a Φ **then return**
10     $f \leftarrow$ def(X)
11   **if not** $downsafe(f)$ **then return**
12     $downsafe(f) \leftarrow$ false
13   **foreach** operand ω of f **do**
14   **if not** $has\_real\_use\ (\omega)$ **then** Reset_downsafe(ω)

11.2.2   The Computational Optimality Criterion

At this point, we have eliminated the unsafe Φs based on the safety criterion. Next, we want to identify all the Φs that are possible candidates for insertion, by disqualifying Φs that cannot be insertion candidates in any computationally optimal placement. An unsafe Φ can still be an insertion candidate if the expression is fully available there, though the inserted computation will itself be fully redundant. We define the *can_be_avail* attribute for the current step, whose purpose is to identify the region where, after appropriate insertions, the computation can become fully available. A Φ is ¬*can_be_avail* if and only if inserting there violates computational optimality. The *can_be_avail* attribute can be viewed as

$$can\_be\_avail(\Phi) = downsafe(\Phi) \cup avail(\Phi).$$

We could compute the *avail* attribute separately using the full availability analysis, which involves propagation in the forward direction with respect to the control-flow graph. But this would have performed some useless computation because we do not need to know its values within the region where the Φs are *downsafe*. Thus, we choose to compute *can_be_avail* directly by initializing a Φ to be ¬*can_be_avail* if the Φ is not *downsafe* and one of its operands is ⊥. In the propagation phase, we propagate ¬*can_be_avail* forward when a ¬*downsafe* Φ has an operand that is defined by a ¬*can_be_avail* Φ and that operand is not marked *has_real_use*.

After *can_be_avail* has been computed, computational optimality could be fulfiled simply by performing insertions at all the *can_be_avail* Φs. In this case, full redundancies would be created among the insertions themselves, but the subsequent full redundancy elimination step would remove any fully redundant inserted or non-inserted computation. This would leave the earliest computations as the optimal code placement.

11.2.3   The Lifetime Optimality Criterion

To fulfil lifetime optimality, we perform a second forward propagation called Later that is derived from the well-understood partial availability analysis. The purpose is to disqualify *can_be_avail* Φs where the computation is partially available based on the original occurrences of X. A Φ is marked *later* if it is not necessary to insert there because a later insertion is possible. In other words, there exists a computationally optimal placement under which X is not available immediately after the Φ. We optimistically consider all the *can_be_avail* Φs to be *later*, except in the following cases: (real) the Φ has an operand defined by a real computation; or (transitive) the Φ has an operand that is *can_be_avail* Φ marked not *later*. Case (real) represents the initialization for our forward propagation of not *later*; all other *can_be_avail* Φs are marked *later*. The Later propagation is based on case (transitive).

The final criterion for performing insertion is to insert at the Φs where *can_be_avail* and ¬*later* hold. We call such Φs *will_be_avail*. At these Φs, insertion is performed at each operand that satisfies either of the following conditions:

> $\Big\{$ it is ⊥; or
> $\Big\{$ *has_real_use* is false and it is defined by a ¬*will_be_avail* Φ

We illustrate our discussion in this section with the example of Fig. 11.5, where the program exhibits partial redundancy that cannot be removed by safe code motion. The two Φs with their computed data-flow attributes are as shown. If insertions were based on *can_be_avail*, $a + b$ would have been inserted at the exits of blocks 4 and 5 due to the Φ in block 6, which would have resulted in

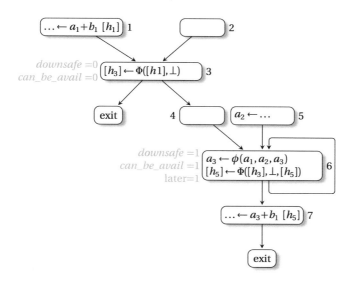

Fig. 11.5 Example to show the need for the *later* attribute

unnecessary code motion increasing register pressure. By considering *later*, no insertion is performed, which is optimal under safe PRE for this example.

11.3 Speculative PRE

If we ignore the safety requirement of PRE discussed in Sect. 11.2, the resulting code motion will involve speculation. Speculative code motion suppresses redundancy in some paths at the expense of another path where the computation is added but result is unused. As long as the paths that are burdened with more computations are executed less frequently than the paths where the redundant computations are avoided, a net gain in program performance can be achieved. Thus, speculative code motion should only be performed when there are clues about the relative execution frequencies of the paths involved.

Without profile data, speculative PRE can be conservatively performed by restricting it to loop-invariant computations. Figure 11.6 shows a loop-invariant computation $a + b$ that occurs in a branch inside the loop. This loop-invariant code motion is speculative because, depending on the branch condition inside the loop, it may be executed zero times, while moving it to the loop header causes it to execute once. This speculative loop-invariant code motion is profitable unless the path inside the loop containing the expression is never taken, which is usually not the case. When performing SSAPRE, marking Φs located at the start of loop bodies as downsafe will effect speculative loop-invariant code motion.

Fig. 11.6 Speculative
loop-invariant code motion

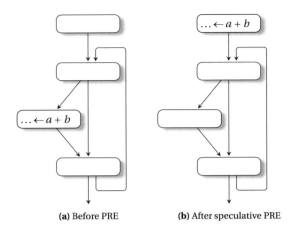

(a) Before PRE   **(b)** After speculative PRE

Computations such as indirect loads and divides are called *dangerous* computations because they may generate a fault. Dangerous computations in general should not be speculated. As an example, if we replace the expression $a + b$ in Fig. 11.6 by a/b and the speculative code motion is performed, it may cause a runtime divide-by-zero fault after the speculation because b can be 0 at the loop header, while it is never 0 in the branch that contains a/b inside the loop body.

Dangerous computations are sometimes protected by tests (or guards) placed in the code by the programmers or automatically generated by language compilers such as those for Java. When such a test occurs in the program, we say the dangerous computation is *safety-dependent* on the control-flow point that establishes its safety. At the points in the program where its safety dependence is satisfied, the dangerous instruction is *fault-safe* and can still be speculated.

We can represent safety dependencies as value dependencies in the form of abstract τ variables. Each successful runtime test defines a τ variable on its fall-through path. During SSAPRE, we attach these τ variables as additional operands to the dangerous computations related to the test. The τ variables are also put into SSA form, so their definitions can be found by following the use-def chains. The definitions of the τ variables have abstract right-hand-side values that are not allowed to be involved in any optimization. Because they are abstract, they are also omitted in the generated code after the SSAPRE phase. A dangerous computation can be defined to have more than one τ operand, depending on its semantics. When all its τ operands have definitions, it means the computation is fault-safe; otherwise, it is unsafe to speculate. By taking the τ operands into consideration, speculative PRE automatically honors the fault-safety of dangerous computations when it performs speculative code motion.

In Fig. 11.7, the program contains a non-zero test for b. We define an additional τ operand for the divide operation in a/b in SSAPRE to provide the information about whether a non-zero test for b is available. At the start of the region guarded by the non-zero test for b, the compiler inserts the definition of τ_1 with the abstract

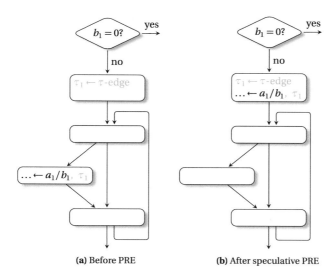

(a) Before PRE   (b) After speculative PRE

Fig. 11.7   Speculative and fault-safe loop-invariant code motion

right-hand-side value τ-edge. Any appearance of a/b in the region guarded by the non-zero test for b will have τ_1 as its τ operand. Having a defined τ operand allows a/b to be freely speculated in the region guarded by the non-zero test, while the definition of τ_1 prevents any hoisting of a/b past the non-zero test.

11.4   Register Promotion via PRE

Variables and most data in programs normally start out residing in memory. It is the compiler's job to promote those memory contents to registers as much as possible to speed up program execution. Load and store instructions have to be generated to transfer contents between memory locations and registers. The compiler also has to deal with the limited number of physical registers and find an allocation that makes the best use of them. Instead of solving these problems all at once, we can tackle them as two smaller problems separately:

1. Register promotion—We assume there is an unlimited number of registers, called *pseudo-registers* (also called symbolic registers, virtual registers, or temporaries). Register promotion will allocate variables to pseudo-registers whenever possible and optimize the placement of the loads and stores that transfer their values between memory and registers.
2. Register allocation (see Chap. 22)—This phase will fit the unlimited number of pseudo-registers to the limited number of *real* or *physical* registers.

In this chapter, we only address the register promotion problem because it can be cast as a redundancy elimination problem.

11.4.1 Register Promotion as Placement Optimization

Variables with no aliases are trivial register promotion candidates. They include the temporaries generated during PRE to hold the values of redundant computations. Variables in the program can also be determined via compiler analysis or by language rules to be alias-free. For these trivial candidates, one can rename them to unique pseudo-registers, and no load or store needs to be generated.

Our register promotion is mainly concerned with scalar variables that have aliases, indirectly accessed memory locations and constants. A scalar variable can have aliases whenever its address is taken, or if it is a global variable, since it can be accessed by function calls. A constant value is a register promotion candidate whenever some operations using it have to refer to it through register operands.

Since the goal of register promotion is to obtain the most efficient placement for loads and stores, register promotion can be modelled as two separate problems: PRE of loads, followed by PRE of stores. In the case of constant values, our use of the term *load* will extend to referring to the operation performed to put the constant value in a register. The PRE of stores does not apply to constants.

From the point of view of redundancy, loads behave like expressions: the later occurrences are the ones to be deleted. For stores, the reverse is true: as illustrated in the examples of Fig. 11.8, the earlier stores are the ones to be deleted. The PRE of stores, also called *partial dead code elimination*, can thus be treated as the dual of the PRE of loads. Thus, performing PRE of stores has the effects of moving stores forward while inserting them as early as possible. Combining the effects of the PRE of loads and stores results in optimal placements of loads and stores while minimizing the live ranges of the pseudo-registers, by virtue of the computational and lifetime optimality of our PRE algorithm.

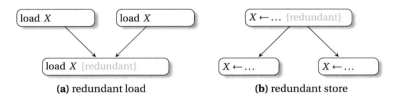

(a) redundant load   (b) redundant store

Fig. 11.8 Duality between load and store redundancies

11.4.2 Load Placement Optimization

PRE applies to any computation, including loads from memory locations or creation of constants. In program representations, loads can be either indirect through a pointer or direct. Indirect loads are automatically covered by the PRE of expressions. Direct loads correspond to scalar variables in the program, and since our input program representation is in HSSA form, the aliasing that affects the scalar variables is completely modelled by the χ and μ functions. In our representation, both direct loads and constants are leaves of the expression trees. When we apply SSAPRE to direct loads, since the hypothetical temporary h can be regarded as the candidate variable itself, the FRG corresponds somewhat to the variable's SSA graph, so the Φ-insertion step and Rename step can be streamlined.

When working on the PRE of memory loads, it is important to also take into account the stores, which we call *l-value* occurrences. A store of the form $X \leftarrow < \mathtt{expr} >$ can be regarded as being made up of the sequence:

$$r \leftarrow < \mathtt{expr} >$$
$$X \leftarrow r$$

Because the pseudo-register r contains the current value of X, any subsequent occurrences of the load of X can reuse the value from r and thus can be regarded as redundant. Figure 11.9 gives examples of loads made redundant by stores.

When we perform the PRE of loads, we thus take the store occurrences into consideration. The Φ-insertion step will insert Φs at the iterated dominance frontiers of store occurrences. In the Rename step, a store occurrence is always given a new h-version, because a store is a definition. Any subsequent load renamed to the same h-version is redundant with respect to the store.

We apply the PRE of loads (LPRE) first, followed by the PRE of stores (STRE). This ordering is based on the fact that LPRE is not affected by the result of STRE, but LPRE creates more opportunities for the SPRE by deleting loads that would otherwise have blocked the movement of stores. In addition, speculation is required for the PRE of loads and stores in order for register promotion to do a decent job in loops.

The example in Fig. 11.10 illustrates what is discussed in this section. During LPRE, $A \leftarrow \ldots$ is regarded as a store occurrence. The hoisting of the load of A to

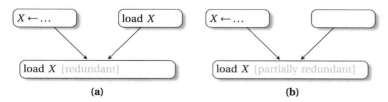

(a)   **(b)**

Fig. 11.9 Redundant loads after stores

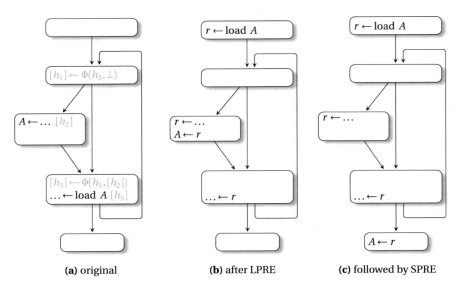

(a) original   **(b)** after LPRE   **(c)** followed by SPRE

Fig. 11.10 Register promotion via load PRE followed by store PRE

the loop header does not involve speculation. The occurrence of $A \leftarrow \ldots$ causes r to be updated by splitting the store into the two statements $r \leftarrow \ldots ; A \leftarrow r$. In the PRE of stores (SPRE), speculation is needed to sink $A \leftarrow \ldots$ to outside the loop because the store occurs in a branch inside the loop. Without performing LPRE first, the load of A inside the loop would have blocked the sinking of $A \leftarrow \ldots$.

11.4.3   Store Placement Optimization

As mentioned earlier, SPRE is the dual of LPRE. Code motion in SPRE will have the effect of moving stores forward with respect to the control-flow graph. Any presence of (aliased) loads has the effect of blocking the movement of stores or rendering the earlier stores non-redundant.

To apply the dual of the SSAPRE algorithm, it is necessary to compute a program representation that is the dual of the SSA form, the *static single use* (SSU) form (see Chap. 13—SSU is a special case of SSI). In SSU, use-def edges are factored at divergence points in the control-flow graph using σ-functions (see Sect. 13.1.4). Each use of a variable establishes a new version (we say the load *uses* the version), and every store reaches exactly one load.

We call our store PRE algorithm SSUPRE, which is made up of the corresponding steps in SSAPRE. The insertion of σ-functions and renaming phases constructs the SSU form for the variable whose store is being optimized. The data-flow analyses consist of UpSafety to compute the *upsafe* (fully available)

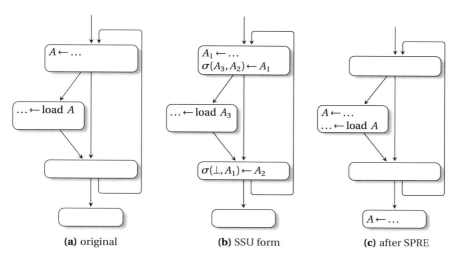

(a) original   **(b)** SSU form   **(c)** after SPRE

Fig. 11.11 Example of program in SSU form and the result of applying SSUPRE

attribute, CanBeAnt to compute the *can_be_ant* attribute, and Earlier to compute the *earlier* attribute. Though store elimination itself does not require the introduction of temporaries, lifetime optimality still needs to be considered for the temporaries introduced in the LPRE phase, which hold the values to the point where the stores are placed. It is desirable not to sink the stores too far down.

Figure 11.11 gives the SSU form and the result of SSUPRE on an example program. The sinking of the store to outside the loop is traded for the insertion of a store in the branch inside the loop. The optimized code no longer exhibits any store redundancy.

11.5   Value-Based Redundancy Elimination

The PRE algorithm we have described so far is not capable of recognizing redundant computations among lexically different expressions that yield the same value. In this section, we discuss redundancy elimination based on value analysis.

11.5.1   Value Numbering

The term *value number* originates from a hash-based method for recognizing when two expressions evaluate to the same value within a basic block. The value number of an expression tree can be regarded as the index of its hashed entry in the hash table. An expression tree is hashed recursively bottom-up, starting with the leaf

$$a \leftarrow \ldots \qquad \text{VN}(a) = v_1$$
$$c \leftarrow \ldots \qquad \text{VN}(c) = v_2$$
$$b \leftarrow c \qquad \text{VN}(b) = \text{VN}(c) = v_2$$
$$c \leftarrow a + c \qquad \text{VN}(c) = \text{VN}(+(\text{VN}(a), \text{VN}(c))) = \text{VN}(+(v_1, v_2)) = v_3$$
$$d \leftarrow a + b \qquad \text{VN}(d) = \text{VN}(+(\text{VN}(a), \text{VN}(b))) = \text{VN}(+(v_1, v_2)) = v_3$$

(a) straight-line code   **(b)** value numbers

Fig. 11.12 Value numbering in a local scope

$$a_1 \leftarrow \ldots \qquad \text{VN}(a_1) = v_1$$
$$c_1 \leftarrow \ldots \qquad \text{VN}(c_1) = v_2$$
$$b_1 \leftarrow c_1 \qquad \text{VN}(b_1) = \text{VN}(c_1) = v_2$$
$$c_2 \leftarrow a_1 + c_1 \qquad \text{VN}(c_2) = \text{VN}(+(\text{VN}(a_1), \text{VN}(c_1))) = \text{VN}(+(v_1, v_2)) = v_3$$
$$d_1 \leftarrow a_1 + b_1 \qquad \text{VN}(d_1) = \text{VN}(+(\text{VN}(a_1), \text{VN}(b_1))) = \text{VN}(+(v_1, v_2)) = v_3$$

(a) processed statements   **(b)** value numbers

Fig. 11.13 Global value numbering on SSA form

nodes. Each internal node is hashed based on its operator and the value numbers of its operands. The local algorithm for value numbering will conduct a scan down the instructions in a basic block, assigning value numbers to the expressions. At an assignment, the assigned variable will be assigned the value number of the right-hand side expression. The assignment will also cause any value number that refers to that variable to be killed. For example, the program code in Fig. 11.12a will result in the value numbers v_1, v_2, and v_3 shown in Fig. 11.12b. Note that variable c is involved with both value numbers v_2 and v_3 because it has been redefined.

SSA form enables value numbering to be easily extended to the global scope, called global value numbering (GVN), because each SSA version of a variable corresponds to at most one static value for the variable. In the example of Fig. 11.13, a traversal along any topological ordering of the SSA graph can be used to assign value numbers to variables. One subtlety is regarding the ϕ-functions. When we value number a ϕ-function, we would like the value numbers for its use operands to have been determined already. One strategy is to perform the global value numbering by visiting the nodes in the control-flow graph in a reverse post-order traversal of the dominator tree. This traversal strategy can minimize the instances when a ϕ-use has an unknown value number, which arises only in the case of back edges from loops. When this arises, we have no choice but to assign a new value number to the variable defined by the ϕ-function. For example, in the following loop:

```
1  i₁ ← 0
2  j₁ ← 0
3  while <cond> do
4  │   i₂ ← φ(i₃, i₁)
5  │   j₂ ← φ(j₃, j₁)
6  │   i₃ ← i₂ + 4
7  └   j₃ ← j₂ + 4
```

When we try to hash a value number for either of the two ϕs, the value numbers for i_3 and j_3 are not yet determined. As a result, we create different value numbers for i_2 and j_2. This makes the above algorithm unable to recognize that i_2 and j_2 can be given the same value number, or that i_3 and j_3 can be given the same value number.

The above hash-based value numbering algorithm can be regarded as pessimistic, because it will not assign the same value number to two different expressions unless it can prove they compute the same value. There exists a different approach (see Sect. 11.6 for references) to performing value numbering that is not hash-based and is optimistic. It does not depend on any traversal over the program's flow of control and so is not affected by the presence of back edges. The algorithm partitions all the expressions in the program into *congruence classes*. Expressions in the same congruence class are considered *equivalent* because they evaluate to the same static value. The algorithm is optimistic because when it starts, it assumes all expressions that have the same operator to be in the same congruence class. Given two expressions within the same congruence class, if their operands at the same operand position belong to different congruence classes, the two expressions may compute to different values and thus should not be in the same congruence class. This is the subdivision criterion. As the algorithm iterates, the congruence classes are subdivided into smaller ones, while the total number of congruence classes increases. The algorithm terminates when no more subdivisions can occur. At this point, the set of congruence classes in this final partition will represent all the values in the program that we care about, and each congruence class is assigned a unique value number.

While such a partition-based algorithm is not obstructed by the presence of back edges, it does have its own deficiencies. Because it has to consider one operand position at a time, it is not able to apply commutativity to detect more equivalences. Since it is not applied bottom-up with respect to the expression tree, it is not able to apply algebraic simplifications while value numbering. To get the best of both the hash-based and the partition-based algorithms, it is possible to apply the two algorithms independently and then combine their results together to shrink the final set of value numbers.

11.5.2 Redundancy Elimination Under Value Numbering

So far, we have discussed finding computations that compute to the same values but have not addressed eliminating the redundancies among them. Two computations that compute to the same value exhibit redundancy only if there is a control-flow path that leads from one to the other.

An obvious approach is to consider PRE for each value number separately. This can be done by introducing, for each value number, a temporary that stores the redundant computations. But value-number-based PRE has to deal with the issue of *how* to generate an insertion. Because the same value can come from different forms of expressions at different points in the program, it is necessary to determine which form to use at each insertion point. If the insertion point is outside the live range of any variable version that can compute that value, then the insertion point has to be disqualified. Due to this complexity, and the expectation that strictly partial redundancy is rare among computations that yield the same value, it seems to be sufficient to perform only full redundancy elimination among computations that have the same value number.

However, it is possible to broaden the scope and consider PRE among lexically identical expressions and value numbers at the same time. In this hybrid approach, it is best to relax our restriction on the style of program representation described in Sect. 11. By not requiring Conventional SSA Form, we can more effectively represent the flow of values among the program variables. By considering the live range of each SSA version to extend from its definition to program exit, we allow its value to be used whenever convenient. The program representation can even be in the form of triplets, in which the result of every operation is immediately stored in a temporary. It will just assign the value number of the right-hand side to the left-hand-side variables. This hybrid approach (GVN-PRE—see below) can be implemented based on an adaptation of the SSAPRE framework. Since each ϕ-function in the input can be viewed as merging different value numbers from the direct predecessor blocks to form a new value number, the Φ-function insertion step will be driven by the presence of ϕs for the program variables. Several FRGs can be formed, each being regarded as a representation of the flow and merging of computed values. Using each individual FRG, PRE can be performed by applying the remaining steps of the SSAPRE algorithm.

11.6 Further Reading

The concept of partial redundancy was first introduced by Morel and Renvoise. In their seminal work [201], Morel and Renvoise showed that global common subexpressions and loop-invariant computations are special cases of partial redundancy, and they formulated PRE as a code placement problem. The PRE algorithm developed by Morel and Renvoise involves bidirectional data-flow analysis, which incurs more overhead than unidirectional data-flow analysis. In addition, their algorithm does not yield optimal results in certain situations. A better placement strategy, called lazy code motion (LCM), was later developed by Knoop et al. [170, 172]. It improved on Morel and Renvoise's results by avoiding unnecessary code movements, by removing the bidirectional nature of the original PRE data-flow analysis and by proving the optimality of their algorithm. Since lazy code motion was introduced, there have been alternative formulations of PRE algorithms that achieve the same optimal results but differ in the formulation approach and implementation details [98, 106, 217, 313].

The above approaches to PRE are all based on encoding program properties in bit-vector forms and the iterative solution of data-flow equations. Since the bit-vector representation uses basic blocks as its granularity, a separate algorithm is needed to detect and suppress local common subexpressions. Chow et al. [73, 164] came up with the first SSA-based approach to perform PRE. Their SSAPRE algorithm is an adaptation of LCM that takes advantage of the use-def information inherent in SSA. It avoids having to encode data-flow information in bit-vector form and eliminates the need for a separate algorithm to suppress local common subexpressions. Their algorithm was the first to make use of SSA to solve data-flow problems for expressions in the program, taking advantage of SSA's sparse representation so that fewer steps are needed to propagate data-flow information. The SSAPRE algorithm thus brings the many desirable characteristics of SSA-based solution techniques to PRE.

In the area of speculative PRE, Murphy et al. [206] introduced the concept of fault-safety and used it in the SSAPRE framework for the speculation of dangerous computations. When execution profile data are available, it is possible to tailor the use of speculation to maximize runtime performance for the execution that matches the profile. Xue and Cai [312] presented a computationally and lifetime optimal algorithm for speculative PRE based on profile data. Their algorithm uses data-flow analysis based on bit-vector and applies minimum cut to flow networks formed out of the control-flow graph to find the optimal code placement. Zhou et al. [317] applied the minimum cut approach to flow networks formed out of the FRG in the SSAPRE framework to achieve the same computational and lifetime optimal code motion. They showed their sparse approach based on SSA results in smaller flow networks, enabling the optimal code placements to be computed more efficiently.

Lo et al. [187] showed that register promotion can be achieved by load placement optimization followed by store placement optimization. Other optimizations can potentially be implemented using the SSAPRE framework, for instance code hoisting, register shrink-wrapping [70], and live range shrinking. Moreover, PRE has traditionally provided the context for integrating additional optimizations into its framework. They include operator strength reduction [171] and linear function test replacement [163].

Hashed-based value numbering originated from Cocke and Schwartz [77], and Rosen et al. [249] extended it to global value numbering based on SSA. The partition-based algorithm was developed by Alpern et al. [6]. Briggs et al. [49] presented refinements to both the hash-based and partition-based algorithms, including applying the hash-based method in a post-order traversal of the dominator tree.

VanDrunen and Hosking proposed A-SSAPRE (anticipation-based SSAPRE) which removes the requirement of Conventional SSA Form and is best for program representations in the form of triplets [296]. Their algorithm determines optimization candidates and constructs FRGs via a depth-first, pre-order traversal over the basic blocks of the program. Within each FRG, non-lexically identical expressions are allowed, as long as there are potential redundancies among them. VanDrunen and Hosking [297] subsequently presented GVN-PRE (Value-based Partial Redundancy Elimination), which is claimed to subsume both PRE and GVN.

Part III
Extensions

Chapter 12
Introduction

Vivek Sarkar and Fabrice Rastello

So far, we have introduced the foundations of SSA form and its use in different program analyses. We now explain the need for extensions to SSA form to enable a larger class of program analyses. The extensions arise from the fact that many analyses need to make finer-grained distinctions between program points and data accesses than what can be achieved by vanilla SSA form. However, these richer flavours of extended SSA-based analyses still retain many of the benefits of SSA form (e.g., sparse data-flow propagation) which distinguish them from classical data-flow frameworks for the same analysis problems.

12.1 Static Single Information Form

The sparseness in vanilla SSA form arises from the observation that information for an unaliased scalar variable can be safely propagated from its (unique) definition to all its reaching uses without examining any intervening program points. As an example, SSA-based constant propagation aims to compute for each single assignment variable, the (usually over-approximated) set of possible values carried by the definition of that variable. For instance, consider an instruction that defines a variable a and uses two variables, b and c. An example is $a = b+c$. In an SSA-form program, constant propagation will determine if a is a constant by looking directly at the definition point of b and at the definition point of c. We say that information

V. Sarkar (✉)
Georgia Institute of Technology, Atlanta, GA, USA
e-mail: vsarkar@gatech.edu

F. Rastello
Inria, Grenoble, France
e-mail: fabrice.rastello@inria.fr

© The Author(s), under exclusive license to Springer Nature Switzerland AG 2022
F. Rastello, F. Bouchez Tichadou (eds.), *SSA-based Compiler Design*,
https://doi.org/10.1007/978-3-030-80515-9_12

is propagated from these two definition points directly to the instruction $a = b + c$. However, there are many analyses for which definition points are not the only source of new information. For example, consider an if-then-else statement that uses a in both the then and else parts. If the branch condition involves a, say $a == 0$, we now have additional information that can distinguish the value of a in the then and else parts, even though both uses have the same reaching definition. Likewise, the use of a reference variable p in certain contexts can be the source of new information for subsequent uses, e.g., the fact that p is non-null because no exception was thrown at the first use.

The goal of Chap. 13 is to present a systematic approach to deal with these additional sources of information while still retaining the space and time benefits of vanilla SSA form. This approach is called static single information (SSI) form, and it involves additional renaming of variables to capture new sources of information. For example, SSI form can provide distinct names for the two uses of a in the then and else parts of the if-then-else statement mentioned above. This additional renaming is also referred to as *live range splitting*, akin to the idea behind splitting live ranges in optimized register allocation. The sparseness of SSI form follows from formalization of the Partitioned Variable Lattice (PVL) problem and the Partitioned Variable Problem (PVP), both of which establish orthogonality among transfer functions for renamed variables. The ϕ-functions inserted at join nodes in vanilla SSA form are complemented by σ-functions in SSI form that can perform additional renaming at branch nodes and interior (instruction) nodes. Information can be propagated in the forward and backward directions using SSI form, enabling analyses such as range analysis (leveraging information from branch conditions) and null-pointer analysis (leveraging information from prior uses) to be performed more precisely than with vanilla SSA form.

12.2   Control Dependencies

So as to expose parallelism and locality, one needs to get rid of the CFG at some points. For loop transformations, software pipelining, there is a need to manipulate a higher degree of abstraction to represent the iteration space of nested loops and to extend data-flow information to this abstraction. One can expose even more parallelism (at the level of instructions) by replacing control flow by control dependencies: the goal is either to express a predicate expression under which a given basic block is to be executed or to select afterwards (using similar predicate expressions) the correct value among a set of eagerly computed ones.

1. Technically, we say that SSA provides data flow (data dependencies). The goal is to enrich it with control dependencies. The program dependence graph (PDG) constitutes the basis of such IR extensions. Gated single assignment (gated SSA, GSA) mentioned below provides an interpretable (data- or demand-driven) IR

that uses this concept. Psi-SSA (ψ-SSA) also mentioned below is a very similar IR but more appropriate to code generation for architectures with predication.

2. Note that such extensions sometimes face difficulties handling loops correctly (need to avoid deadlock between the loop predicate and the computation of the loop body, replicate the behaviour of infinite loops, etc.). However, we believe that, as we will illustrate further, loop carried control dependencies complicate the recognition of possible loop transformations: it is usually better to represent loops and their corresponding iteration space using a dedicated abstraction.

12.3   Gated SSA Forms

As already mentioned, one of the strengths of SSA form is its associated data-flow graph (DFG), the SSA graph that is used to propagate directly the information along the def-use chains. This is what makes data-flow analysis sparse. By combining the SSA graph with the control-flow graph, static analysis can be made context-sensitive. This can be done in a more natural and powerful way by incorporating in a unified representation both the data-flow and control-flow information.

The program dependence graph (PDG) adds to the data dependence edges (SSA graph as the data dependence graph—DDG) the control dependence edges (control dependence graph—CDG). As already mentioned, one of the main motivations for the development of the PDG was to aid automatic parallelization of instructions across multiple basic blocks. However, in practice, it also exposes the relationship between the control predicates and their related control-dependent instructions, thus allowing us to propagate the associated information. A natural way to represent this relationship is through the use of gating functions that are used in some extensions such as the gated SSA (GSA) or the value state dependence graph (VSDG) . Gating functions are directly interpretable versions of ϕ-nodes. As an example, $\phi_{if}(P, v_1, v_2)$ can be interpreted as a function that selects the value v_1 if predicated P evaluates to true and the value v_2 otherwise. PDG, GSA, and VSDG are described in Chap. 14.

12.4   Psi-SSA Form

ψ-SSA form (Chap. 15) addresses the need for modelling Static Single Assignment form in predicated operations. A predicated operation is an alternate representation of a fine-grained control-flow structure, often obtained by using the well-known *if-conversion* transformation (see Chap. 20). A key advantage of using predicated operations in a compiler's intermediate representation is that it can often enable more optimizations by creating larger basic blocks compared to approaches in which predicated operations are modelled as explicit control-flow graphs. From an SSA-form perspective, the challenge is that a predicated operation may or may not update its definition operand, depending on the value of the predicate guarding that assignment. This challenge is addressed in ψ-SSA form by introducing ψ-functions that perform merge functions for predicated operations, analogous to the merge performed by ϕ-functions at join points in vanilla SSA form.

In general, a ψ-function has the form, $a_0 = \psi (p_1?a_1, ..., p_i?a_i, ..., p_n?a_n)$, where each input argument a_i is associated with a predicate p_i as in a nested if-then-else expression for which the value returned is the rightmost argument whose predicate is true. Observe that the semantic of a ψ-function imposes the logical disjunction of its predicates to evaluate to true. A number of algebraic transformations can be performed on ψ-functions, including ψ-inlining, ψ-reduction, ψ-projection, ψ-permutation, and ψ-promotion. Chapter 15 also includes an algorithm for transforming a program out of Psi-SSA form that extends the standard algorithm for destruction of vanilla SSA form.

12.5 Hashed SSA Form

The motivation for SSI form arose from the need to perform additional renaming to distinguish among different uses of an SSA variable. Hashed SSA (HSSA) form introduced in Chap. 16 addresses another important requirement, viz., the need to model aliasing among variables. For example, a static use or definition of indirect memory access $*p$ in the C language could represent the use or definition of multiple local variables whose addresses can be taken and may potentially be assigned to p along different control-flow paths. To represent aliasing of local variables, HSSA form extends vanilla SSA form with *MayUse* (μ) and *MayDef* (χ) functions to capture the fact that a single static use or definition could potentially impact multiple variables. Note that MayDef functions can result in the creation of new names (versions) of variables, compared to vanilla SSA form. HSSA form does not take a position on the accuracy of alias analysis that it represents. It is capable of representing the output of any alias analysis performed as a pre-pass to HSSA construction. As summarized above, a major concern with HSSA form is that its size could be quadratic in the size of the vanilla SSA form, since each use or definition can now be augmented by a set of MayUse's and MayDef's, respectively. A heuristic approach to dealing with this problem is to group together all variable versions that have no "real" occurrence in the program, i.e., do not appear in a real instruction outside of a ϕ, μ, or χ function. These versions are grouped together into a single version called the *zero version* of the variable.

In addition to aliasing of locals, it is important to handle the possibility of aliasing among heap-allocated variables. For example, $*p$ and $*q$ may refer to the same location in the heap, even if no aliasing occurs among local variables. HSSA form addresses this possibility by introducing a *virtual variable* for each address expression used in an indirect memory operation and renaming virtual variables with ϕ-functions as in SSA form. Further, the alias analysis pre-pass is expected to provide information on which virtual variables may potentially be aliased, thereby leading to the insertion of μ or χ functions for virtual variables as well. Global value numbering is used to increase the effectiveness of the virtual variable approach, since all indirect memory accesses with the same address expression can be merged into a single virtual variable (with SSA renaming as usual). In fact, the Hashed SSA name in HSSA form comes from the use of hashing in most value-numbering algorithms.

12.6 Array SSA Form

In contrast to HSSA form, Array SSA form (Chap. 17) takes an alternate approach to modelling aliasing of indirect memory operations by focusing on aliasing in arrays as its foundation. The aliasing problem for arrays is manifest in the fact that accesses to elements $A[i]$ and $A[j]$ of array A refer to the same location when $i = j$. This aliasing can occur with just local array variables, even in the absence of pointers and heap-allocated data structures. Consider a program with a definition of $A[i]$ followed by a definition of $A[j]$. The vanilla SSA approach can be used to rename these two definitions to (say) $A_1[i]$ and $A_2[j]$. The challenge with arrays arises when there is a subsequent use of $A[k]$. For scalar variables, the reaching definition for this use can be uniquely identified in vanilla SSA form. However, for array variables, the reaching definition depends on the subscript values. In this example, the reaching definition for $A[k]$ will be A_2 or A_1 if $k == j$ or $k == i$ (or a prior definition A_0 if $k \neq j$ and $k \neq i$). To provide $A[k]$ with a single reaching definition, Array SSA form introduces a definition-Φ ($d\Phi$) operator that represents the merge of $A_2[j]$ with the prevailing value of array A prior to A_2. The result of this $d\Phi$ operator is given a new name, A_3 (say), which serves as the single definition that reaches use $A[k]$ (which can then be renamed to $A_3[k]$). This extension enables sparse data-flow propagation algorithms developed for vanilla SSA form to be applied to array variables, as illustrated by the algorithm for *sparse constant propagation of array elements* presented in this chapter. The accuracy of analyses for Array SSA form depends on the accuracy with which pairs of array subscripts can be recognized as being *definitely same* ($\mathcal{DS}$) or *definitely different* ($\mathcal{DD}$).

To model heap-allocated objects, Array SSA form builds on the observation that all indirect memory operations can be modelled as accesses to elements of abstract arrays that represent disjoint subsets of the heap. For modern object-oriented languages such as Java, type information can be used to obtain a partitioning of the heap into disjoint subsets, e.g., instances of field x are guaranteed to be disjoint from instances of field y. In such cases, the set of instances of field x can be modelled as a logical array (map) $\mathcal{H}^x$ that is indexed by the object reference (key). The problem of resolving aliases among field accesses $p.x$ and $q.x$ then becomes equivalent to the problem of resolving aliases among array accesses $\mathcal{H}^x[p]$ and $\mathcal{H}^x[q]$, thereby enabling Array SSA form to be used for analysis of objects as in the algorithm for *redundant load elimination* among object fields presented in this chapter. For weakly typed languages such as C, the entire heap can be modelled as a single heap array. As in HSSA form, an alias analysis pre-pass can be performed to improve the accuracy of *definitely same* and *definitely different* information. In particular, global value numbering is used to establish *definitely same* relationships, analogous to its use in establishing equality of address expressions in HSSA form. In contrast to HSSA form, the size of Array SSA form is guaranteed to be linearly proportional to the size of a comparable scalar SSA form for the original program (if all array variables were treated as scalars for the purpose of the size comparison).

12.7 Memory-Based Data Flow

SSA provides data flow/dependencies between *scalar* variables. Execution order of side effect instructions must also be respected. Indirect memory access can be considered very conservatively as such and lead to (sometimes called state, see Chap. 14) dependence edges. Too conservative dependencies annihilate the potential of optimizations. Alias analysis is the first step towards more precise dependence information. Representing this information efficiently in the IR is important. One could simply add a dependence (or a flow arc) between two "consecutive" instructions that may or must alias. Then, in the same spirit as SSA ϕ-nodes aim at combining the information as early as possible (as opposed to standard def-use chains, see the discussion in Chap. 2), similar nodes can be used for memory dependencies. Consider the following sequential C-like code involving pointers, where p and q may alias:

$$*p \leftarrow \ldots; \quad *q \leftarrow \ldots; \quad \ldots \leftarrow *p; \quad \ldots \leftarrow *q;$$

Without the use of ϕ-nodes, the amount of def-use chains required to link the assignments to their uses would be quadratic (4 here). Hence, the usefulness of generalizing SSA and its ϕ-node for scalars to handle memory access for sparse analyses. HSSA (see Chap. 16) and Array-SSA (see the Chap. 17) are two different implementations of this idea. One has to admit that if this early combination is well suited for analysis or interpretation, the introduction of a ϕ-function might add a control dependence to an instruction that would not exist otherwise. In other words, only simple loop carried dependencies can be expressed this way. Let us illustrate this point using a simple example:

for i **do**
$$\quad \lfloor \quad A[i] \leftarrow f(A[i-2])$$

Here, the computation at iteration indexed by $i + 2$ accesses the value computed at iteration indexed by i. Suppose we know that $f(x) > x$ and that prior to entering the loop $A[*] \geq 0$. Consider the following SSA-like representation:

for i **do**
$$\quad \lfloor \quad A_2 \leftarrow \phi(A_0, A_1)$$
$$\quad \quad A_2 \leftarrow \phi(j = i : A_1[j] \leftarrow f(A_2[j-2]), j \neq i : A_1[j] \leftarrow A_2[j])$$

This would easily allow for the propagation of information that $A_2 \geq 0$. On the other hand, by adding this ϕ-node, it becomes difficult to devise that iteration i and $i + 1$ can be executed in parallel: the ϕ-node adds a loop carried dependence. If you are interested in performing loop transformations that are more sophisticated than just exposing fully parallel loops (such as loop interchange, loop tiling, or multidimensional software pipelining), then (dynamic) single assignment forms should be your friend. There exist many formalisms including Kahn process networks (KPN) or Fuzzy Data-flow Analysis (FADA) that implement this idea. But each time restrictions apply. This is part of the huge research area of automatic parallelization outside of the scope of this book. For further details, we refer the reader to the corresponding Encyclopedia of Parallel Computing [216].

Chapter 13
Static Single Information Form

Fernando Magno Quintão Pereira and Fabrice Rastello

The objective of a data-flow analysis is to discover facts that are true about a program. We call such facts *information*. Using the notation introduced in Chap. 8, information is an element in the data-flow lattice. For example, the information that concerns liveness analysis is the set of variables alive at a certain program point. Similarly to liveness analysis, many other classical data-flow approaches bind information to pairs formed by a variable and a program point. However, if an invariant occurs for a variable v at any program point where v is alive, then we can associate this invariant directly with v. If the intermediate representation of a program guarantees this correspondence between information and variable for every variable, then we say that the program representation provides the *Static Single Information* (SSI) property.

In Chap. 8 we have shown how the SSA form allows us to solve sparse forward data-flow problems such as constant propagation. In the particular case of constant propagation, the SSA form lets us assign to each variable the invariant—or information—of being constant or not. The SSA intermediate representation gives us this invariant because it splits the live ranges of variables in such a way that each variable name is defined only once. Now we will show that live range splitting can also provide the SSI property not only to forward but also to backward data-flow analyses.

Different data-flow analyses might extract information from different program facts. Therefore, a program representation may provide the SSI property to some data-flow analyses but not to all of them. For instance, the SSA form naturally

F. M. Q. Pereira (✉)
Federal University of Minas Gerais, Belo Horizonte, Brazil
e-mail: fernando@dcc.ufmg.br

F. Rastello
Inria, Grenoble, France
e-mail: fabrice.rastello@inria.fr

© The Author(s), under exclusive license to Springer Nature Switzerland AG 2022   165
F. Rastello, F. Bouchez Tichadou (eds.), *SSA-based Compiler Design*,
https://doi.org/10.1007/978-3-030-80515-9_13

provides the SSI property to the reaching definition analysis. Indeed, the SSA form provides the static single information property to any data-flow analysis that obtains information at the definition sites of variables. These analyses and transformations include copy and constant propagation, as illustrated in Chap. 8. However, for a data-flow analysis that derives information from the use sites of variables, such as the class inference analysis that we will describe in Sect. 13.1.6, the information associated with a variable might not be unique along its entire live range even under SSA: In that case the SSA form does not provide the SSI property.

There are extensions of the SSA form that provide the SSI property to more data-flow analyses than the original SSA does. Two classic examples—detailed later—are the *Extended-SSA* (e-SSA) form and the *Static Single Information* (SSI) form. The e-SSA form provides the SSI property to analyses that take information from the definition site of variables, and also from conditional tests where these variables are used. The SSI form provides the static single information property to data-flow analyses that extract information from the definition sites of variables and from the last use sites (which we define later). These different intermediate representations rely on a common strategy to achieve the SSI property: *live range splitting*. In this chapter we show how to use live range splitting to build program representations that provide the static single information property to different types of data-flow analyses.

13.1 Static Single Information

The goal of this section is to define the notion of *Static Single Information*, and to explain how it supports the sparse data-flow analyses discussed in Chap. 8. With this aim in mind, we revisit the concept of sparse analysis in Sect. 13.1.1. There is a special class of data-flow problems, which we call *Partitioned Lattice per Variable* (PLV), which fits into the sparse data-flow framework of this chapter very well. We will look more carefully into these problems in Sect. 13.1.2. The intermediate program representations discussed in this chapter provide the static single information property—formalized in Sect. 13.1.3—to any PLV problem. In Sect. 13.1.5 we give algorithms to solve sparsely any data-flow problem that contains the SSI property. This sparse framework is very broad: Many well-known data-flow problems are partitioned lattice, as we will see in the examples in Sect. 13.1.6.

13.1.1 Sparse Analysis

Traditionally, non-relational data-flow analyses bind information to pairs formed by a variable and a program point. Consider for example the problem of *range analysis*, i.e., estimating the interval of values that an integer variable may assume throughout

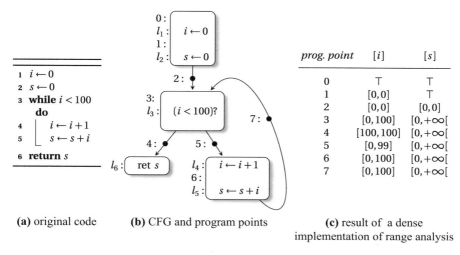

(a) original code   **(b)** CFG and program points   **(c)** result of a dense
                                                                        implementation of range analysis

Fig. 13.1 An example of a dense data-flow analysis that finds the range of possible values associated with each variable at each program point

the execution of a program. A traditional implementation of this analysis would find, for each pair (v, p) of a variable v and a program point p, the interval of possible values that v might assume at p (see the example in Fig. 13.1). In this case, we call a program point any region between two consecutive instructions and denote as $[v]$ the abstract information associated with variable v. Because this approach keeps information at each program point, we call it *dense*, in contrast to the sparse analyses seen in Chap. 8, Sect. 8.2.

The dense approach might result in a large quantity of redundant information during the data-flow analysis. For instance, if we denote $[v]^p$ the abstract state of variable v at program point p, we have for instance in our example $[i]^1 = [i]^2$, $[s]^5 = [s]^6$ and $[i]^6 = [i]^7$ (see Fig. 13.1). This redundancy happens because some transfer functions are identities: In range analysis, an instruction that neither defines nor uses any variable is associated with an identity transfer function. Similarly, the transfer function that updates the abstract state of i at program point 2 is an identity, because the instruction immediately before 2 does not add any new information to the abstract state of i, $[i]^2$ is updated with the information that flows directly from the direct predecessor point 1.

The goal of *sparse* data-flow analysis is to shortcut the identity transfer functions, a task that we accomplish by grouping contiguous program points bound to identities into larger regions. Solving a data-flow analysis sparsely has many advantages over doing it densely: Because we do not need to keep bitvectors associated with each program point, the sparse solution tends to be more economical in terms of space and time. Going back to our example, a given variable v may be mapped to the same interval along many consecutive program points. Furthermore, if the information associated with a variable is invariant along its entire live range, then we can bind this information to the variable itself. In other words, we can replace all the constraint variables $[v]^p$ by a single constraint variable $[v]$, for each variable v and every $p \in \text{live}(v)$.

Although not every data-flow problem can be easily solved sparsely, many of them can as they fit into the family of PLV problems described in the next section.

13.1.2 Partitioned Lattice per Variable (PLV) Problems

The non-relational data-flow analysis problems we are interested in are the ones that bind information to pairs of program variables and program points. We refer to this class of problems as *Partitioned Lattice per Variable* problems and formally describe them as follows.

Definition 13.1 (PLV) Let $\mathcal{V} = \{v_1, \ldots, v_n\}$ be the set of program variables. Let us consider, without loss of generality, a forward data-flow analysis that searches for a maximum. This data-flow analysis can be written as an equation system that associates each program point p, with n element of a lattice $\mathcal{L}$, given by the following equation:

$$x^p = \bigwedge_{s \in \text{directpreds}(p)} F^{s \to p}(x^s),$$

where x^p denotes the abstract state associated with program point p, and $F^{s \to p}$ is the transfer function from direct predecessor s to p. The analysis can alternatively be written as a constraint system that binds to each program point p and each $s \in$ directpreds(p) the equation $x^p = x^p \wedge F^{s \to p}(x^s)$ or, equivalently, the in equation

$$x^p \sqsubseteq F^{s \to p}(x^s).$$

The corresponding Maximum Fixed Point (MFP) problem is said to be a *Partitioned Lattice per Variable Problem* iff $\mathcal{L}$ can be decomposed into the product of $\mathcal{L}_{v_1} \times \cdots \times \mathcal{L}_{v_n}$ where each $\mathcal{L}_{v_i}$ is the lattice associated with program variable v_i. In other words x^s can be written as $([v_1]^s, \ldots, [v_n]^s)$ where $[v]^s$ denotes the abstract state associated with variable v and program point s. $F^{s \to p}$ can thus be decomposed into the product of $F_{v_1}^{s \to p} \times \cdots \times F_{v_n}^{s \to p}$ and the constraint system decomposed into the inequalities $[v_i]^p \sqsubseteq F_{v_i}^{s \to p}([v_1]^s, \ldots, [v_n]^s)$.

Going back to range analysis, if we denote as $\mathcal{I}$ the lattice of integer intervals, then the overall lattice can be written as $\mathcal{L} = \mathcal{I}^n$, where n is the number of variables. Note that the class of PLV problems includes a smaller class of problems called *Partitioned Variable Problems* (PVP). These analyses, which include live variables reaching definitions and forward/backward printing, can be decomposed into a set of sparse data-flow problems—usually one per variable—each independent of the others.

Note that not all data-flow analyses are PLV, for instance problems dealing with relational information, such as "$i < j$?", which needs to hold information on *pairs* of variables.

13.1.3   The Static Single Information Property

If the information associated with a variable is invariant along its entire live range, then we can bind this information to the variable itself. In other words, we can replace all the constraint variables $[v]^p$ by a single constraint variable $[v]$, for each variable v and every $p \in$ live(v). Consider the problem of range analysis again. There are two types of control-flow points associated with non-identity transfer functions: definitions and conditionals. (1) At the definition point of variable v, F_v simplifies to a function that depends only on some $[u]$ where each u is an argument of the instruction defining v; (2) At the conditional tests that use a variable v, F_v can be simplified to a function that uses $[v]$ and possibly other variables that appear in the test. The other program points are associated with an identity transfer function and can thus be ignored: $[v]^p = [v]^p \wedge F_v^{s \to p}([v_1]^s, \ldots, [v_n]^s)$ simplifies to $[v]^p = [v]^p \wedge [v]^p$ i.e., $[v]^p = [v]^p$. This gives the intuition on why a propagation engine along the def-use chains of an SSA form program can be used to solve the constant propagation problem in an equivalent, yet "sparser," manner.

A program representation that fulfils the Static Single Information (SSI) property allows us to attach the information to variables, instead of program points, and needs to fulfil the following four properties: *Split* forces the information related to a variable to be invariant along its entire live range; *Info* forces this information to be irrelevant outside the live range of the variable; *Link* forces the def-use chains to reach the points where information is available for a transfer function to be evaluated; finally, *Version* provides a one-to-one mapping between variable names and live ranges.

We now give a formal definition of the SSI and the four properties.

Property 1 (SSI)   STATIC SINGLE INFORMATION: Consider a forward (resp. backward) monotone PLV problem E_{dense} stated as a set of constraints

$$[v]^p \sqsubseteq F_v^{s \to p}([v_1]^s, \ldots, [v_n]^s)$$

for every variable v, each program point p, and each $s \in$ directpreds(p) (resp. $s \in$ directsuccs(p)). A program representation fulfils the Static Single Information property if and only if it fulfils the following four properties:

Split   Let s be the unique direct predecessor (resp. direct successor) of a program point where a variable v is live and such that $F_v^{s \to p} \neq \lambda x.\bot$ is non-trivial, i.e., is not the simple projection on $\mathscr{L}_v$, then s should contain a definition (resp. last use) of v; for $(v, p) \in variables \times progPoints$, let (Y_v^p) be a maximum solution to E_{dense}. Each node p that has several direct predecessors (resp. direct successors), and for which $F_v^{s \to p}(Y_{v_1}^s, \ldots, Y_{v_n}^s)$ has different values on its incoming edges ($s \to p$) (resp. outgoing edges ($p \to s$)), should have a ϕ-function at the entry of p (resp. σ-function at the exit of p) for v as defined in the next section.

Info   Each program point p such that $v \notin$ live-out(p) (resp. $v \notin$ live-in(p)) should be bound to an undefined transfer function, i.e., $F_v^p = \lambda x.\bot$.

Link   Each instruction *inst* for which F_v^{inst} depends on some $[u]^s$ should contain a use (resp. definition) of u live-in (resp. live-out) at *inst*.

Version   For each variable v, live(v) is a connected component of the CFG.

We must split live ranges using special instructions to provide the SSI properties. A naive way would be to split them between each pair of consecutive instructions, then we would automatically provide these properties, as the newly created variables would be live at only one program point. However, this strategy would lead to the creation of many trivial program regions, and we would lose sparsity. We provide a sparser way to split live ranges that fit Property 1 in Sect. 13.2. We may also have to extend the live range of a variable to cover every program point where the information is relevant; we accomplish this last task by inserting pseudo-uses and pseudo-definitions of this variable.

13.1.4 Special Instructions Used to Split Live Ranges

We perform live range splitting via special instructions: the σ-functions and parallel copies that, together with ϕ-functions, create new definitions of variables. These notations are important elements of the propagation engine described in the section that follows. In short, a σ-function (for a branch point) is the dual of a ϕ-function (for a join point), and a parallel copy is a copy that *must* be done in parallel with another instruction. Each of these special instructions, ϕ-function, σ-functions, and parallel copies, split live ranges at different kinds of program points: interior nodes, branches, and joins.

Interior nodes are program points that have a unique direct predecessor and a unique direct successor. At these points we perform live range splitting via copies. If the program point already contains another instruction, then this copy *must* be done *in parallel* with the existing instruction. The notation,

$$inst \parallel v_1' = v_1 \parallel \ldots \parallel v_m' = v_m$$

denotes m copies $v_i' = v_i$ performed in parallel with instruction *inst*. This means that all the uses of *inst* plus all right-hand variables v_i are read simultaneously, then *inst* is computed, then all definitions of *inst* plus all left-hand variables v_i' are written simultaneously. For a usage example of parallel copies, we will see later in this chapter an example of null-pointer analysis: Fig. 13.4.

We call *joins* the program points that have one direct successor and multiple direct predecessors. For instance, two different definitions of the same variable v

might be associated with two different constants, hence providing two different pieces of information about v. To avoid these definitions reaching the same use of v, we merge them at the earliest program point where they meet. We do it via our well-known ϕ-functions.

In backward analyses the information that emerges from different uses of a variable may reach the same *branch point*, which is a program point with a unique direct predecessor and multiple direct successors. To ensure Property 1, the use that reaches the definition of a variable must be unique, in the same way that in an SSA form program the definition that reaches a use is unique. We ensure this property via special instructions called σ-functions. The σ-functions are the dual of ϕ-functions, performing a parallel assignment depending on the execution path taken. The assignment

$$(l^1 : v_1^1, \ldots, l^q : v_1^q) = \sigma(v_1) \parallel \ldots \parallel (l^1 : v_m^1, \ldots, l^q : v_m^q) = \sigma(v_m)$$

represents m σ-functions that assign to each variable v_i^j the value in v_i if control flows into block l^j. As with ϕ-functions, these assignments happen in parallel, i.e., the m σ-functions encapsulate m parallel copies. Also, note that variables live in different branch targets are given different names by the σ-function.

13.1.5   Propagating Information Forward and Backward

Let us consider a unidirectional forward (resp. backward) PLV problem E_{dense}^{ssi} stated as a set of equations $[v]^p \sqsubseteq F_v^{s \to p}([v_1]^s, \ldots, [v_n]^s)$ (or equivalently $[v]^p = [v]^p \wedge F_v^{s \to p}([v_1]^s, \ldots, [v_n]^s)$ for every variable v, each program point p, and each $s \in \text{directpreds}(p)$ (resp. $s \in \text{directsuccs}(p)$). To simplify the discussion, any ϕ-function (resp. σ-function) is seen as a set of copies, one per direct predecessor (resp. direct successor), which leads to many constraints. In other words, a ϕ-function such as $p : a = \phi(a_1 : l^1, \ldots, a_m : l^m)$ gives us n constraints such as

$$[a]^p \sqsubseteq F_a^{l^j \to p}([a_1]^{l^j}, \ldots, [a_n]^{l^j})$$

which usually simplifies into $[a]^p \sqsubseteq [a_j]^{l^j}$. This last can be written equivalently into the classical meet

$$[a]^p \sqsubseteq \bigwedge_{l^j \in \text{directpreds}(p)} [a_j]^{l^j}$$

Algorithm 13.1: Backward propagation engine under SSI

---

1   *worklist* ← ∅
2   **foreach** $v \in$ vars **do** $[v] \leftarrow \top$
3   **foreach** $i \in$ insts **do** push(*worklist*, i)
4   **while** *worklist* ≠ ∅ **do**
5      $i \leftarrow$ pop(*worklist*)
6   **foreach** $v \in i.uses$ **do**
7         $[v]_{new} \leftarrow [v] \wedge G_v^i([i.defs])$
8   **if** $[v] \neq [v]_{new}$ **then**
9   *worklist* ← *worklist* ∪ *v.defs*
10           $[v] \leftarrow [v]_{new}$

---

used in Chap. 8. Similarly, a σ-function $(l^1 : a_1, \ldots, l^m : a_m) = \sigma(p : a)$ after program point p yields n constraints such as

$$[a_j]^{l_j} \sqsubseteq F_v^{p \to l^j}([a_1]^p, \ldots, [a_n]^p)$$

which usually simplifies into $[a_j]^{l_j} \sqsubseteq [a]^p$. Given a program that fulfils the SSI property for E_{dense}^{ssi} and the set of transfer functions F_v^s, we show here how to build an equivalent sparse constrained system.

Definition 13.2 (SSI Constrained System) Consider that a program in SSI form gives us a constraint system that associates with each variable v the constraints $[v]^p \sqsubseteq F_v^{s \to p}([v_1]^s, \ldots, [v_n]^s)$. We define a system of sparse equations E_{sparse}^{ssi} as follows:

- For each instruction i that defines (resp. uses) a variable v, let $a \ldots z$ be the set of used (resp. defined) variables. Because of the *Link* property, $F_v^{s \to p}$ (that we will denote F_v^i from now) depends only on some $[a]^s \ldots [z]^s$. Thus, there exists a function G_v^i defined as the restriction of F_v^i on $\mathscr{L}_a \times \cdots \times \mathscr{L}_z$, i.e., informally, "$G_v^i([a], \ldots, [z]) = F_v^i([v_1], \ldots, [v_n])$."
- The sparse constrained system associates the constraint $[v] \sqsubseteq G_v^i([a], \ldots, [z])$ with each variable v, for each definition (resp. use) point i of v, where $a, \ldots, z$ are used (resp. defined) at i.

The propagation engine discussed in Chap. 8 sends information forwards along the def-use chains naturally formed by the SSA form program. If a given program fulfils the SSI property for a backward analysis, we can use a very similar propagation algorithm to communicate information backwards, such as the worklist Algorithm 13.1. A slightly modified version, presented in Algorithm 13.2, propagates information forwards. If necessary, these algorithms can be made control-flow sensitive, like Algorithm 8.1 in Chap. 8.

Still, we should highlight a quite important subtlety that appears in line 7 of Algorithms 13.1 and 13.2: $[v]$ appears on the right-hand side of the assignment for

---

Algorithm 13.2: Forward propagation engine under SSI

---

1   *worklist* ← ∅
2   **foreach** $v \in$ vars **do** $[v] \leftarrow \top$
3   **foreach** $i \in$ insts **do** push(*worklist*, i)
4   **while** *worklist* $\neq \emptyset$ **do**
5   $\quad$ $i \leftarrow$ pop(*worklist*)
6   $\quad$ **foreach** $v \in i.defs$ **do**
7   $\quad\quad$ $[v]_{new} \leftarrow G_v^i([i.uses])$
8   $\quad\quad$ **if** $[v] \neq [v]_{new}$ **then**
9   $\quad\quad\quad$ *worklist* ← *worklist* ∪ $v.uses$
10 $\quad\quad\quad$ $[v] \leftarrow [v]_{new}$

Algorithm 13.1 while it does not for Algorithm 13.2. This stems from the asymmetry of our SSI form that ensures (for practical purposes only, as we will explain soon) the Static Single Assignment property but not the Static Single Use (SSU) property. If we have several uses of the same variable, then the sparse backward constraint system will have several inequations—one per variable use—with the same left-hand side. Technically this is the reason why we manipulate a constraint system (system with inequations) and not an equation system as in Chap. 8. Both systems can be solved[1] using a scheme known as *chaotic iteration* such as the worklist algorithm we provide here. The slight and important difference for a constraint system as opposed to an equation system is that one needs to meet $G_v^i(\ldots)$ with the old value of $[v]$ to ensure the monotonicity of the consecutive values taken by $[v]$. It would still be possible to enforce the SSU property, in addition to the SSA property, of our intermediate representation, at the expense of adding more ϕ-functions and σ-functions. However, this guarantee is not necessary to every sparse analysis. The dead-code elimination problem illustrates this point well: For a program under SSA form, replacing G_v^i in Algorithm 13.1 by the property "i is a useful instruction or one of the variables it defines is marked as useful" leads to the standard SSA-based dead-code elimination algorithm. The sparse constraint system does have several equations (one per variable use) for the same left-hand side (one for each variable). It is not necessary to enforce the SSU property in this instance of dead-code elimination, and doing so would lead to a less efficient solution in terms of compilation time and memory consumption. In other words, a code under SSA form fulfils the SSI property for dead-code elimination.

13.1.6   *Examples of Sparse Data-Flow Analyses*

As we have mentioned before, many data-flow analyses can be classified as PLV problems. In this section we present some meaningful examples.

---

[1] In an ideal world, with monotone framework and lattice of finite height.

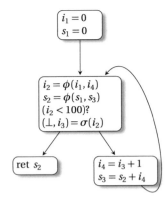

$$[i_1] = [i_1] \cup [0,0] \qquad\qquad = [0,0]$$
$$[s_1] = [s_1] \cup [0,0] \qquad\qquad = [0,0]$$
$$[i_2] = [i_2] \cup [i_1] \cup [i_4] \qquad\quad = [0,100]$$
$$[s_2] = [s_2] \cup [s_1] \cup [s_3] \qquad\quad = [0,+\infty[$$
$$[i_3] = [i_3] \cup ([i_2] \cap]\!-\!\infty,99]) = [0,99]$$
$$[i_4] = [i_4] \cup ([i_3]+1) \qquad\quad = [1,100]$$
$$[s_3] = [s_3] \cup ([s_2]+[i_4]) \qquad\;\; = [1,+\infty[$$

(a) Live range splitting used by a
sparse implementation of range analysis

(b) sparse constraint system & solution

Fig. 13.2 Live range splitting on Fig. 13.1 and a solution to this instance of the range analysis
problem

Range Analysis Revisited

We start this section by revisiting the initial example of data-flow analysis of this
chapter, given in Fig. 13.1. A range analysis acquires information from either the
points where variables are defined, or from the points where variables are tested. In
the original figure we know that i must be bound to the interval $[0, 0]$ immediately
after instruction l_1. Similarly, we know that this variable is upper bounded by 100
when arriving at l_4, due to the conditional test that happens before. Therefore, in
order to achieve the SSI property, we should split the live ranges of variables at their
definition points, or at the conditionals where they are used. Figure 13.2 shows on
the left the original example after live range splitting. In order to ensure the SSI
property in this example, the live range of variable i must be split at its definition,
and at the conditional test. The live range of s, on the other hand, must be split only
at its definition point, as it is not used in the conditional. Splitting at conditionals
is done via σ-functions. The representation that we obtain by splitting live ranges
at definitions and conditionals is called the Extended Static Single Assignment (e-
SSA) form. Figure 13.2 also shows on the right the result of the range analysis
on this intermediate representation. This solution assigns to each variable a unique
range interval.

Class Inference

Some dynamically typed languages, such as Python, JavaScript, Ruby, or Lua,
represent objects as tables containing methods and fields. It is possible to improve
the execution of programs written in these languages if we can replace these simple
tables by actual classes with virtual tables. A class inference engine tries to assign
a class to a variable v based on the ways that v is defined and used. Figure 13.3
illustrates this optimization on a Python program (a). Our objective is to infer the
correct suite of methods for each object bound to variable v. Figure 13.3b shows
the results of a dense implementation of this analysis. Because type inference is

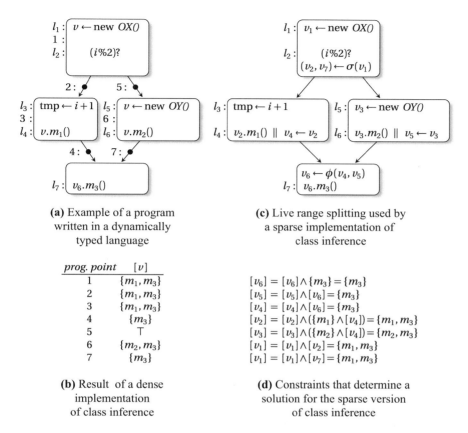

(a) Example of a program
written in a dynamically
typed language

prog. point	$[v]$
1	$\{m_1, m_3\}$
2	$\{m_1, m_3\}$
3	$\{m_1, m_3\}$
4	$\{m_3\}$
5	$\top$
6	$\{m_2, m_3\}$
7	$\{m_3\}$

(b) Result of a dense
implementation
of class inference

(c) Live range splitting used by
a sparse implementation of
class inference

$$[v_6] = [v_6] \wedge \{m_3\} = \{m_3\}$$
$$[v_5] = [v_5] \wedge [v_6] = \{m_3\}$$
$$[v_4] = [v_4] \wedge [v_6] = \{m_3\}$$
$$[v_2] = [v_2] \wedge (\{m_1\} \wedge [v_4]) = \{m_1, m_3\}$$
$$[v_3] = [v_3] \wedge (\{m_2\} \wedge [v_4]) = \{m_2, m_3\}$$
$$[v_1] = [v_1] \wedge [v_2] = \{m_1, m_3\}$$
$$[v_1] = [v_1] \wedge [v_7] = \{m_1, m_3\}$$

(d) Constraints that determine a
solution for the sparse version
of class inference

Fig. 13.3 Class inference analysis as an example of backward data-flow analysis that takes information from the uses of variables

a backward analysis that extracts information from use sites, we split live ranges using parallel copies at these program points and rely on σ-functions to merge them back, as shown in Fig. 13.3c. The use-def chains that we derive from the program representation lead naturally to a constraint system, shown in Fig. 13.3d, where $[v_j]$ denotes the set of methods associated with variable v_j. A fixed point to this constraint system is a solution to our data-flow problem. This instance of class inference is a Partitioned Variable Problem (PVP),[2] because the data-flow information associated with a variable v can be computed independently from the other variables.

Null-Pointer Analysis

The objective of null-pointer analysis is to determine which references may hold null values. This analysis allows compilers to remove redundant null-exception

---

[2] Actually, class inference is no longer a PVP as soon as we want to propagate the information through copies.

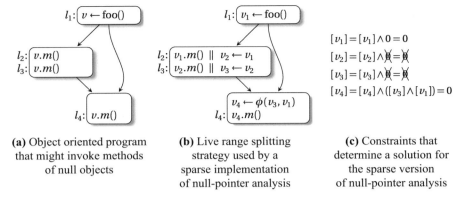

(a) Object oriented program that might invoke methods of null objects

(b) Live range splitting strategy used by a sparse implementation of null-pointer analysis

(c) Constraints that determine a solution for the sparse version of null-pointer analysis

Fig. 13.4 Null-pointer analysis as an example of forward data-flow analysis that takes information from the definitions and uses of variables (0 represents the fact that the pointer is possibly null, ⦰ if it cannot be) **(a)** Object oriented program that might invoke methods of null objects **(b)** Live range splitting strategy used by a sparse implementation of null-pointer analysis **(c)** Constraints that determine a solution for the sparse version of null-pointer analysis

tests and helps developers find null-pointer dereferences. Figure 13.4 illustrates this analysis. Because information is produced not only at definition but also at use sites, we split live ranges after each variable is used, as shown in Fig. 13.4b. For instance, we know that v_2 cannot be null, otherwise an exception would have been thrown during the invocation $v_1.m()$; hence the call $v_2.m()$ cannot result in a null-pointer dereference exception. On the other hand, we notice in Fig. 13.4a that the state of v_4 is the meet of the state of v_3, definitely not-null, and the state of v_1, possibly null, and we must conservatively assume that v_4 may be null.

13.2 Construction and Destruction of the Intermediate Program Representation

In the previous section we have seen how the static single information property gives the compiler the opportunity to solve a data-flow problem sparsely. However, we have not yet seen how to convert a program to a format that provides the SSI property. This is a task that we address in this section, via the three-step algorithm from Sect. 13.2.2.

13.2.1 Splitting Strategy

A *live range splitting strategy* $\mathscr{P}_v = I_\uparrow \cup I_\downarrow$ over a variable v consists of a set of "oriented" program points. We let $I_\downarrow$ denote a set of points i with forward direction.

Client	Splitting strategy $\mathcal{P}$
Alias analysis, reaching defs., cond. constant propagation	$Defs_{\downarrow}$
Partial Redundancy Elimination	$Defs_{\downarrow} \bigcup LastUses_{\uparrow}$
ABCD, taint analysis, range analysis	$Defs_{\downarrow} \bigcup Out(Conds)_{\downarrow}$
Stephenson's bitwidth analysis	$Defs_{\downarrow} \bigcup Out(Conds)_{\downarrow} \bigcup Uses_{\uparrow}$
Mahlke's bitwidth analysis	$Defs_{\downarrow} \bigcup Uses_{\uparrow}$
An's type inference, Class inference	$Uses_{\uparrow}$
Hochstadt's type inference	$Uses_{\uparrow} \bigcup Out(Conds)_{\uparrow}$
Null-pointer analysis	$Defs_{\downarrow} \bigcup Uses_{\downarrow}$

Fig. 13.5 Live range splitting strategies for different data-flow analyses. *Defs* (resp. *Uses*) denotes the set of instructions that define (resp. use) the variable; *Conds* denotes the set of instructions that apply a conditional test on a variable; Out(*Conds*) denotes the exits of the corresponding basic blocks; *LastUses* denotes the set of instructions where a variable is used, and after which it is no longer live

Similarly, we let $I_{\uparrow}$ denote a set of points i with backward direction. The live range of v must be split at least at every point in $\mathcal{P}_v$. Going back to the examples from Sect. 13.1.6, we have the live range splitting strategies enumerated below. The list in Fig. 13.5 gives further examples of live range splitting strategies. Corresponding references are given in the last section of this chapter.

- Range analysis is a forward analysis that takes information from points where variables are defined and conditional tests that use these variables. For instance, in Fig. 13.1, we have $\mathcal{P}_i = \{l_1, \text{Out}(l_3), l_4\}_{\downarrow}$ where $\text{Out}(l_i)$ is the exit of l_i (i.e., the program point immediately after l_i), and $\mathcal{P}_s = \{l_2, l_5\}_{\downarrow}$.
- Class inference is a backward analysis that takes information from the uses of variables; thus, for each variable, the live range splitting strategy is characterized by the set $Uses_{\uparrow}$ where $Uses$ is the set of use points. For instance, in Fig. 13.3, we have $\mathcal{P}_v = \{l_4, l_6, l_7\}_{\uparrow}$.
- Null-pointer analysis takes information from definitions and uses and propagates this information forwardly. For instance, in Fig. 13.4, we have $\mathcal{P}_v = \{l_1, l_2, l_3, l_4\}_{\downarrow}$.

The algorithm SSIfy in Fig. 13.6 implements a live range splitting strategy in three steps. Firstly, it splits live ranges, inserting new definitions of variables into the program code. Secondly, it renames these newly created definitions; hence, ensuring that the live ranges of two different re-definitions of the same variable do not overlap. Finally, it removes dead and non-initialized definitions from the program code. We describe each of these phases in the rest of this section.

```
1  Function SSIfy(var v, Splitting_Strategy 𝒫ᵥ)
2  │  split(v, 𝒫ᵥ)
3  │  rename(v)
4  └  clean(v)
```

Fig. 13.6 Split the live ranges of v to convert it to SSI form

13.2.2 Splitting Live Ranges

In order to implement $\mathscr{P}_v$ we must split the live ranges of v at each program point listed by $\mathscr{P}_v$. However, these points are not the only ones where splitting might be necessary. As we have pointed out in Sect. 13.1.4, we might have, for the same original variable, many different sources of information reaching a common program point. For instance, in Fig. 13.1, there exist two definitions of variable i, l_1 and l_4, which reach the use of i at l_3. The information that flows forward from l_1 and l_4 collides at l_3, the loop entry. Hence the live range of i has to be split immediately before l_3—at $In(l_3)$—leading, in our example, to a new definition, i_1. In general, the set of program points where information collides can be easily characterized by the notion of join sets and iterated dominance frontier (DF^+) seen in Chap. 4. Similarly, split sets created by the backward propagation of information can be over-approximated by the notion of *iterated post-dominance frontier* (pDF^+), which is the dual of DF^+. That is, the post-dominance frontier is the dominance frontier in a CFG where the directions of edges have been reversed. Note that, just as the notion of dominance requires the existence of a unique entry node that can reach every CFG node, the notion of post-dominance requires the existence of a unique exit node reachable by any CFG node. For control-flow graphs that contain several exit nodes or loops with no exit, we can ensure the single-exit property by creating a dummy common exit node and inserting some never-taken exit edges into the program.

Figure 13.7 shows the algorithm that we use to create new definitions of variables. This algorithm has three main phases. First, in lines 2–7 we create new definitions to split the live ranges of variables due to backward collisions of information. These new definitions are created at the iterated post-dominance frontier of points at which information originates. If a program point is a join node, then each of its direct predecessors will contain the live range of a different definition of v, as we ensure in lines 5–6 of our algorithm. Note that these new definitions are not placed parallel to an instruction, but in the region immediately after it, which we denote as "Out(...)." In lines 8–13 we perform the inverse operation: We create new definitions of variables due to the forward collision of information. Our starting points $S_\downarrow$, in this case, also include the original definitions of v, as we see in line 9, because we want to stay in SSA form in order to have access to a fast liveness check as described in Chap. 9. Finally, in lines 14–20 we

---

```
1  Function split (var v, Splitting_Strategy 𝒫ᵥ = I↓ ∪ I↑)
        ▷ compute the set of split nodes
2      S↑ ← ∅
3      foreach i ∈ I↑ do
4          if i.is_join then
5              foreach e ∈ incoming_edges(i) do
6                  S↑ ← S↑ ∪ Out(pDF⁺(e))
7          else S↑ ← S↑ ∪ Out(pDF⁺(i))
8      S↓ ← ∅
9      foreach i ∈ S↑ ∪ Defs(v) ∪ I↓ do
10         if i.is_branch then
11             foreach e ∈ outgoing_edges(i) do
12                 S↓ ← S↓ ∪ In(DF⁺(e))
13         else S↓ ← S↓ ∪ In(DF⁺(i))
14     S ← 𝒫ᵥ ∪ S↑ ∪ S↓
        ▷ Split live-range of v by inserting φ, σ, and copies
15     foreach i ∈ S do
16         if i does not already contain any definition of v then
17             if i.is_join then insert "v ← φ(v, ..., v)" at i
18             else
19                 if i.is_branch then insert "(v, ..., v) ← σ(v)" at i
20                 else insert a copy "v ← v" at i
```

Fig. 13.7 Live range splitting. In(l) denotes a program point immediately before l, and Out(l) a program point immediately after l

actually insert the new definitions of v. These new definitions might be created by σ functions (due to $\mathscr{P}_v$ or to the splitting in lines 2–7); by ϕ-functions (due to $\mathscr{P}_v$ or to the splitting in lines 8–13); or by parallel copies.

13.2.3   Variable Renaming

The `rename` algorithm in Fig. 13.8 builds def-use and use-def chains for a program after live range splitting. This algorithm is similar to the classical algorithm used to rename variables during the SSA construction that we saw in Chap. 3. To rename a variable v we traverse the program's dominance tree, from top to bottom, stacking each new definition of v that we find. The definition currently on the top of the stack is used to replace all the uses of v that we find during the traversal. If the stack is empty, this means that the variable is not defined at this point. The renaming process replaces the uses of undefined variables by $\bot$ (see comment of function `stack.set_use`). We have two methods, `stack.set_use` and

```
 1  Function rename(var v)
        ▷ Compute use-def & def-use chains.
 2      stack ← ∅
 3      foreach CFG node n in dominance order do
 4          if ∃v ← φ(v : l¹,..., v : l�q) in In(n) then
 5              ⌊ stack.set_def(v ← φ(v : l¹,..., v : l�q))

 6          foreach instruction u in n that uses v do
 7              ⌊ stack.set_use(u)

 8          if ∃ instruction d in n that defines v then
 9              ⌊ stack.set_def(d)

10          foreach instruction (...) ← σ(v) in Out(n) do
11              ⌊ stack.set_use((...) ← σ(v))

12          if ∃(v : l¹,..., v : lq) ← σ(v) in Out(n) then
13              foreach v : lⁱ ← v in (v : l¹,..., v : lq) ← σ(v) do
14                  ⌊ stack.set_def(v : lⁱ ← v)

15          foreach m in direct-successors(n) do
16              if ∃v ← φ(..., v : lⁿ,...) in In(m) then
17                  ⌊ stack.set_use(v ← v : lⁿ)
```

```
 1  Function stack.set_use(instruction inst)
        ▷ We consider here that stack.peek() = ⊥ if stack.isempty(), and that
          Def(⊥) = entry
 2      while Def(stack.peek()) does not dominate inst do
 3          ⌊ stack.pop()
 4      vᵢ ← stack.peek()
 5      replace the uses of v by vᵢ in inst
 6      if vᵢ ≠ ⊥ then set Uses(vᵢ) = Uses(vᵢ)∪ inst
```

```
 1  Function stack.set_def(instruction inst)
 2      let vᵢ be a fresh version of v
 3      replace the defs of v by vᵢ in inst
 4      set Def(vᵢ) = inst
 5      stack.push(vᵢ)
```

Fig. 13.8 Versioning

`stack.set_def`, that build the chains of relations between variables. Note that sometimes we must rename a single use inside a ϕ-function, as in lines 16–17 of the algorithm. For simplicity we consider this single use as a simple assignment when calling `stack.set_use`, as can be seen in line 17. Similarly, if we must rename a single definition inside a σ-function, then we treat it as a simple assignment, like we do in lines 12–14 of the algorithm.

```
 1  Function clean(var v)
 2      let web = {v_i | v_i is a version of v}
 3      defined ← ∅
 4      active ← {inst | inst actual instruction and web ∩ inst.defs ≠ ∅}
 5      while ∃inst ∈ active | web ∩ inst.defs\defined ≠ ∅ do
 6          foreach v_i ∈ web ∩ inst.defs\defined do
 7              active ← active ∪ Uses(v_i)
 8              defined ← defined ∪ {v_i}

 9      used ← ∅
10      active ← {inst | inst actual instruction and web ∩ inst.uses ≠ ∅}
11      while ∃inst ∈ active | inst.uses\used ≠ ∅ do
12          foreach v_i ∈ web ∩ inst.uses\used do
13              active ← active ∪ Def(v_i)
14              used ← used ∪ {v_i}

15      let live = defined ∩ used
16      foreach non actual inst ∈ Def(web) do
17          foreach v_i operand of inst | v_i ∉ live do
18              replace v_i by ⊥
19          if inst.defs = {⊥} or inst.uses = {⊥} then
20              remove inst
```

Fig. 13.9 Dead and undefined code elimination. Original instructions not inserted by `split` are called *actual* instructions. *inst*.defs denotes the (set) of variable(s) defined by *inst*, and *inst*.uses denotes the set of variables used by *inst*

13.2.4 Dead and Undefined Code Elimination

Just like Algorithm 3.7, the algorithm in Fig. 13.9 eliminates ϕ-functions and parallel copies that define variables not actually used in the code. By way of symmetry, it also eliminates σ-functions and parallel copies that use variables not actually defined in the code. We mean by "actual" instructions those that already existed in the program before we transformed it with `split`. In line 2, "web" is fixed to the set of versions of v, so as to restrict the cleaning process to variable v, as we see in the first two loops. The "active" set is initialized to actual instructions, line 4. Then, during the first loop in lines 5–8, we augment it with ϕ-functions, σ-functions, and copies that can reach actual definitions through use-def chains. The corresponding version of v is hence marked as *defined* (line 8). The next loop, lines 11–14, performs a similar process, this time to add to the active set instructions that can reach actual uses through def-use chains. The corresponding version of v is then marked as *used* (line 14). Each non-live variable, i.e., either undefined or dead (non-used), hence not in the "live" set (line 15) is replaced by $\bot$ in all ϕ, σ, or copy functions where it appears by the loop, lines 15–18. Finally, all useless ϕ, σ, or copy functions are removed by lines 19–20.

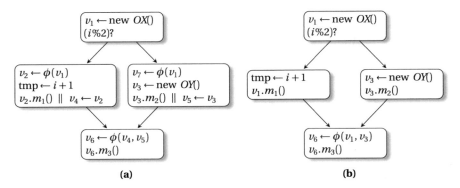

Fig. 13.10 (**a**) Implementing σ-functions via single arity ϕ-functions; (**b**) getting rid of copies and σ-functions

13.2.5 Implementation Details

Implementing σ-Functions

The most straightforward way to implement σ-functions, in a compiler that already supports the SSA form, is to represent them by ϕ-functions. In this case, the σ-functions can be implemented as single arity ϕ-functions. As an example, Fig. 13.10a shows how we would represent the σ-functions of Fig. 13.3d. If l is a branch point with n direct successors that would contain a σ-function (l^1 : $v_1, \ldots, l^n : v_n$) $\leftarrow \sigma(v)$, then, for each direct successor l^j of l, we insert at the beginning of l^j an instruction $v_j \leftarrow \phi(l^j : v)$. Note that l^j may already contain a ϕ-function for v. This happens when the control-flow edge $l \rightarrow l^j$ is *critical*: A critical edge links a basic block with several direct successors to a basic block with several direct predecessors. If l^j already contains a ϕ-function $v' \leftarrow \phi(\ldots, v_j, \ldots)$, then we rename v_j to v.

SSI Destruction

Traditional instruction sets do not provide ϕ-functions or σ-functions. Thus, before producing an executable program, the compiler must implement these instructions. We have already seen in Chap. 3 how to replace ϕ-functions with actual assembly instructions; however, now we must also replace σ-functions and parallel copies. A simple way to eliminate all the σ-functions and parallel copies is via copy-propagation. In this case, we copy-propagate the variables that these special instructions define. As an example, Fig. 13.10b shows the result of copy folding applied on Fig. 13.10a.

13.3   Further Reading

The monotone data-flow framework is an old ally of compiler writers. Since the work of pioneers like Prosser [234], Allen [3, 4], Kildall [166], Kam [158], and Hecht [143], data-flow analyses such as reaching definitions, available expressions, and liveness analysis have made their way into the implementation of virtually every important compiler. Many compiler textbooks describe the theoretical basis of the notions of lattice, monotone data-flow framework, and fixed points. For a comprehensive overview of these concepts, including algorithms and formal proofs, we refer the interested reader to Nielson et al.'s book [208] on static program analysis.

The original description of the intermediate program representation known as Static Single Information form was given by Ananian in his Master's thesis [8]. The notation for σ-functions that we use in this chapter was borrowed from Ananian's work. The SSI program representation was subsequently revisited by Jeremy Singer in his PhD thesis [261]. Singer proposed new algorithms to convert programs to SSI form, and also showed how this program representation could be used to handle truly bidirectional data-flow analyses. We have not discussed bidirectional data-flow problems, but the interested reader can find examples of such analyses in Khedker et al.'s work [165]. Working on top of Ananian's and Singer's work, Boissinot et al. [37] have proposed a new algorithm to convert a program to SSI form. Boissinot et al. have also separated the SSI program representation into two flavours, which they call *weak* and *strong*. Tavares et al. [282] have extended the literature on SSI representations, defining building algorithms and giving formal proofs that these algorithms are correct. The presentation that we use in this chapter is mostly based on Tavares et al.'s work.

There exist other intermediate program representations that, like the SSI form, make it possible to solve some data-flow problems sparsely. Well-known among these representations is the *Extended Static Single Assignment* form, introduced by Bodik *et al.* to provide a fast algorithm to eliminate array bound checks in the context of a JIT compiler [32]. Another important representation, which supports data-flow analyses that acquire information at use sites, is the *Static Single Use* form (SSU). As uses and definitions are not fully symmetric (the live range can "traverse" a use while it cannot traverse a definition), there are different variants of SSU [125, 187, 228]. For instance, the "strict" SSU form enforces that each definition reaches a single use, whereas SSI and other variations of SSU allow two consecutive uses of a variable on the same path. All these program representations are very effective, having seen use in a number of implementations of flow analyses; however, they only fit specific data-flow problems.

The notion of *Partitioned Variable Problem* (PVP) was introduced by Zadeck, in his PhD dissertation [316]. Zadeck proposed fast ways to build data structures that allow one to solve these problems efficiently. He also discussed a number of data-flow analyses that are partitioned variable problems. There are data-flow analyses that do not meet the Partitioned Lattice per Variable property. Notable examples include abstract interpretation problems on relational domains, such as Polyhedrons [86], Octagons [199], and Pentagons [188].

In terms of data structures, the first, and best known method proposed to support sparse data-flow analyses is Choi et al.'s *Sparse Evaluation Graph* (SEG) [67]. The nodes of this graph represent program regions where information produced by the data-flow analysis might change. Choi et al.'s ideas have been further expanded, for example by Johnson et al.'s *Quick Propagation Graphs* [153], or Ramalingam's *Compact Evaluation Graphs* [237]. Nowadays we have efficient algorithms that build such data structures [154, 224, 225]. These data structures work best when applied on partitioned variable problems.

As opposed to those approaches, the solution promoted by this chapter consists in an intermediate representation (IR) based evaluation graph, and has advantages and disadvantages when compared to the data structure approach. The intermediate representation based approach has two disadvantages, which we have already discussed in the context of the standard SSA form. First it has to be maintained and at some point destructed. Second, because it increases the number of variables, it might add some overhead to analyses and transformations that do not require it. On the other hand, IR based solutions to sparse data-flow analyses have many advantages over data structure based approaches. For instance, an IR allows concrete or abstract interpretation. Solving any coupled data-flow analysis problem along with a SEG was mentioned by Choi et al. [67] as an open problem. However, as illustrated by the conditional constant propagation problem described in Chap. 8, coupled data-flow analysis can be solved naturally in IR based evaluation graphs. Last, SSI is compatible with SSA extensions such as gated SSA described in Chap. 14, which allows demand-driven interpretation.

The data-flow analyses discussed in this chapter are well-known in the literature. Class inference was used by Chambers et al. in order to compile Self programs more efficiently [63]. Nanda and Sinha have used a variant of null-pointer analysis to find which call sites may cause errors due to the dereference of null objects [207]. Ananian [8], and later Singer [261], have shown how to use the SSI representation to do partial redundancy elimination sparsely. In addition to being used to eliminate redundant array bound checks [32], the e-SSA form has been used to solve Taint Analysis [248], and range analysis [123, 277]. Stephenson et al. [273] described a bitwidth analysis that is both forward and backward, taking information from definitions, uses, and conditional tests. For another example of bidirectional bitwidth analysis, see Mahlke et al.'s algorithm [194]. The type inference analysis that we mentioned in Fig. 13.5 was taken from Hochstadt et al.'s work [284].

Chapter 14
Graphs and Gating Functions

James Stanier and Fabrice Rastello

Many compilers represent the input program as some form of graph in order to support analysis and transformation. Over time a cornucopia of program graphs have been presented in the literature and subsequentially implemented in real compilers. Many of these graphs use SSA concepts as the core principle of their representation, ranging from literal translations of SSA into graph form to more abstract graphs which are implicitly in SSA form. We aim to introduce a selection of program graphs that use these SSA concepts, and examine how they may be useful to a compiler writer.

A well-known graph representation is the control-flow graph (CFG) which we encountered at the beginning of the book while being introduced to the core concept of SSA. The CFG models control flow in a program, but the graphs that we will study instead model *data flow*. This is useful as a large number of compiler optimizations are based on data-flow analysis. In fact, all graphs that we consider in this chapter are all data-flow graphs.

In this chapter, we will look at a number of SSA-based graph representations. An introduction to each graph will be given, along with diagrams to show how sample programs look when translated into that particular graph. Additionally, we will describe the techniques that each graph was created to solve, with references to the literature for further research.

For this chapter, we assume that the reader already has familiarity with SSA (see Chap. 1) and the applications that it is used for.

J. Stanier (✉)
Brandwatch, Cumbria, UK

F. Rastello
Inria, Grenoble, France
e-mail: fabrice.rastello@inria.fr

F. Rastello, F. Bouchez Tichadou (eds.), *SSA-based Compiler Design*,
https://doi.org/10.1007/978-3-030-80515-9_14

14.1 Data-Flow Graphs

Since all of the graphs in this chapter are data-flow graphs, let us define them. A data-flow graph (DFG) is a directed graph $G = (V, E)$ where the edges E represent the flow of data from the result of one instruction to the input of another. An instruction executes once all of its input data values have been computed. When an instruction executes, it produces a new data value which is propagated to other connected instructions.

Whereas the CFG imposes a total ordering on instructions—the same ordering that the programmer wrote them in—the DFG has no such concept of ordering; it just models the flow of data. This means that it typically needs a companion representation such as the CFG to ensure that optimized programs are still correct. However, with access to both the CFG and DFG, optimizations such as dead code elimination, constant folding, and common subexpression elimination can be performed effectively. But this comes at a price: keeping both graphs updated during optimization can be costly and complicated.

14.2 The SSA Graph

We begin our exploration with a graph that is a literal representation of SSA: the SSA graph. The SSA graph can be constructed from a program in SSA form by explicitly adding use-def chains. To demonstrate what the graph looks like, we present some sample code in Fig. 14.1 which is then translated into an SSA graph.

An SSA graph consists of vertices that represent instructions (such as + and print) or ϕ-functions, and directed edges that connect uses to definitions of values. The outgoing edges of a vertex represent the arguments required for that instruction, and the ingoing edge(s) to a vertex represent the propagation of the instruction's result(s) after they have been computed. We call these types of graphs *demand-based* representations. This is because in order to compute an instruction, we must first *demand* the results of the operands.

Although the textual representation of SSA is much easier for a human to read, the primary benefit of representing the input program in graph form is that the compiler writer is able to apply a wide array of graph-based optimizations by using standard graph traversal and transformation techniques.

In the literature, the SSA graph has been used to detect induction variables in loops, for performing instruction selection (see Chap. 19), operator strength reduction, rematerialization, and has been combined with an extended SSA language to support compilation in a parallelizing compiler. The reader should note that the exact specification of what constitutes an SSA graph changes from paper to paper. The essence of the intermediate representation (IR) has been presented here, as each author tends to make small modifications for their particular implementation.

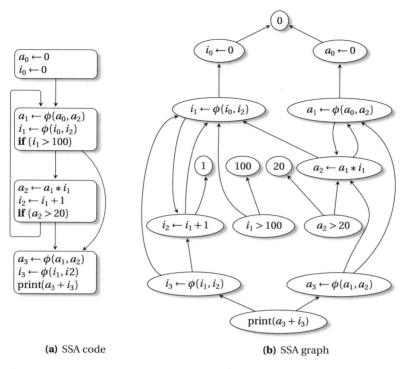

(a) SSA code   **(b)** SSA graph

Fig. 14.1 Some SSA code translated into an SSA graph. Note how edges *demand* the input values for a node

14.2.1 Finding Induction Variables with the SSA Graph

We illustrate the usefulness of the SSA graph through a basic induction variable (IV) recognition technique. A more sophisticated technique is developed in Chap. 10. Given that a program is represented as an SSA graph, the task of finding induction variables is simplified. A *basic linear induction variable i* is a variable that appears only in the form:

```
1  i = 10
2  while <cond> do
3      . . .
4      i = i + k
5      . . .
```

where k is a constant or loop-invariant. A simple IV recognition algorithm is based on the observation that each basic linear induction variable will belong to a non-trivial strongly connected component (SCC) in the SSA graph. SCCs can be easily discovered in linear time using any depth-first search traversal. Each such SCC must conform to the following constraints:

- The SCC contains only one ϕ-function at the header of the loop.
- Every component is either $i = \phi(i_0, ..., i_n)$ or $i = i_k \oplus n$, where $\oplus$ is addition or subtraction, and n is loop-invariant.

This technique can be expanded to detect a variety of other classes of induction variables, such as wraparound variables, non-linear induction variables, and nested induction variables. Scans and reductions also show a similar SSA graph pattern and can be detected using the same approach.

14.3 Program Dependence Graph

The program dependence graph (PDG) represents both control and data dependencies together in one graph. The PDG was developed to support optimizations requiring reordering of instructions and graph rewriting for parallelism, as the strict ordering of the CFG is relaxed and complemented by the presence of data dependence information. The PDG is a directed graph $G = (V, E)$ where nodes V are statements, predicate expressions, or region nodes, and edges E represent either control or data dependencies. Thus, the set of all edges E has two distinct subsets: the control dependence subgraph E_C and the data dependence subgraph E_D.

Statement nodes represent instructions in the program. Predicate nodes test a conditional statement and have true and false edges to represent the choice taken on evaluation of the predicate. Region nodes group control dependencies with identical source and label together. If the control dependence for a region node is satisfied, then it follows that all of its children can be executed. Thus, if a region node has three different control-independent statements as immediate children, then those statements could potentially be executed in parallel. Diagrammatically, rectangular nodes represent statements, diamond nodes predicates, and circular nodes are region nodes. Dashed edges represent control dependence, and solid edges represent data dependence. Loops in the PDG are represented by back edges in the control dependence subgraph. We show example code translated into a PDG in Fig. 14.2.

Building a PDG is a multi-stage process involving:

- Construction of the post-dominator tree
- Use of the post-dominator tree to generate the control dependence subgraph
- Insertion of region nodes
- Construction of DAGs for each basic block which are then joined to create the data dependence subgraph

Let us explore this construction process in more detail.

An ENTRY node is added with one edge labeled true pointing to the CFG entry node, and another labeled false going to the CFG exit node.

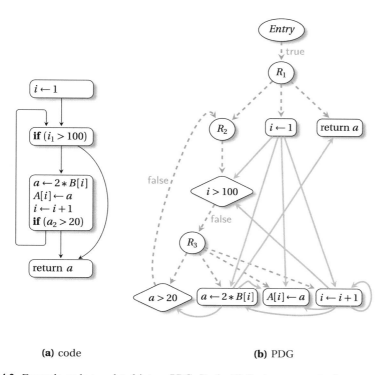

(a) code   (b) PDG

Fig. 14.2 Example code translated into a PDG. Dashed/full edges, respectively, represent control/data dependencies

Before constructing the rest of the control dependence subgraph E_C, let us define control dependence. A node w is said to be control dependent on edge (u, v) if w post-dominates v and w does not strictly post-dominate u. Control dependence between nodes is equivalent to the post-dominance frontier on the reversed CFG. To compute the control dependence subgraph, the post-dominator tree is constructed for the CFG. Then, the control dependence edges from u to w are labeled with the boolean value taken by the predicate computed in u when branching on edge (u, v). Then, let S consist of the set of all edges (A, B) in the CFG such that B is not an ancestor of A in the post-dominator tree. Each of these edges has an associated label true or false. Then, each edge in S is considered in turn. Given (A, B), the post-dominator tree is traversed backwards from B until we reach A's parent, marking all nodes visited (including B) as control dependent on A with the label of S.

Next, region nodes are added to the PDG. Each region node summarizes a set of control conditions and "groups" all nodes with the same set of control conditions together. Region nodes are also inserted so that predicate nodes will only have two direct successors. To begin with, an unpruned PDG is created by checking, for each node of the CFG, which control region it depends on. This is done by traversing the post-dominator tree in post-order and mapping sets of control dependencies to region nodes. For each node N visited in the post-dominator tree, the map is checked for an existing region node with the same set CD of control dependencies. If none exists, a new region node R is created with these control dependencies and entered into the map. R is made to be the only control dependence direct predecessor of N. Next, the intersection INT of CD is computed for each immediate child of N in the post-dominator tree. If $INT = CD$, then the corresponding dependencies are removed from the child and replaced with a single dependence on the child's control direct predecessor. Then, a pass over the graph is made to make sure that each predicate node has a unique direct successor for each boolean value. If more than one exists, the corresponding edges are replaced by a single edge to a freshly created region node that itself points to the direct successor nodes.

Finally, the data dependence subgraph is generated. This begins with the construction of DAGs for each basic block where each upwards reaching leaf is called a *merge node*. Data-flow analysis is used to compute reaching definitions. All individual DAGs are then connected together: edges are added from definition nodes to the corresponding merge nodes that may be reached. The resulting graph is the data dependence subgraph, and PDG construction is complete.

The PDG has been used for generating code for parallel architectures and has also been used in order to perform accurate program slicing and testing.

14.3.1 Detecting Parallelism with the PDG

An instruction scheduling algorithm running on a CFG lacks the necessary dataflow information to make decisions about parallelization. It requires additional code transformations such as loop unrolling or if-conversion (see Chap. 20) in order to expose any instruction-level parallelism.

However, the structure of the PDG can give the instruction scheduler this information for free. Any node of a CFG loop that is not contained in a strongly connected component of the PDG (using *both* control and data dependence edges) can be parallelized.

In the example in Fig. 14.2, since the instruction A[i]=a in the loop does not form a strongly connected component in the PDG, it can be vectorized provided that variable a has a private scope. On the other hand, because of the circuit involving the test on a, the instruction a=2*B[i] cannot.

14.4   Gating Functions and GSA

In SSA form, ϕ-functions are used to identify points where variable definitions converge. However, they cannot be directly *interpreted*, as they do not specify the condition that determines which of the variable definitions to choose. By this logic, we cannot directly interpret the SSA graph. Being able to interpret our IR is a useful property as it gives the compiler writer more information when implementing optimizations and also reduces the complexity of performing code generation. Gated single assignment form (GSA—sometimes called gated SSA) is an extension of SSA with *gating functions*. These gating functions are directly interpretable versions of ϕ-nodes and replace ϕ-nodes in the representation. We usually distinguish the three following forms of gating functions:

- The ϕ_{if} function explicitly represents the condition that determines which ϕ value to select. A ϕ_{if} function of the form $\phi_{if}(p, v_1, v_2)$ has p as a predicate, and v_1 and v_2 as the values to be selected if the predicate evaluates to true or false, respectively. This can be read simply as *if-then-else*.
- The ϕ_{entry} function is inserted at loop headers to select the initial and loop carried values. A ϕ_{entry} function of the form $\phi_{entry}(v_{init}, v_{iter})$, has v_{init} as the initial input value for the loop, and v_{iter} as the iterative input. We replace ϕ-functions at loop headers with ϕ_{entry} functions.
- The ϕ_{exit} function determines the value of a variable when a loop terminates. A ϕ_{exit} function of the form $\phi_{exit}(p, v_{exit})$ has P as predicate and v_{exit} as the definition reaching beyond the loop.

It is easiest to understand these gating functions by means of an example. Figure 14.3 shows how our earlier code in Fig. 14.2 translates into GSA form. Here,

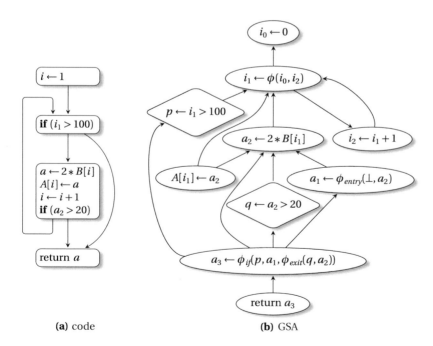

(a) code   (b) GSA

Fig. 14.3 A graph representation of our sample code in (demand-based) GSA form

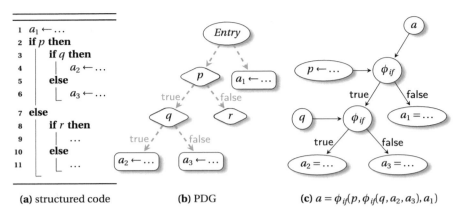

(a) structured code   (b) PDG   (c) $a = \phi_{if}(p, \phi_{if}(q, a_2, a_3), a_1)$

Fig. 14.4 (**a**) A structured code; (**b**) the PDG (with region nodes omitted); (**c**) the DAG representation of the nested gated ϕ_{if}

we can see the use of both ϕ_{entry} and ϕ_{exit} gating functions. At the header of our sample loop, the ϕ-function has been replaced by a ϕ_{entry} function which determines between the initial and iterative value of i. After the loop has finished executing, the nested ϕ_{exit} function selects the correct live-out version of a.

This example shows several interesting points. First, the semantics of both the ϕ_{exit} and ϕ_{if} are strict in their gate: here a_1 or $\phi_{exit}(q, a_2)$ is not evaluated before p is known.[1] Similarly, a ϕ_{if} function that results from the nested if-then-else code of Fig. 14.4 would be itself nested as $a = \phi_{if}(p, \phi_{if}(q, a_2, a_3), a_1)$. Second, this representation of the program does not allow for an interpreter to decide whether an instruction with a side effect (such as $A[i_1] = a_2$ in our running example) has to be executed or not. Finally, computing the values of gates is highly related to the simplification of path expressions: in our running example, a_2 should be selected when the path $\neg p$ followed by q (denoted $\neg p.q$) is taken, while a_1 should be selected when the path p is taken; for our nested if-then-else example, a_1 should be selected either when the path $\neg p.r$ is taken or when the path $\neg p.\neg r$ is taken, which simplifies to $\neg p$. Diverse approaches can be used to generate the correct nested ϕ_{if} or ϕ_{exit} gating functions.

The most natural way uses a data-flow analysis that computes, for each program point and each variable, its unique reaching definition and the associated set of reaching paths. This set of paths is abstracted using a *path expression*. If the code is not already under SSA, and if at a merge point of the CFG, its direct predecessor basic blocks are reached by different variables, a ϕ-function is inserted. The gate of each operand is set to the path expression of its corresponding incoming edge. If a unique variable reaches all the direct predecessor basic blocks, the corresponding

---

[1] As opposed to the ψ-function described in Chap. 15 that would use syntax such as $a_3 = \phi((p \wedge \neg q)?a_1, (\neg p \wedge q)?a_2)$ instead.

path expressions are merged. Of course, a classical path compression technique can be used to minimize the number of visited edges. One can observe the similarities with the ϕ-function placement algorithm described in Sect. 4.4.

There also exists a relationship between the control dependencies and the gates: from a code already under strict and conventional SSA form, one can derive the gates of a ϕ_{if} function from the control dependencies of its operands. This relationship is illustrated by Fig. 14.4 in the simple case of a structured code.

These gating functions are important as the concept will form components of the value state dependence graph later. GSA has seen a number of uses in the literature including analysis and transformations based on data flow. With the diversity of applications (see Chaps. 10 and 23), many variants of GSA have been proposed. Those variations concern the correct handling of loops in addition to the computation and representation of gates.

By using gating functions, it becomes possible to construct IRs based solely on data dependencies. These IRs are sparse in nature compared to the CFG, making them good for analysis and transformation. This is also a more attractive proposition than generating and maintaining both a CFG and DFG, which can be complex and prone to human error. One approach has been to combine both of these into one representation, as is done in the PDG. Alternatively, we can utilize gating functions along with a data-flow graph for an effective way of representing whole program information using data-flow information.

14.4.1   Backward Symbolic Analysis with GSA

GSA is useful for performing symbolic analysis. Traditionally, symbolic analysis is performed by forward propagation of expressions through a program. However, complete forward substitution is expensive and can result in a large quantity of unused information and complicated expressions. Instead, *backward*, demand-driven substitutions can be performed using GSA which only substitutes *needed* information. Consider the following program:

```
1  JMAX ← EXPR
2  if p then
3  │    J ← JMAX − 1
4  else
5  └    J ← JMAX
6  assert (J ≤ JMAX)
```

If forward substitutions were to be used in order to determine whether the assertion is correct, then the symbolic value of J must be discovered, starting at the top of the program in statement at line 1. Forward propagation through this program results in statement at line 6 being

$$\textbf{assert } ((\textbf{if } p \textbf{ then } EXPR - 1 \textbf{ else } EXPR) \leq EXPR)$$

and thus the **assert** () statement evaluates to true. In real, non-trivial programs, these expressions can get unnecessarily long and complicated.

Using GSA instead allows for backwards, demand-driven substitutions. The program above has the following GSA form:

---

1 $JMAX_1 \leftarrow EXPR$
2 **if** p **then**
3 $\quad | \quad J_1 \leftarrow JMAX_1 - 1$
4 **else**
5 $\quad \lfloor \quad J_2 \leftarrow JMAX_1$
6 $J_3 \leftarrow \phi_{if}(p, J_1, J_2)$
7 **assert** $(J_3 \leq JMAX_1)$

Using this backward substitution technique, we start at statement on line 7 and follow the SSA links of the variables from J_3. This allows for skipping of any intermediate statements that do not affect variables in this statement. Thus, the substitution steps are

$$J_3 = \phi_{if}(p, J_1, J_2)$$
$$= \phi_{if}(p, JMAX_1 - 1, JMAX_1)$$

The backward substitution then stops because enough information has been found, avoiding the redundant substitution of $JMAX_1$ by $EXPR$. In non-trivial programs, this can greatly reduce the number of redundant substitutions, making symbolic analysis significantly cheaper.

14.5 Value State Dependence Graph

The gating functions defined in the previous section were used in the development of a sparse data-flow graph IR called the value state dependence graph (VSDG). The VSDG is a directed graph consisting of operation nodes, loop, and merge nodes together with value and state dependency edges. Cycles are permitted but must satisfy various restrictions. A VSDG represents a single procedure: this matches the classical CFG. An example VSDG is shown in Fig. 14.5.

14.5.1 Definition of the VSDG

A VSDG is a labeled directed graph $G = (N, E_V, E_S, \ell, N_0, N_\infty)$ consisting of nodes N (with unique entry node N_0 and exit node N_∞), value dependency edges $E_V \subseteq N \times N$, and state dependency edges $E_S \subseteq N \times N$. The labelling function ℓ associates each node with an operator.

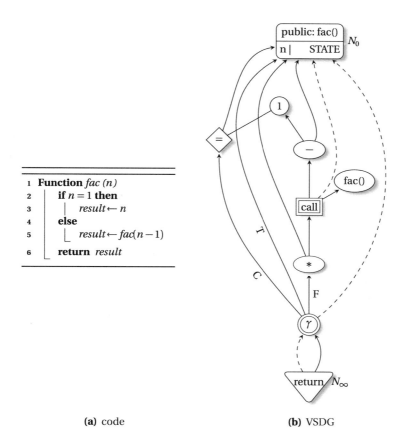

```
1  Function fac (n)
2      if n = 1 then
3          result ← n
4      else
5          result ← fac(n − 1)
6      return result
```

(a) code   **(b)** VSDG

Fig. 14.5 A recursive factorial function, whose VSDG illustrates the key graph components—value dependency edges (solid lines), state dependency edges (dashed lines), a `const` node, a `call` node, a γ-node, a conditional node, and the function entry and exit nodes

The VSDG corresponds to a reducible program, e.g., there are no cycles in the VSDG except those mediated by θ-nodes (loop).

Value dependency (E_V) indicates the flow of values between nodes. State dependency (E_S) represents two things; the first is essentially a sequential dependency required by the original program, e.g., a given `load` instruction may be required to follow a given `store` instruction without being re-ordered, and a `return` node in general must wait for an earlier loop to terminate even though there might be no value dependency between the loop and the `return` node. The second purpose is that state dependency edges can be added incrementally until the VSDG corresponds to a unique CFG. Such state dependency edges are called *serializing* edges .

The VSDG is implicitly represented in SSA form: a given operator node, n, will have zero or more E_V-consumers using its value. Note that, in implementation terms, a single register can hold the produced value for consumption at all consumers; it is therefore useful to talk about the idea of an output *port* for n being allocated a specific register, r, to abbreviate the idea of r being used for each edge (n_1, n_2), where $n_2 \in \text{directsucc}(n_1)$.

There are four main classes of VSDG nodes: value nodes (representing pure arithmetic) ,γ-nodes (conditionals), θ-nodes (loops) , and state nodes (side effects). The majority of nodes in a VSDG generates a value based on some computation (add, subtract, etc.) applied to their dependent values (constant nodes, which have no dependent nodes, are a special case).

γ-Nodes

The γ-node is similar to the ϕ_{if} gating function in being dependent on a control predicate, rather than the control-independent nature of SSA ϕ-functions. A γ-node $\gamma(C : p,\ T : v_{\text{true}},\ F : v_{\text{false}})$ evaluates the condition dependency p and returns the value of v_{true} if p is true, otherwise v_{false}. We generally treat γ-nodes as single-valued nodes (contrast θ-nodes, which are treated as tuples), with the effect that two separate γ-nodes with the same condition can be later combined into a tuple using a single test. Figure 14.6 illustrates two γ-nodes that can be combined in this way. Here, we use a pair of values (2-tuple)·for prots T and F. We also see how two syntactically different programs can map to the same structure in the VSDG.

θ-Nodes

The θ-node models the iterative behaviour of loops, modelling loop state with the notion of an *internal value* which may be updated on each iteration of the loop. It has five specific elements that represent dependencies at various stages of computation.

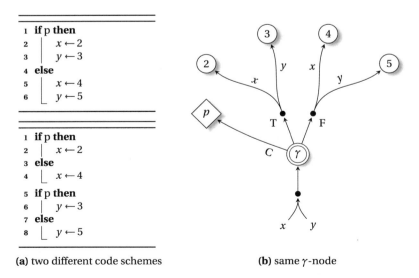

(a) two different code schemes   **(b)** same γ-node

Fig. 14.6 Two different code schemes map to the same γ-node structure

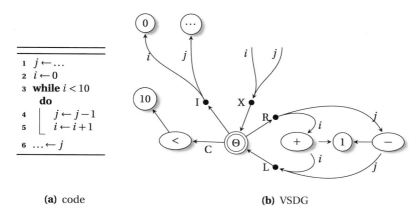

```
1  j ← ...
2  i ← 0
3  while i < 10
      do
4    |  j ← j − 1
5    |  i ← i + 1
6  ... ← j
```

(a) code   **(b)** VSDG

Fig. 14.7 An example showing a `for` loop. Evaluating **X** triggers it to evaluate the **I** value (outputting the value **L**). While **C** evaluates to true, it evaluates the **R** value (which in this case also uses the θ-node's **L** value). When **C** is false, it returns the final internal value through **X**. As i is not used after the loop, there is no dependency on i at **X**

The θ-node corresponds to a merge of the ϕ_{entry} and ϕ_{exit} nodes in gated SSA. A θ-node $\theta(C : p, \ I : v_{init}, \ R : v_{return}, \ L : v_{iter}, \ X : v_{exit})$ sets its internal value to initial value v_{init} and then, while condition value p holds true, sets v_{iter} to the current internal value and updates the internal value with the repeat value v_{return}. When p evaluates to false, computation ceases, and the last internal value is returned through v_{exit}.

A loop that updates k variables will have: a single condition p, initial values $v_{init}^1, \ldots, v_{init}^k$, loop iterations $v_{iter}^1, \ldots, v_{iter}^k$, loop returns $v_{return^1}, \ldots, v_{return}^k$, and loop exits $v_{exit}^1, \ldots, v_{exit}^k$. The example in Fig. 14.7 also shows a pair (2-tuple) of values being used on ports I, R, L, X, one for each loop-variant value.

The θ-node directly implements pre-test loops (`while`, `for`); post-test loops (`do...while`, `repeat...until`) are synthesized from a pre-test loop preceded by a duplicate of the loop body. At first, this may seem to cause unnecessary duplication of code, but it has two important benefits: (1) it exposes the first loop body iteration to optimization in post-test loops (cf. loop-peeling) and (2) it normalizes all loops to one loop structure, which both reduces the cost of optimization and increases the likelihood of two schematically dissimilar loops being isomorphic in the VSDG.

State Nodes

Loads and stores compute a value and state. The `call` node takes both the name of the function to call and a list of arguments and returns a list of results; it is treated as a state node as the function body may read or update state.

We maintain the simplicity of the VSDG by imposing the restriction that *all* functions have *one* return node (the exit node N_∞), which returns at least one result (which will be a state value in the case of `void` functions). To ensure that function calls and definitions are able to be allocated registers easily, we suppose that the number of arguments to, and results from, a function is smaller than the number of physical registers—further arguments can be passed via a stack as usual.

Note also that the VSDG does not force loop-invariant code into or out of loop bodies, but rather allows later phases to determine, by adding serializing edges, such placement of loop-invariant nodes for later phases.

14.5.2 Dead Node Elimination with the VSDG

By representing a program as a VSDG, many optimizations become trivial. For example, consider dead node elimination (Fig. 14.1). This combines both dead code elimination and unreachable code elimination. Dead code generates VSDG nodes for which there is no value or state dependency path from the `return` node, i.e., the result of the function does not in any way depend on the results of the dead nodes. Unreachable code generates VSDG nodes that are either dead or become dead after some other optimization. Thus, a *dead node* is a node that is not post-dominated by the exit node N_∞. To perform dead node elimination, only two passes are required over the VSDG resulting in linear runtime complexity: one pass to identify all of the live nodes, and a second pass to delete the unmarked (i.e., dead) nodes. It is safe because all nodes that are deleted are guaranteed never to be reachable from the `return` node.

Algorithm 14.1: Dead node elimination on the VSDG

---

1 **Input:** A VSDG $G(N, E_V, E_S, N_\infty)$ with zero or more dead nodes
2 **Output:** A VSDG with no dead nodes
3 *WalkAndMark*(N_∞, G)
4 *DeleteMarked*(G)

5 **Function** *WalkAndMark*(n, G)
6   **if** n is marked **then return**
7   mark n
8   **foreach** node $m \in N \wedge (n, m) \in (E_V \cup E_S)$ **do**
9   *WalkAndMark*(m)

10 **Function** *DeleteMarked*(G)
11   **foreach** node $n \in N$ **do**
12   **if** n is unmarked **then** *delete*(n)
13

14.6 Further Reading

A compiler's intermediate representation can be a graph, and many different graphs exist in the literature. We can represent the control flow of a program as a control-flow graph (CFG) [3], where straight-line instructions are contained within basic blocks and edges show where the flow of control may be transferred to once leaving that block. A CFG is traditionally used to convert a program into SSA form [90]. We can also represent programs as a type of data-flow graph (DFG) [96, 97], and SSA can be represented in this way as an SSA graph [84]. An example was given that used the SSA graph to detect a variety of induction variables in loops [127, 306]. It has also been used for performing instruction selection techniques [108, 254], operator strength reduction [84], and rematerialization [48] and has been combined with an extended SSA language to aid compilation in a parallelizing compiler [274].

The program dependence graph (PDG) as defined by Ferrante et al. [118] represents control and data dependencies in one graph. We choose to report the definition of Bilardi and Pingali [29]. Section 14.3 mentions possible abstractions to represent data dependencies for dynamically allocated objects. Among others, the book of Darte et al. [93] provides a good overview of such representations. The PDG has been used for program slicing [214], testing [20] and widely for parallelization [21, 117, 119, 259].

Gating functions can be used to create directly interpretable ϕ-functions. These are used in gated single assignment form. Alpern et al. [6] presented a precursor of GSA for structured code, to detect equality of variables. This chapter adopts their notations, i.e., a ϕ_{if} for an if-then-else construction, a ϕ_{entry} for the entry of a loop, and a ϕ_{exit} for its exit. The original usage of GSA was by Ballance et al. [215] as an intermediate stage in the construction of the Program Dependence Web IR. Further, GSA papers replaced ϕ_{if} by γ, ϕ_{entry} by μ, and ϕ_{exit} by η. Havlak [141] presented an algorithm for construction of a simpler version of GSA—Thinned GSA—which is constructed from a CFG in SSA form. The construction technique sketched in this chapter is developed in more detail in [289]. GSA has been used for a number of analyses and transformations based on data flow. The example given of how to perform backward demand-driven symbolic analysis using GSA has been borrowed from [290]. If-conversion (see Chap. 20) converts control dependencies into data dependencies. To avoid the potential loss of information related to the lowering of ϕ-functions into conditional moves or select instructions, gating ψ-functions (see Chap. 15) can be used.

We then described the value state dependence graph (VSDG) [151], which is an improvement on a previous, unmentioned graph, the Value Dependence Graph [304]. It uses the concept of gating functions, data dependencies, and state to model a program. We gave an example of how to perform dead node elimination on the VSDG. Detailed semantics of the VSDG are available [151], as well as semantics of a related IR: the gated data dependence graph [292]. Further study has taken place on the problem of generating code from the VSDG [178, 269, 291], and it has also been used to perform a combined register allocation and code motion algorithm [152].

Chapter 15
Psi-SSA Form

François de Ferrière

In the SSA representation, each definition of a variable is given a unique name, and new pseudo-definitions are introduced on ϕ-functions to merge values coming from different control-flow paths. An example is given in Fig. 15.1b. Each definition is an unconditional definition, and the value of a variable is the value of the expression on the unique assignment to this variable. This essential property of the SSA representation no longer holds when definitions may be conditionally executed. When a variable is defined by a predicated operation, the value of the variable will or will not be modified depending on the value of a guard register. As a result, the value of the variable after the predicated operation is either the value of the expression on the assignment if the predicate is true, or the value the variable had before this operation if the predicate is false. This is represented in Fig. 15.1c where we use the notation $p \; ? \; a = \text{op}$ to indicate that an operation $a = \text{op}$ is executed only if predicate p is true , and is ignored otherwise. We will also use the notation $\overline{p}$ to refer to the complement of predicate p. The goal of the ψ-SSA form advocated in this chapter is to express these conditional definitions while keeping the Static Single Assignment property.

15.1 Definition and Construction

Predicated operations are used to convert control-flow regions into straight-line code. Predicated operations may be used by the intermediate representation in an early stage of the compilation process as a result of inlining intrinsic functions. Later

F. de Ferrière (✉)
STMicroelectronics, Grenoble, France
e-mail: francois.de-ferriere@st.com

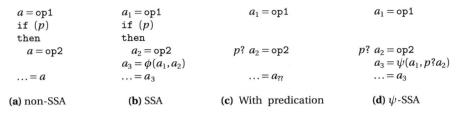

Fig. 15.1 SSA representation

on, the compiler may also generate predicated operations through if-conversion optimizations as described in Chap. 20.

In Fig. 15.1c, the use of a on the last instruction refers to the variable a_1 if p is false, or to the variable a_2 if p is true. These multiple reaching definitions on the use of a cannot be represented by the standard SSA representation. One possible representation would be to use the gated SSA form, presented in Chap. 14. In such a representation, the ϕ-function would be augmented with the predicate p to tell which value between a_1 and a_2 is to be considered. However, gated SSA is a completely different intermediate representation where the control flow is no longer represented. This representation is better suited to program interpretation than to optimizations at code-generation level as addressed in this chapter. Another possible representation would be to add a reference to a_1 on the definition of a_2. In this case, $p ? a_2 = \text{op2} \mid a_1$ would have the following semantic: a_2 takes the value computed by op2 if p is true, or holds the value of a_1 if p is false. The use of a on the last instruction of Fig. 15.1c would now refer to the variable a_2, which holds the correct value. The drawback of this representation is that it adds dependencies between operations (here a flow dependence from op1 to op2), which would prevent code reordering for scheduling.

Our solution is presented in Fig. 15.1d. The ϕ-function of the SSA code with control flow is "replaced" by a ψ-function on the corresponding predicated code, with information on the predicate associated with each argument. This representation is adapted to code optimization and code generation on a low-level intermediate representation. A ψ-function $a_0 = \psi(p_1?a_1, \ldots, p_i?a_i, \ldots, p_n?a_n)$ defines one variable, a_0, and takes a variable number of arguments a_i; each argument a_i is associated with a predicate p_i. In the notation, the predicate p_i will be omitted if $p_i \equiv$ true.

A ψ-function has the following properties:

- *It is an operation*: A ψ-function is a regular operation. It can occur at any location in a basic block where a regular operation is valid. Each argument a_i, and each predicate p_i, must be dominated by its definition.
- *It is predicated*: A ψ-function is a predicated operation, under the predicate $\bigcup_{k=1}^{n} p_k$, although this predicate is not explicit in the representation.

Fig. 15.2 ψ-SSA with non-disjoint predicates

```
if (p)
then
    a₁ = 1
else
    a₂ = -1
x₁ = φ(a₁, a₂)
if (q)
then
    a₃ = 0
x₂ = φ(x₁, a₃)
```

(a) control-flow code

$p?\ a_1 = 1$

$\overline{p}?\ a_2 = -1$
$\quad x_1 = \psi(p?a_1, \overline{p}?a_2)$

$q?\ a_3 = 0$
$\quad x_2 = \psi(p?a_1, \overline{p}?a_2, q?a_3)$

(b) Predicated code

- *It has an ordered list of arguments*: The order of the arguments in a ψ-function is significant. A ψ-function is evaluated from left to right. The value of a ψ-function is the value of the rightmost argument whose predicate evaluates to true.
- *Rule on predicates*: The predicate p_i associated with the argument a_i in a ψ-function must be included in or equal to the predicate on the definition of the variable a_i. In other words, for the code $q\ ?\ a_i = \mathrm{op};\ a_0 = \psi(\ldots,\ p_i?a_i, \ldots)$, we must have $p_i \subseteq q$ (or $p_i \Rightarrow q$).

A ψ-function can represent cases where variables are defined on arbitrary independent predicates such as p and q in the example of Fig. 15.2: For this example, during the SSA construction, a unique variable a was renamed into the variables a_1, a_2, and a_3, and the variables x_1 and x_2 were introduced to merge values coming from different control-flow paths. In the control-flow version of the code, there is a control dependence between the basic block that defines x_1 and the operation that defines a_3, which means that the definition of a_3 must be executed after the value for x_1 has been computed. In the predicated form of this example, there are no longer any control dependencies between the definitions of a_1, a_2, and a_3. A compiler transformation can now freely move these definitions independently of each other, which may allow more optimizations to be performed on this code. However, the semantics of the original code requires that the definition of a_3 occurs after the definitions of a_1 and a_2. The order of the arguments in a ψ-function gives information on the original order of the definitions. We take the convention that the order of the arguments in a ψ-function is, from left to right, equal to the original order of their definitions, from top to bottom, in the control-flow dominance tree of the program in a non-SSA representation. This information is needed to maintain the correct semantics of the code during transformations of the ψ-SSA representation and to revert the code back to a non-ψ-SSA representation.

The construction of the ψ-SSA representation is a small modification to the standard algorithm to build an SSA representation (see Sect. 3.1). The insertion of ψ-functions is performed during the SSA renaming phase. During this phase, basic blocks are processed in their dominance order, and operations in each basic block are scanned from top to bottom. On an operation, for each predicated definition of a variable, a new ψ-function will be inserted just after the operation. Consider the definition of a variable x under predicate p_2 ($p_2\ ?\ x = \mathrm{op}$); suppose that x_1 is the current version of x before proceeding to op and that x_1 is defined through

p_2? $x = \mathrm{op}$ p_2? $x = \mathrm{op}$ p_2? $x_2 = \mathrm{op}$ p_2? $x_2 = \mathrm{op}$

$\qquad\qquad\qquad\quad x = \psi(p_1?x_1, p_2?x) \qquad\quad x = \psi(p_1?x_1, p_2?x) \qquad\quad x_3 = \psi(p_1?x_1, p_2?x_2)$

(a) Initial   **(b)** ψ-insertion   **(c)** op-renaming   **(d)** ψ-renaming

Fig. 15.3 Construction and renaming of ψ-SSA

predicate p_1 (possibly true); after renaming x into a freshly created version, say x_2, a ψ-function of the form $x = \psi(p_1?x_1, p_2?x)$ is inserted right after op. Then the renaming of this new operation proceeds. The first argument of the ψ-function is already renamed and thus is not modified. The second argument is renamed into the current version of x, that is, x_2. On the definition of the ψ-function, the variable x is given a new name, x_3, which becomes the current version for further references to the x variable. This insertion and renaming of a ψ-function is shown in Fig. 15.3.

ψ-functions can also be introduced into an SSA representation by applying an if-conversion transformation, such as the one described in Chap. 20. Local transformations of control-flow patterns can also require the replacement of ϕ-functions by ψ-functions.

15.2 SSA Algorithms

With this definition of the ψ-SSA representation, implicit data-flow links to predicated operations are now explicitly expressed through ψ-functions. The usual algorithms that perform optimizations or transformations on the SSA representation can now be easily adapted to the ψ-SSA representation, without compromising the efficiency of the transformations performed. Actually, within the ψ-SSA representation, predicated definitions behave exactly the same as non-predicated ones for optimizations on the SSA representation. Only the ψ-functions have to be treated in a specific way. As an example, the classical constant propagation algorithm under SSA can be easily adapted to the ψ-SSA representation. In this algorithm, the only modification concerns ψ-functions, which have to be handled using the same rules as with the ϕ-functions. Other algorithms such as dead code elimination (see Chap. 3), global value numbering, partial redundancy elimination (see Chap. 11), and induction variable analysis (see Chap. 10) are examples of algorithms that can easily be adapted to this representation with little effort.

15.3 Psi-SSA Algorithms

In addition to standard algorithms that can be applied to ψ-functions and predicated code, a number of specific transformations can be performed on the ψ-functions, namely ψ-inlining, ψ-reduction, ψ-projection, ψ-permutation, and ψ-promotion.

$$a_1 = \text{op1}$$
$$p_2? \; a_2 = \text{op2}$$
$$x_1 = \psi(a_1, p_2?a_2)$$
$$p_3? \; a_3 = \text{op3}$$
$$x_2 = \psi(p_1?x_1, p_3?a_3)$$

$$a_1 = \text{op1}$$
$$p_2? \; a_2 = \text{op2}$$
$$x_1 = \psi(a_1, p_2?a_2) \; // \; \text{dead}$$
$$p_3? \; a_3 = \text{op3}$$
$$x_2 = \psi(p_1?a_1, p_1 \wedge p_2?a_2, p_3?a_3)$$

Fig. 15.4 ψ-Inlining of the definition of x_1

$$a_1 = \text{op1}$$
$$p_2? \; a_2 = \text{op2}$$
$$\overline{p_2}? \; a_3 = \text{op3}$$
$$x_2 = \psi(a_1, p_2?a_2, \overline{p_2}?a_3)$$

$$a_1 = \text{op1}$$
$$p_2? \; a_2 = \text{op2}$$
$$\overline{p_2}? \; a_3 = \text{op3}$$
$$x_2 = \psi(p_2?a_2, \overline{p_2}?a_3)$$

Fig. 15.5 ψ-Reduction. The first argument a_1 of the ψ-function can safely be removed

For a ψ-function $a_0 = \psi(p_1?a_1, \ldots, p_i?a_i, \ldots, p_n?a_n)$, these transformations are defined as follows:

ψ-**Inlining** recursively replaces in a ψ- function an argument a_i that is defined on another ψ- function by the arguments of this other ψ-function. The predicate p_i associated with argument a_i will be distributed with an and operation over the predicates associated with the inlined arguments. This is shown in Fig. 15.4.

ψ-**Reduction** removes from a ψ-function an argument a_i whose value will always be overridden by arguments on its right in the argument list. An argument a_i associated with predicate p_i can be removed if $p_i \subseteq \bigcup_{k=i+1}^{n} p_k$. This can be illustrated by the example of Fig. 15.5.

ψ-**Projection** creates from a ψ-function a new ψ-function on a restricted predicate, say p. In this new ψ-function, an argument a_i initially guarded by p_i will be guarded by the conjunction $p_i \wedge p$. If p_i is known to be disjoint with p, a_i actually contributes no value to the ψ-function and thus can be removed. ψ-projection on predicate p is usually performed when the result of a ψ-function is used in an operation predicated by p. This is illustrated in Fig. 15.6.

ψ-**Permutation** changes the order of the arguments in a ψ-function. In a ψ-function, the order of the arguments is significant. Two arguments in a ψ-function can be permuted if the intersection of their associated predicate in the ψ-function is empty. An example of such a permutation is shown in Fig. 15.7.

ψ-**Promotion** changes one of the predicates used in a ψ-function to a larger predicate. The promotion must obey the following condition so that the semantics of the ψ-function is not altered by the transformation: consider an operation $a_0 = \psi(p_1?x_1, \ldots, p_i?x_i, \ldots, p_n?x_n)$ promoted into $a_0 =$

Fig. 15.6 ψ-Projection of x_2 on p_2. Second argument a_3 can be removed

$$p_2? \; a_2 = \text{op2}$$
$$\overline{p_2}? \; a_3 = \text{op3}$$
$$x_2 = \psi(p_2?a_2, \overline{p_2}?a_3)$$
$$p_2? \; y_1 = x_2$$

$$p_2? \; a_2 = \text{op2}$$
$$\overline{p_2}? \; a_3 = \text{op3}$$
$$x_2 = \psi(p_2?a_2, \overline{p_2}?a_3)$$
$$x_3 = \psi(p_2?a_2)$$
$$p_2? \; y_1 = x_3$$

Fig. 15.7 ψ-Permutation of arguments a_2 and a_3

$$\overline{p_2}? \; a_3 = \text{op3}$$
$$p_2? \; a_2 = \text{op2}$$
$$x_2 = \psi(p_2?a_2, \overline{p_2}?a_3)$$

$$\overline{p_2}? \; a_3 = \text{op3}$$
$$p_2? \; a_2 = \text{op2}$$
$$x_2 = \psi(\overline{p_2}?a_3, p_2?a_2)$$

```
if (p)
then
    a₁ = ADD i₁,1;          a₁ = ADD i₁,1;          a₁ = ADD i₁,1;
else
    a₂ = ADD i₁,2;          a₂ = ADD i₁,2;          a₂ = ADD i₁,2;
    x = φ(a₁,a₂)            x = ψ(p?a₁,p̄?a₂)        x = ψ(a₁,p̄?a₂)

    (a) Control flow         (b) ψ-SSA            (c) after ψ-promotion
```

Fig. 15.8 ψ-SSA for partial predication. ψ-Promotion of argument a_1

$\psi(p_1?x_1, \ldots, p_i'?x_i, \ldots, p_n?x_n)$ with $p_i \subseteq p_i'$, and then p_i' must fulfil

$$\left(p_i' \setminus \bigcup_{k=i}^{n} p_k\right) \cap \bigcup_{k=1}^{i-1} p_k = \emptyset, \tag{15.1}$$

where $p_i' \setminus \bigcup_{k=i}^{n} p_k$ corresponds to the possible increase of the predicate of the ψ-function, $\bigcup_{k=1}^{n} p_k$. This promotion must also satisfy the properties of ψ-functions, and, in particular, that the predicate associated with a variable in a ψ-function must be included in or equal to the predicate on the definition of that variable (which itself can be a ψ-function). A simple ψ-promotion is illustrated in Fig. 15.8c.

The ψ-SSA representation can be used on a partially predicated architecture, where only a subset of the instructions supports a predicate operand. Figure 15.8 shows an example where some code with control-flow edges was transformed into a linear sequence of instructions. Taking the example of an architecture where the ADD operation cannot be predicated, the ADD operation must be speculated under the true predicate. On an architecture where the ADD operation can be predicated, it may also be profitable to perform speculation in order to reduce the number of predicates on predicated code and to reduce the number of operations to compute these predicates. Once speculation has been performed on the definition of a variable used in a ψ-function, the predicate associated with this argument can be promoted, provided that the semantics of the ψ-function is maintained (Eq. 15.1).

Usually, the first argument of a ψ-function can be promoted under the true predicate. Also, when disjoint conditions are computed, one of them can be promoted to include the other conditions, usually reducing the number of predicates. A side effect of this transformation is that it may increase the number of copy instructions to be generated during the ψ-SSA destruction phase, as will be explained in the following section.

15.4   Psi-SSA Destruction

The SSA destruction phase reverts an SSA representation into a non-SSA representation. This phase must be adapted to the ψ-SSA representation. This algorithm uses ψ-ϕ-webs to create a conventional ψ-SSA representation. The notion of ϕ-

webs is extended to ϕ and ψ operations so as to derive the notion of conventional ψ-SSA (ψ-C-SSA) form. A ψ-ϕ-web is a non-empty, minimal set of variables such that if two variables are referenced on the same ϕ or ψ-function, then they are in the same ψ-ϕ-web. The property of the ψ-C-SSA form is that the renaming into a single variable of all variables that belong to the same ψ-ϕ-web, and the removal of the ψ and ϕ functions, results in a program with the same semantics as the original program.

Now, consider Fig. 15.9 to illustrate the transformations that must be performed to convert a program from a ψ-SSA form into a program in ψ-C-SSA form.

Looking at the first example (Fig. 15.9a), the dominance order of the definitions for the variables a and b differs from their order from left to right in the ψ-function. Such code may appear after a code motion algorithm has moved the definitions for a and b relatively to each other. Here, the renaming of the variables a, b, and x into a single variable will not restore the semantics of the original program. The order in which the definitions of the variables a, b, and x occur must be corrected. This is done through the introduction of the variable c that is defined as a predicated copy of the variable b, after the definition of a. Now, the renaming of the variables a, c, and x into a single variable will result in the correct behaviour.

In Fig. 15.9d, the definition of the variable b has been speculated. However, according to the semantics of the ψ-function, the variable x will only be assigned the value of b when p is true. A new variable c must be defined as a predicated copy

$p?\ b = \dots$	$p?\ b = \dots$	$p?\ b = \dots$
$a = \dots$	$a = \dots$	$x = \dots$
	$p?\ c = b$	$p?\ x = b$
$x = \psi(a, p?b)$	$x = \psi(a, p?c)$	
(a) ψ-T-SSA form	**(b)** ψ-C-SSA form	**(c)** non-SSA form
$a = \dots$	$a = \dots$	$x = \dots$
$b = \dots$	$b = \dots$	$b = \dots$
	$p?\ c = b$	$p?\ x = b$
$x = \psi(a, p?b)$	$x = \psi(a, p?c)$	
(d) ψ-T-SSA form	**(e)** ψ-C-SSA form	**(f)** non-SSA form
$a = \dots$	$a = \dots$	$x = \dots$
	$d = a$	$y = x$
$p?\ b = \dots$	$p?\ b = \dots$	$p?\ x = \dots$
$q?\ c = \dots$	$q?\ c = \dots$	$q?\ y = \dots$
$x = \psi(a, p?b)$	$x = \psi(a, p?b)$	
$y = \psi(a, q?c)$	$y = \psi(d, q?c)$	
(g) ψ-T-SSA form	**(h)** ψ-C-SSA form	**(i)** non-SSA form

Fig. 15.9 Non-conventional ψ-SSA (ψ-T-SSA) form, ψ-C-SSA forms, and non-SSA form after destruction

of the variable b, after the definition of b and p; in the ψ-function, variable b is then replaced by variable c. The renaming of variables a, c, and x into a single variable will now follow the correct behaviour.

In Fig. 15.9g, the renaming of the variables a, b, c, x, and y into a single variable will not give the correct semantics. In fact, the value of a used in the second ψ-function would be overridden by the definition of b before the definition of the variable c. Such code will occur after copy folding has been applied on a ψ-SSA representation. We see that the value of a has to be preserved before the definition of b. This is done through the definition of a new variable (d here), resulting in the code given in Fig. 15.9h. Now, the variables a, b, and x can be renamed into a single variable, and the variables d, c, and y will be renamed into another variable, resulting in a program in a non-SSA form with the correct behaviour.

We will now present an algorithm that will transform a program from a ψ-SSA form into its ψ-C-SSA form. This algorithm comprises three parts:

- *Psi-normalize*: This phase puts all ψ-functions in what we call a *normalized* form.
- *Psi-web*: This phase grows ψ-webs from ψ-functions and introduces repair code where needed such that each ψ-web is interference-free.
- *Phi-web*: This phase is the standard SSA destruction algorithm (e.g., see Chap. 21) with the additional constraint that all variables in a ψ-web must be coalesced together. This can be done using the pinning mechanism presented in Chap. 21.

We now detail the implementation of each of the first two parts.

15.4.1   Psi-Normalize

We define the notion of *normalized-ψ*. The normalized form of a ψ-function has two characteristics:

- The order of the arguments in a normalized-ψ-function is, from left to right, equal to the order of their definitions, from top to bottom, in the control-flow dominance tree.
- The predicate associated with each argument in a normalized-ψ-function is equal to the predicate used on the unique definition of this argument.

These two characteristics correspond respectively to the two cases presented in Fig. 15.9a, d. When some arguments of a ψ-function are also defined by ψ-functions, the normalized-ψ characteristics must hold on a virtual ψ-function where ψ-inlining has been performed on these arguments.

When ψ-functions are created during the construction of the ψ-SSA representation, they are naturally built in their normalized form. Later, transformations are applied to the ψ-SSA representation. Predicated definitions may be moved relatively to each other. Also, operation speculation and copy folding may enlarge the domain of the predicate used on the definition of a variable. These transformations may cause some ψ-functions to be in a non-normalized form.

PSI-Normalize Implementation

Each ψ-function is processed independently. An analysis of the ψ-functions in a top-down traversal of the dominator tree reduces the amount of repair code that is inserted during this pass. We only detail the algorithm for such a traversal.

For a ψ-function $a_0 = \psi(p_1?a_1, \ldots, p_i?a_i, \ldots, p_n?a_n)$, the argument list is processed from left to right. For each argument a_i, the predicate p_i associated with this argument in the ψ-function and the predicate used on the definition of this argument are compared. If they are not equal, a new variable a_i' is introduced and is initialized at the highest point in the dominator tree after the definitions of a_i and p_i. a_i' is defined by the operation $p_i ? a_i' = a_i$. Then a_i is replaced by a_i' in the ψ-function.

Next, we consider the dominance order of the definition for a_i in respect of the definition for a_{i-1}. When a_i is defined on a ψ-function, we recursively look for the definition of the first argument of this ψ-function, until a definition on a non-ψ-function is found. If the definition we found for a_i dominates the definition for a_{i-1}, a correction is needed. If the predicates p_{i-1} and p_i are disjoint, a ψ-permutation can be applied between a_{i-1} and a_i, so as to reflect into the ψ-function the actual dominance order of the definitions of a_{i-1} and a_i. If ψ-permutation cannot be applied, a new variable a_i' is created for repair. a_i' is defined by the operation $p_i?a_i' = a_i$. This copy operation is inserted at the highest point that is dominated by the definitions of a_{i-1} and a_i.[1] Then a_i is replaced in the ψ-function by a_i'.

The algorithm continues with the argument a_{i+1}, until all arguments of the ψ-function are processed. When all arguments are processed, the ψ is in its normalized form. When all ψ functions are processed, the function will contain only normalized-ψ-functions.

15.4.2   Psi-web

The role of the psi-web phase is to repair the ψ-functions that are part of a non-interference-free ψ-web. This case corresponds to the example presented in Fig. 15.9g. In the same way as there is a specific point of use for arguments on ϕ-functions for liveness analysis (e.g., see Sect. 21.2), we give a definition of the actual point of use of arguments on normalized ψ-functions for liveness analysis. With this definition, liveness analysis is computed accurately, and an interference graph can be built. The cases where repair code is needed can be easily and accurately detected by observing that variables in a ψ-function interfere.

---

[1] When a_i is defined by a ψ-function, its definition may appear after the definition for a_{i-1}, although the non-ψ definition for a_i appears before the definition for a_{i-1}.

Fig. 15.10 ψ-Functions and
C conditional operations
equivalence

$$a = \text{op1}$$
$$p? \; b = \text{op2}$$
$$q? \; c = \text{op3}$$
$$x = \psi(a, p?b, q?c)$$

(a) ψ-SSA form

$$a = \text{op1}$$
$$b = p \; ? \; \text{op2} \; : \; a$$
$$c = q \; ? \; \text{op3} \; : \; b$$
$$x = c$$

(b) conditional form

Liveness and Interferences in Psi-SSA

Consider the code in Fig. 15.10b. Instead of using a representation with ψ-functions, predicated definitions have been modified to make a reference to the value the predicated definition will have in the event that the predicate evaluates to false. In this example, we use the notation of the select operator $x = cond \; ? \; exp1 \; : \; exp2$ that assigns $exp1$ to x if $cond$ is true and $exp2$ otherwise. Each of the predicated definitions makes explicit use of the variable immediately to its left in the argument list of the original ψ-function from Fig. 15.10a. We can see that a renaming of the variables a, b, c, and x into a single representative name will still compute the same value for the variable x. Note that this transformation can only be performed on normalized ψ-functions, since the definition of an argument must be dominated by the definition of the argument immediately to its left in the argument list of the ψ-function, and the same predicate must be used on the definition of an argument and with this argument in the ψ operation. Using this equivalence for the representation of a ψ-function, we now give a definition of the point of use for the arguments of a ψ-function.

Definition 15.1 (Use Points) Let $a_0 = \psi(p_1?a_1, \ldots, p_i?a_i, \ldots, p_n?a_n)$ be a normalized ψ-function. For $i < n$, the point of use of argument a_i occurs at the operation that defines a_{i+1}. The point of use for the last argument a_n occurs at the ψ-function itself.

Given this definition of point of use of ψ-function arguments, and using the usual point of use of ϕ-function arguments, a traditional liveness analysis can be run. Then an interference graph can be built to collect the interferences between variables involved in ψ or ϕ-functions. For the construction of the interference graph, an interference between two variables that are defined on disjoint predicates can be ignored.

Repairing Interferences on ψ-Functions

We now present an algorithm that resolves the interferences as it builds the ψ-webs. A pseudo-code is given in Algorithm 15.1. First, the ψ-webs are initialized with a single variable per ψ-web. Next, ψ-functions are processed one at a time, in no specific order, merging when non-interfering the ψ-webs of its operands together. Two ψ-webs interfere if at least one variable in the first ψ-web interferes with at least one variable in the other one. The arguments of the ψ-function, say $a_0 = \psi(p_1?a_1, \ldots, p_i?a_i, \ldots, p_n?a_n)$, are processed from right (a_n) to left (a_1). If the ψ-web that contains a_i does not interfere with the ψ-web that contains a_0, they are merged together. Otherwise, repair code is needed. A new variable, a_i', is created and is initialized with a predicated copy $p_i \; ? \; a_i' = a_i$, inserted just above the definition for a_{i+1}, or just above the ψ-function in case of the last argument. The current argument a_i in the ψ-function is replaced by the new variable a_i'. The interference graph is updated. This can be done by considering the set of variables, say U, that a_i interferes with. For each $u \in U$, if u is in the merged ψ-web, it should not interfere with a_i'; if the definition of u dominates the definition of a_i, it is live through the definition of a_i, thus it should be made interfering with a_i'; last, if the definition of a_i dominates the definition of b, it should be made interfering only if this definition is within the live range of a_i' (see Chap. 9).

Algorithm 15.1: ψ-Webs merging during the processing of a ψ-function $a_0 = \psi(p_1?a_1, \ldots, p_i?a_i, \ldots, p_n?a_n)$

---

1 **let** $psiWeb$ be the web containing a_0
2 **foreach** a_i **in** $[a_n, a_{n-1}, \ldots, a_1]$ **do**
3   **let** $opndWeb$ be the web containing a_i
4   **if** $opndWeb \neq psiWeb$ **then**
5   **if** $IGraph.$interfere($psiWeb$, $opndWeb$) **then**
6   **let** a_i' be a freshly created variable
7   **let** C_i be a new predicated copy p_i ? $a_i' \leftarrow a_i$
8   **let** op be the operation that defines a_{i+1}, $a_i.def.op$, or the psi operation
9   **while** op is a ψ-function **do**
10   replace op by the operation that defines its first argument
11   append C_i right before op
12   replace a_i by a_i' in the ψ-function
13          $opndWeb \leftarrow \{a_i'\}$
14   **foreach** u **in** $IGraph.$interferenceSet(a_i) **do**
15   **if** $u \notin psiWeb$ **then**
16   **if** $u.def.op$dominates $a_i.def.op \bigvee a_i' \in$ livein($u.def.op$) **then**
17                   $IGraph.$addInterference(a_i', a)

18    $psiWeb \leftarrow psiWeb \cup opndWeb$

Consider the code in Fig. 15.11 to see how this algorithm works. The liveness on the ψ-function creates a live range for variable a that extends down to the definition

$p?\ a = \ldots$	$p?\ a = \ldots$	$p?\ x = \ldots$
$q?\ b = \ldots$	$q?\ b = \ldots$	$q?\ b = \ldots$
	$q?\ b' = b$	$q?\ x = b$
$r?\ c = \ldots$	$r?\ c = \ldots$	$r?\ x = \ldots$
$x = \psi(p?a, q?b, r?c)$	$x = \psi(p?a, q?b', r?c)$	
$s?\ d = b+1$	$s?\ d = b+1$	$s?\ d = b+1$
(a) before processing the ψ-function	**(b)** after processing the ψ-function	**(c)** after actual coalescing

Fig. 15.11 Elimination of ψ live interference

of b, but not further down. Thus, the variable a does not interfere with the variables b, c, or x. The live range for variable b extends down to its use in the definition of variable d. This live range interferes with the variables c and x. The live range for variable c extends down to its use in the ψ-function that defines the variable x. At the beginning of the processing on the ψ-function $x = \psi(p?a, q?b, r?c)$, ψ-webs are singletons $\{a\}$, $\{b\}$, $\{c\}$, $\{x\}$, $\{d\}$. The argument list is processed from right to left, i.e., starting with variable c. $\{c\}$ does not interfere with $\{x\}$, and they can be merged together, resulting in $psiWeb = \{x, c\}$. Next, variable b is processed. Since it interferes with both x and c, repair code is needed. A variable b' is created and is initialized just below the definition for b, as a predicated copy of b. The interference graph is updated conservatively, with no changes. $psiWeb$ now becomes $\{x, b', c\}$. Then variable a is processed, and as no interference is encountered, $\{a\}$ is merged to $psiWeb$. The final code after SSA destruction is shown in Fig. 15.11c.

15.5 Further Reading

In this chapter, we have mainly described the ψ-SSA representation and have detailed specific transformations that can be performed thanks to this representation. More details on the implementation of the ψ-SSA algorithms, and figures on the benefits of this representation, can be found in [275] and [104].

We mentioned in this chapter that a number of classical SSA-based algorithms can be easily adapted to the ψ-SSA representation, usually by just adapting the rules on the ϕ-functions to the ψ-functions. Among these algorithms, we can mention the constant propagation algorithm described in [303], dead code elimination [202], global value numbering [76], partial redundancy elimination [73], and induction variable analysis [306], which have already been implemented into a ψ-SSA framework.

There are also other SSA representations that can handle predicated instructions, of which the Predicated SSA representation [58]. This representation is targeted at very low-level optimization to improve operation scheduling in the presence of predicated instructions. Another representation is the gated SSA form, presented in Chap. 14.

The ψ-SSA destruction algorithm presented in this chapter is inspired from the SSA destruction algorithm of Sreedhar et al. [267], which introduces repair code when needed as it grows ϕ-webs from ϕ-functions. The phi-web phase mentioned in this chapter to complete the ψ-SSA destruction algorithm can use exactly the same approach by simply initializing ψ-ϕ-webs by ψ-webs.

Chapter 16
Hashed SSA Form: HSSA

Massimiliano Mantione and Fred Chow

Hashed SSA (or in short, HSSA) is an SSA extension that can effectively represent how aliasing relations affect a program in SSA form. It works equally well for aliasing among scalar variables and, more generally, for indirect load and store operations on arbitrary memory locations. Thus, all common SSA-based optimizations can be applied uniformly to any storage area no matter how they are represented in the program.

It should be noted that only the representation of aliasing is discussed here. HSSA relies on a separate alias analysis pass that runs before its creation. Depending on the actual alias analysis algorithm used, the HSSA representation will reflect the accuracy produced by the alias analysis pass.

This chapter starts by defining notations to model the effects of aliasing for scalar variables in SSA form. We then introduce a technique that can reduce the overhead linked to the SSA representation by avoiding an explosion in the number of SSA versions for aliased variables. Next, we introduce the concept of *virtual variables* to model indirect memory operations as if they were scalar variables, effectively allowing indirect memory operations to be put into SSA form together with scalar variables. Finally, we apply global value numbering (GVN) to the program to derive Hashed SSA form[1] as the effective SSA representation of all storage entities in the program.

---

[1] The name *Hashed* SSA comes from the use of hashing in value numbering.

M. Mantione (✉)
WorkWave, Gorgonzola, Milan, Italy
e-mail: massimiliano.mantione@gmail.com

F. Chow
Huawei, Fremont, CA, USA
e-mail: fchow99@comcast.net

© The Author(s), under exclusive license to Springer Nature Switzerland AG 2022
F. Rastello, F. Bouchez Tichadou (eds.), *SSA-based Compiler Design*,
https://doi.org/10.1007/978-3-030-80515-9_16

16.1   SSA and Aliasing: μ- and χ-Functions

Aliasing occurs in a program when a storage location (that contains a value) referred to in the program code can potentially be accessed through a different means; this can occur under one of the following four conditions:

- Two or more storage locations partially overlap. For example, in the C *union* construct, a storage location can be accessed via different field names.
- A variable is pointed to by a pointer. In this case, the variable can be accessed in two ways: *directly*, through the variable name, and *indirectly*, through the pointer that holds its address.
- The address of a variable is passed in a procedure call. This enables the called procedure to access the variable indirectly.
- The variable is declared in the global scope. This allows the variable to potentially be accessed in any function call.

We seek to model the *effects* of aliasing on a program in SSA form based on the results of the alias analysis performed. To characterize the effects of aliasing, we distinguish between two types of definitions of a variable: *MustDef* and *MayDef*. A MustDef must redefine the variable and thus blocks the references to its previous definitions from that point on. A MayDef only potentially redefines the variable and so does not prevent previous definitions of the same variable from being referenced later in the program.[2] We represent MayDef through the use of χ-functions. On the use side, in addition to real uses of the variable, which are *MustUses*, there are *MayUses* that arise in places in the program where there are potential references to the variable. We represent MayUse through the use of μ-functions. The semantics of μ and χ operators can be illustrated through the C-like example in Fig. 16.1, where $*p$ represents an indirect access through pointer p. The argument of the μ-operator is the potentially used variable. The argument of the χ-operator is the potentially assigned variable itself, to express the fact that the variable's original value will *flow through* if the MayDef does not modify the variable.

The use of μ- and χ-functions does not alter the complexity of transforming a program into SSA form. All that is necessary is a pre-pass that inserts them in the program. Ideally, μ- and χ-functions should be placed *parallel* to the instruction that led to their insertion. Parallel instructions are represented in Fig. 16.1 using the notation introduced in Sect. 13.1.4. Nonetheless, practical implementations may choose to insert μ- and χ-functions before or after the instructions that involve aliasing. In particular, μ-functions can be inserted immediately *before* the involved statement or expression and χ-operators immediately *after* the statement. This distinction allows us to model call effects correctly: the called function appears to potentially use the values of variables before the call, and the potentially modified values appear after the call.

---

[2] MustDefs are often referred to as Killing Defs and MayDefs as Preserving or Non-killing Defs in the literature.

1 $i = 2$	1 $i = 2$	1 $i_1 = 2$
2 **if** j **then**	2 **if** j **then**	2 **if** j_1 **then**
3 $\mid$ ...	3 $\mid$ ...	3 $\mid$ ...
4 **else**	4 **else**	4 **else**
5 $\mid$ $f()$	5 $\mid$ $f() \parallel \mu(i)$	5 $\mid$ $f() \parallel \mu(i_1)$
6 $\mid$ $*p = 3$	6 $\mid$ $*p = 3 \parallel i = \chi(i)$	6 $\mid$ $*p = 3 \parallel i_2 = \chi(i_1)$
7	7	7 $i_3 = \phi(i_1, i_2)$
8 $i = 4$	8 $i = 4$	8 $i_4 = 4$
9 return	9 return $\parallel \mu(i)$	9 return $\parallel \mu(i_4)$
(a) Initial C code	**(b)** After μ and χ insertion	**(c)** After ϕ insertion and versioning

Fig. 16.1 Program example where $*p$ might alias i, and function f might indirectly use i but not alter it

Thanks to the systematic insertion of μ-functions and χ-functions, an assignment to any scalar variable can be safely considered dead if it is not marked live by a standard SSA-based dead-code elimination. In our running example of Fig. 16.1c, the potential use of the assigned value of i outside the function, represented through the μ-function at the return, allows us to conclude that the assignment to i_4 is not dead.

16.2   Introducing "Zero Versions" to Limit Version Explosion

While it is true that μ and χ insertion does not alter the complexity of SSA construction, applying it to a production compiler as described in the previous section may make working with code in SSA form inefficient in programs that exhibit a lot of aliasing. Each χ-function introduces a new version, and it may in turn cause new ϕ-functions to be inserted. This can make the number of distinct variable versions needlessly large.

The major issue is that the SSA versions introduced by χ-operators are useless for most optimizations that deal with variable values. χ definitions add uncertainty to the analysis of variable values: the actual value of a variable after a χ definition could be its original value, or it could be the one indirectly assigned by the χ.

Our solution to this problem is to factor all variable versions that are considered *useless* together, so that SSA versions are not wasted. We assign number 0 to this special variable version and call it *zero version*.

Our notion of useless versions relies on the concept of *real occurrence* of a variable, which is an actual definition or use of a variable in the *original* program. From this point of view, in the SSA form, variable occurrences in μ-, χ-, and ϕ-functions are not regarded as real occurrences. In our example in Fig. 16.1, i_2 has no real occurrence, while i_1, i_3, and i_4 do have. The idea is that variable versions that have no real occurrence do not play important roles in the optimization of the program. Once the program is converted back from SSA form, these variables will no longer show up. Since they do not directly appear in the code and their values are usually unknown, we can dispense with the cost of distinguishing among them.

For these reasons, we assign the zero version to versions of variables that have no real occurrence and whose values are derived from at least one χ-function through zero or more intervening ϕ-functions. An equivalent recursive definition is as follows:

- The result of a χ has zero version if it has no real occurrence.
- If the operand of a ϕ is zero version, then the result of the ϕ is zero version if it has no real occurrence.

For a program in full SSA form, Algorithm 16.1 determines which variable versions are to be made zero version under the above definition. The algorithm assumes that only use-def edges (and *not* def-use edges) are available. A *HasRealOcc* flag is associated with each original SSA version, being set to true whenever it has a real occurrence in the program. This can be done during the initial SSA construction. A list *NonZeroPhiList*, initially empty, is also associated with each original program variable.

Algorithm 16.1: Zero-version detection based on SSA use-def chains

---

```
 1  foreach variable v do
 2      foreach version vᵢ of v do
 3          if ¬vᵢ.HasRealOcc ∧ vᵢ.def.operator = χ then
 4              vᵢ.version ← 0
 5          if ¬vᵢ.HasRealOcc ∧ vᵢ.def.operator = φ then
 6              let V ← vᵢ.def.operands
 7              if ∀vⱼ ∈ V, vⱼ.HasRealOcc then vᵢ.HasRealOcc ← true
 8              else if ∃vⱼ ∈ V, vⱼ.version = 0 then vᵢ.version ← 0
 9              else v.NonZeroPhiList.add(vᵢ)

10      changes ← true
11      while changes do
12          changes ← false
13          foreach vᵢ ∈ v.NonZeroPhiList do
14              let V = vᵢ.def.operands
15              if ∀vⱼ ∈ V, vⱼ.HasRealOcc then
16                  vᵢ.HasRealOcc ← true
17                  v.NonZeroPhiList.remove(vᵢ)
18                  changes ← true
19              else if ∃vⱼ ∈ V, vⱼ.version = 0 then
20                  vᵢ.version ← 0
21                  v.NonZeroPhiList.remove(vᵢ)
22                  changes ← true
```

The loop from lines 2 to 12 of Algorithm 16.1 can be regarded as the initialization pass, processing each variable version once. The while loop from lines 14 to 25 is the propagation pass, with time bound by the length of the longest chain of contiguous ϕ assignments. This bound can easily be reduced to the deepest loop nesting depth of the program by traversing the versions based on a topological order of the forward control flow graph. All in all, zero-version detection in the presence of μ- and χ-functions does not significantly change the complexity of SSA construction, while the corresponding reduction in the number of variable versions can reduce the overhead in the subsequent SSA-based optimization phases.

Because zero versions can have multiple static assignments, they do not have fixed or known values. Thus, two zero versions of the same variable cannot be assumed to have identical values. The occurrence of a zero version breaks use-def chains. But since the results of χ-functions have unknown values, zero versioning does not affect the performance of optimizations that propagate known values, such as constant propagation, because they cannot be propagated across points of MayDefs in any case. Optimizations that operate on real occurrences, such as equivalencing and redundancy detection, are also unaffected. In performing dead store elimination, zero versions have to be assumed live. Since zero versions can only have uses represented by μ-functions, the chances of the deletion of stores associated with χ-functions with zero versions are small. But if some later optimizations delete the code that contains a μ-function, zero versioning could prevent its defining χ-function from being recognized as dead. Overall, zero versioning should only cause a small loss of effectiveness in the subsequent SSA-based optimization phases.

16.3   SSA for Indirect Memory Operations: Virtual Variables

The techniques described in the previous sections only apply to scalar variables in the program and not to arbitrary memory locations accessed indirectly. As an example, in Fig. 16.1, μ-, χ-, and ϕ-functions have been introduced to keep track of i's use-defs, but $*p$ is not considered as an SSA variable. Thus, even though we can apply SSA-based optimizations to scalar variables when they are affected by aliasing, indirect memory access operations are not targeted in our optimizations.

This situation is far from ideal, because code written in current mainstream imperative languages (like C++, Java, or C#) typically contains many operations on data stored in unfixed memory locations. For instance, in C, a two-dimensional vector could be represented as a struct declared as "`typedef struct {double x; double y;} point;`" Then, we can have a piece of code that computes the modulus of the vector p written as "`m = (p->x * p->x) + (p->y * p->y);`" As x is accessed twice with both accesses yielding the same value, the second access could be optimized away, and the same goes for y. The problem is that x and y are not scalar variables: p is a pointer variable, while "`p->x`" and "`p->y`" are indirect memory operations. Representing this code snippet in SSA form tells us that the value of p never changes, but it reveals nothing about the values stored in the locations "p->x" and "p->y." It is worth noting that operations on array elements suffer from the same problem.

The purpose of HSSA is to put indirect memory operations in SSA form just like scalar variables, so we can apply all SSA-based optimizations to them uniformly. In our discussion, we use the C dereferencing operator *dereference* to denote indirection from a pointer. This operator can be placed on either the left- or the right-hand side of the assignment operator. Appearances on the right-hand side represent indirect loads, while those on the left-hand side represent indirect stores. Examples include:

- ∗p: read memory at address p
- ∗(p+4): read memory at address $p + 4$ (as in reading the field of an object at offset 4)
- ∗∗p: double indirection
- ∗p=: indirect store

As noted above, indirect memory operations cannot be handled by the regular SSA construction algorithm because they are *operations*, while SSA construction works on *variables* only. In HSSA, we represent the target locations of indirect memory operations using *virtual variables*. A virtual variable is an abstraction of a memory area and appears under HSSA thanks to the insertion of μ-, χ-, and ϕ-functions. Indeed, like any other variable, they can also have aliases. For the same initial C code shown in Fig. 16.2a, we show two different scenarios after μ-function and χ-function insertion. In Fig. 16.2b, the two virtual variables introduced, v^* and w^*, are associated with the memory locations pointed to by $\ast p$ and $\ast q$, respectively. As a result, v^* and w^* both alias with all the indirect memory operations (of lines 3, 5, 7, and 8). In Fig. 16.2c, x^* is associated with the memory location pointed to by b and y^* is associated with the memory location pointed to by $b + 1$. Assuming alias analysis determines that there is no alias between b and $b + 1$, only x^* aliases with the indirect memory operations of line 3, and only y^* aliases with the indirect memory operations of lines 5, 7, and 8.

1 $p = b$	1 $p = b$	1 $p = b$
2 $q = b$	2 $q = b$	2 $q = b$
3 $\ast p = \ldots$	3 $\ast p = \cdots \parallel v^* = \chi(v^*) \parallel w^* = \chi(w^*)$	3 $\ast p = \cdots \parallel x^* = \chi(x^*)$
4 $p = p + 1$	4 $p = p + 1$	4 $p = p + 1$
5 $\cdots = \ast p$	5 $\cdots = \ast p \parallel \mu(v^*) \parallel \mu(w^*)$	5 $\cdots = \ast p \parallel \mu(y^*)$
6 $q = q + 1$	6 $q = q + 1$	6 $q = q + 1$
7 $\ast p = \ldots$	7 $\ast p = \cdots \parallel v^* = \chi(v^*) \parallel w^* = \chi(w^*)$	7 $\ast p = \cdots \parallel y^* = \chi(y^*)$
8 $\cdots = \ast q$	8 $\cdots = \ast q \parallel \mu(v^*) \parallel \mu(w^*)$	8 $\cdots = \ast q \parallel \mu(y^*)$
(a) Initial C code	**(b)** v and w alias w/ ops 3,5,7, and 8	**(c)** x alias w/ op 2; y with 5,7, and 8

Fig. 16.2 Some virtual variables and their insertion depending on how they alias with operands

It can be seen that the behaviour of a virtual variable in annotating the SSA representation is dictated by its definition. The only discipline imposed by HSSA is that each indirect memory operand must be associated with at least one virtual variable. At one extreme, there could be one virtual variable for each indirect memory operation. On the other hand, it may be desirable to cut down the number of virtual variables by making each virtual variable represent more forms of indirect memory operations. Called *assignment factoring*, this has the effect of replacing multiple use-def chains belonging to different virtual variables with one use-def chain that encompasses more nodes and thus more versions. At the other extreme, the most factored HSSA form would define only one single virtual variable that represents all indirect memory operations in the program.

In practice, it is best not to use assignment factoring among memory operations that do not alias among themselves. This provides high accuracy in the SSA representation without incurring additional representation overhead, because the total number of SSA versions is unaffected. On the other hand, among memory operations that alias among themselves, using multiple virtual variables would result in greater representation overhead. Zero versioning can also be applied to virtual variables to help reduce the number of versions. Appearances of virtual variables in the μ- and χ-functions next to their defined memory operations are regarded as *real* occurrences in the algorithm that determines zero versions for them.

Virtual variables can be instrumented into the program during μ-function and χ-function insertion, in the same pass as for scalar variables. During SSA construction, virtual variables are handled just like scalar variables. In the resulting SSA form, the use-def relationships of the virtual variables will represent the use-def relationships among the memory access operations in the program. At this point, we are ready to complete the construction of the HSSA form by applying global value numbering (GVN).

16.4   Using GVN to Form HSSA

In the previous sections, we have laid the foundations for dealing with aliasing and indirect memory operations in SSA form: we introduced μ- and χ-operators to model aliases, applied zero versioning to keep the number of SSA versions acceptable, and defined virtual variables as a way to apply SSA to storage locations accessed indirectly. However, HSSA is incomplete unless *global value numbering* is applied to handle scalar variables and indirect storage locations uniformly (see Chap. 11).

Value numbering works by assigning a unique number to every expression in the program with the idea that expressions identified by the same number are guaranteed to compute to the same value. The value number is obtained using a hash function applied to each node in an expression tree. For an internal node in the tree, the value number is a hash function of the operator and the value numbers of all its immediate operands. The SSA form enables value numbering to be applied on the global scope, taking advantage of the property that the same SSA version of a variable must store the same value regardless of where it appears.

GVN enables us to easily determine when two address expressions compute the same address when building HSSA. We can then construct the SSA form among indirect memory operations whose address expressions have the same value number. Because a memory location accessed indirectly may be assigned different values at different points in the program, having the same value number for address expressions is not a sufficient condition for the read of the memory location to have the same value number. This is where we make use of virtual variables. In HSSA, for two reads of indirect memory locations to have the same value number, apart from their address expressions having identical value numbers, an additional condition is that they must have the same SSA version for their virtual variable. If they are associated with different versions of their virtual variable, they will be assigned different value numbers, because their reads may return different values. This enables the GVN in HSSA to maintain the consistency whereby nodes with the same value number must yield the same value. Thus, in HSSA, indirect memory operations can be handled in the same rank as scalar variables, and they can benefit transparently from any SSA-based optimizations applied to the program. For instance, in the vector modulus computation described above, every occurrence of the expression "p->x" will always have the same GVN and is therefore guaranteed to return the same value, allowing the compiler to emit code that avoids redundant memory reads (the same holds for "p->y").

Using the same code example from Fig. 16.2, we can see in Fig. 16.3 that p_2 and q_2 have the same value number h_7, while p_1's value number is h_1 and is different. The loads at lines 5 and 8 cannot be considered redundant between each other because the versions for v (v_1^* then v_2^*) are different. On the other hand, the load at line 8 can be safely avoided by reusing the value computed in line 7, as both their versions for v (v_2^*) and their address expressions' value numbers are identical. As a last example, if all the associated virtual variable versions for an indirect memory store (defined in parallel by χ-functions) are found to be dead, then it can be safely eliminated.[3] In other words, HSSA transparently extends SSA's dead store elimination algorithm to indirect memory stores.

Another advantage of HSSA in this context is that it enables uniform treatment of indirect memory operations regardless of the levels of indirections (as in the "**p" expression in C which represents a double indirection). This happens naturally because each node of the expression is identified by its value number, and the fact that it is used as an address in another expression does not cause any additional complications.

Having explained why HSSA uses GVN, we are ready to explain how the HSSA intermediate representation is structured. A program in HSSA keeps its CFG

---

[3] Note that any virtual variable that aliases with a memory region live out of the compiled procedure is considered to alias with the return instruction of the procedure and as a consequence will lead to a live μ-function.

(a) Initial C code

```
1  p = b
2  q = b
3  *p = 3
4  p = p + 1
5  ··· = *p
6  q = q + 1
7  *p = 4
8  ··· = *q
```

(b) with one virtual variable

```
1  p₁ = b
2  q₁ = b
3  *p₁ = 3 ‖ v₁* = χ(v₀*)
4  p₂ = p₁ + 1
5  ··· = *p₂ ‖ μ(v₁*)
6  q₂ = q₁ + 1
7  *p₂ = 4 ‖ v₂* = χ(v₁*)
8  ··· = *q₂ ‖ μ(v₂*)
```

$$
\begin{aligned}
&1\quad p_1 = b\\
&2\quad q_1 = b\\
&3\quad *p_1 = 3 \parallel v_1^* = \chi(v_0^*)\\
&4\quad p_2 = p_1 + 1\\
&5\quad \cdots = *p_2 \parallel \mu(v_1^*)\\
&6\quad q_2 = q_1 + 1\\
&7\quad *p_2 = 4 \parallel v_2^* = \chi(v_1^*)\\
&8\quad \cdots = *q_2 \parallel \mu(v_2^*)
\end{aligned}
$$

(c) HSSA statements

$$
\begin{aligned}
&1\quad h_1 = h_0\\
&2\quad h_2 = h_0\\
&3\quad h_5 = h_3 \parallel h_4 = \chi(\mathsf{VN}(v_0^*))\\
&4\quad h_8 = h_7\\
&5\quad \cdots = h_{10} \parallel \mu(h_4)\\
&6\quad h_{11} = h_7\\
&7\quad h_{14} = h_{12} \parallel h_{13} = \chi(h_4)\\
&8\quad \cdots = h_{14} \parallel \mu(h_{13})
\end{aligned}
$$

key	hash	value
b	h_0	b
p_1	h_1	h_0
q_1	h_2	h_0
3	h_3	const(3)
v_1^*	h_4	v_1^*
$\mathrm{ivar}(*p_1, v_1^*) \rightsquigarrow \mathrm{ivar}(*h_0, v_1^*)$	h_5	h_3
1	h_6	const(1)
$p_1 + 1 \rightsquigarrow +(h_0, h_6)$	h_7	$+(h_0, h_6)$
p_2	h_8	h_7
$\mathrm{ivar}(*p_2, v_1^*) \rightsquigarrow \mathrm{ivar}(*h_7, v_1^*)$	h_{10}	$\mathrm{ivar}(*h_7, v_1^*)$
q_2	h_{11}	h_7
4	h_{12}	const(4)
v_2^*	h_{13}	v_2^*
$\mathrm{ivar}(*p_2, v_2^*) \rightsquigarrow \mathrm{ivar}(*h_7, v_2^*)$	h_{14}	h_{12}

(d) Hash Table

Fig. 16.3 Some code after variable versioning, its corresponding HSSA form along with its hash table entries. $q_1 + 1$ that simplifies into $+(h_0, h_6)$ will be hashed to h_7, and $\mathrm{ivar}(*q_2, v_2^*)$ that simplifies into $\mathrm{ivar}(*h_7, v_2^*)$ will be hashed to h_{14}

structure, with basic blocks made up of a sequence of statements (assignments, procedure calls, etc.), but the expression operands of each statement and the left-hand side of assignments are replaced by their corresponding entries in the hash table. Constants, variables (both scalar and virtual), and expressions all find their entries in the hash table. As the variables are in SSA form, each SSA version is assigned one and only one value number. An expression is hashed using its operator and the value numbers of its operands. An indirect memory operation is regarded as both expression and variable. Because it is not a leaf, it is hashed based on its operator and the value number of its address operand. Because it has the property of a variable, the value number of its associated virtual variable version is also included in the hashing. This is illustrated in lines 3, 5, 7, and 8 of the example code in Fig. 16.3. Such entries are referred to as *ivar* nodes, for *indirect variables*, to denote their operational semantics and to distinguish them from the regular scalar variables.

From the code generation point of view, ivar nodes are operations and not variables. The compiler back end, when processing them, will emit indirect memory accesses to the addresses specified as their operand. On the other hand, the virtual variables have no real counterpart in the program's code generation. In this sense, virtual variables can be regarded as a tool to associate aliasing effects with indirect memory operations and construct use-def relationships in the program that involves the indirect memory operations. Using virtual variables, indirect memory operations are no longer relegated to second-class citizens in SSA-based optimization phases.

As a supplementary note, in HSSA, because values and variables in particular can be uniquely identified using value numbers, the reference to SSA versions has been rendered redundant. In fact, to further ensure uniform treatment among scalar and indirect variables in the implementation, the use of SSA version numbers should be omitted in the representation.

16.5   Building HSSA

We now put together all the topics discussed in this chapter by detailing the steps to build HSSA starting from some non-SSA representations of the program. The first task is to construct the SSA form for the scalar and virtual variables, and this is displayed in Algorithm 16.2.

Algorithm 16.2: SSA form construction

---

1. Perform alias analysis and assign a virtual variable to all indirect memory operations.
2. Insert μ-functions and χ-functions for scalar and virtual variables.
3. Insert ϕ-functions as in standard SSA construction, including the χ's as additional assignment statements.
4. Perform SSA renaming on all scalar and virtual variables as in standard SSA construction.

At the end of Algorithm 16.2, all scalar and virtual variables have SSA information, but the indirect memory operations are only "annotated" with virtual variables and cannot be regarded as being in SSA form.

Next, we perform a round of dead store elimination based on the constructed SSA form and then run our zero-version detection algorithm to detect zero versions in the scalar and virtual variables. These correspond to Steps 5 and 6 in Algorithm 16.3, respectively.

Algorithm 16.3: Detecting zero versions

---

5. Perform dead store elimination on the scalar and virtual variables based on their SSA form.
6. Initialize *HasRealOcc* and *NonZeroPhiList* as for Algorithm 16.1, then run Algorithm 16.1 (Zero-version detection).

At this point, however, the number of unique SSA versions have diminished because of the application of zero versioning. The final task is to build the HSSA representation of the program by applying GVN. The steps are displayed in Algorithm 16.4.

Algorithm 16.4: Applying GVN

---

7. Hash a unique value number and create the hash table *var* node for each scalar and virtual variable version that is live. Only one node needs to be created for the zero version of each variable.
8. Conduct a preorder traversal of the dominator tree of the control flow graph, applying GVN to code in each basic block:

   a. Hash expressions bottom up into the hash table, reusing existing hash table nodes that have identical value numbers; for *var* nodes, use corresponding nodes created in step 7 according to their SSA versions.
   b. Two *ivar* nodes have the same value number if the following two conditions are both satisfied:

      • their address expressions have the same value number, and
      • their virtual variables have the same versions.

   c. For each assignment statement (ϕ and χ included), represent its left-hand side by creating a link from the statement to the *var* or *ivar* node in the hash table; at the same time, make the *var* or *ivar* node point back to its defining statement.
   d. Update all ϕ, μ and χ operands and results to refer to entries in the hash table.

At the end of Algorithm 16.4, HSSA form is complete, and every value in the program code is represented by a reference to a node in the hash table.

The purpose of the preorder traversal processing order in Algorithm 16.4 is not strictly required to ensure the correctness of the HSSA, because we already have SSA version information for the scalar and virtual variables. Because this processing order ensures that we always visit definitions before their uses, it streamlines the implementation and also makes it easier to perform additional optimizations like copy propagation during the program traversal. It is also possible to go up the use-def chains for virtual variables and analyse the address expressions of their associated indirect memory operations to determine more accurate alias relations among *ivar* nodes that share the same virtual variable.

In HSSA form, expression trees are converted to directed acyclic graphs (DAGs), which is more memory efficient than ordinary program representation because of node sharing. Many optimizations can run faster on HSSA because they only need to be applied once on the shared use nodes. Optimization implementations can also take advantage of the fact that it is trivial to check if two expressions compute the same value in HSSA.

16.6 Further Reading

The earliest attempts at accommodating may-alias information into SSA form are represented by the work of Cytron and Gershbein [88], where they defined may-alias sets of the form *MayAlias*(p, S), which gives the variable names aliased with $*p$ at statement S in the program. Calls to an *IsAlias*(p, v) function are then inserted into the program at points in the program where modifications due to aliasing occur. The *IsAlias*(p, v) function contains code that models runtime logic and returns appropriate values based on the pointer values. To address the high cost of this representation, Cytron and Gershbein proposed an incremental approach for including may-alias information into SSA form in their work.

The use of assignment factoring to represent MayDefs was first proposed by Choi et al. [68]. The same factoring was referred to as *location factored SSA form* by Steensgaard [272]. He also used assignment factoring to reduce the number of SSA variables in his SSA form.

The idea of value numbering can be traced to Cocke and Schwartz [77]. While the application of global value numbering (GVN) in this chapter is more for the purpose of representation than optimization, GVN has mostly been discussed in the context of redundancy elimination [76, 249]. Chapter 11 covers redundancy elimination in depth and also contains a discussion of GVN.

HSSA was first proposed by Chow et al. [72] and first implemented in the Open64 compiler [7, 64, 65]. All the SSA-based optimizations in the Open64 compiler were implemented based on HSSA. Their copy propagation and dead store elimination work uniformly on both direct and indirect memory operations. Their redundancy elimination covers indirect memory references as well as other expressions. Other SSA-based optimizations in the Open64 compiler include loop induction variable canonicalization [185], strength reduction [163], and register promotion [187]. Induction variable recognition is discussed in Chap. 10. Strength reduction and register promotion are also discussed in Chap. 11.

The most effective way to optimize indirect memory operations is to promote them to scalars when possible. This optimization is called *indirection removal*. In the Open64 compiler, it depends on the ability of copy propagation to convert the address expressions of indirect memory operations to address constants. The indirect memory operations can then be folded to direct loads and stores. Subsequent register promotion will promote the scalars to registers. If the scalars are free of aliasing, they will not be allocated storage in memory.

Lapkowski and Hendren proposed the use of Extended SSA Numbering to capture the semantics of aliases and pointer accesses [175]. Their SSA number idea borrows from SSA's version numbering, in the sense that a new number is used to annotate the variable whenever it could assume a new value. ϕ-nodes are not represented, and not all SSA numbers need to have explicit definitions. SSA numbers for pointer references are "extended" to two numbers, the primary one for the pointer variable and the secondary one for the pointed-to memory location. But because it is not SSA form, it does not exhibit many of the nice properties of SSA, like single definitions and built-in use-defs. It cannot benefit from the many SSA-based optimization algorithms either.

The GCC compiler originally uses different representation schemes between unaliased scalar variables and aliased memory-residing objects. To avoid compromising the compiler's optimization when dealing with memory operations, Novillo proposed a unified approach for representing both scalars and arbitrary memory expressions in their SSA form [211]. They do not use Hashed SSA, but their approach to representing aliasing in its SSA form is very similar to ours. They define the virtual operators VDEF and VUSE, which correspond to our χ- and μ-functions. They started out by creating symbol names for any memory-residing program entities. This resulted in an explosion in the number of VDEFs and VUSEs inserted. They then used assignment factoring to cut down the number of these virtual operators, in which memory symbols are partitioned into groups. The virtual operators were then inserted on a per-group basis, thus reducing compilation time. Since the reduced representation precision has a negative effect on optimizations, they provided different partitioning schemes to reduce the impact on optimizations.

One class of indirectly accessed memory objects is array, in which each element is addressed via an index or subscript expression. HSSA distinguishes among different elements of an array only to the extent of determining if the address expressions compute to the same value or not. When two address expressions have the same value, the two indirect memory references are definitely the same. But when the two address expressions cannot be determined to be the same, they may still be the same. Thus, HSSA cannot provide *definitely different* information. In contrast, the array SSA form can enable more accurate program analysis among accesses to array elements by incorporating the indices into the SSA representation. The array SSA form can capture element-level use-defs, whereas HSSA cannot. In addition, the heap memory storage area can be modelled using abstract arrays that represent disjoint subsets of the heap, with pointers to the heap being treated like indices. Array SSA is covered in detail in Chap. 17. When array references are affine expressions of loop indices, the appropriate representation for dependencies is the so-called *dependence relations* in the polyhedral model [116].

Chapter 17
Array SSA Form

Vivek Sarkar, Kathleen Knobe, and Stephen Fink

In this chapter, we introduce an Array SSA form that captures element-level data-flow information for array variables and coincides with standard SSA form when applied to scalar variables. Any program with arbitrary control-flow structures and arbitrary array subscript expressions can be automatically converted to this Array SSA form, thereby making it applicable to structures, heap objects, and any other data structure that can be modelled as a logical array. A key extension over standard SSA form is the introduction of a *definition-*Φ function (denoted Φ_{def}) that is capable of merging values from distinct array definitions on an element-by-element basis. There are several potential applications of Array SSA form in compiler analysis and optimization of sequential and parallel programs. In this chapter, we focus on sequential programs and use *constant propagation* as an exemplar of a program analysis that can be extended to array variables using Array SSA form and *redundant load elimination* as an exemplar of a program optimization that can be extended to heap objects using Array SSA form.

The rest of the chapter is organized as follows. Section 17.1 introduces *full Array SSA form* for runtime evaluation and *partial Array SSA form* for static analysis. Section 17.2 extends the scalar SSA constant propagation algorithm to enable constant propagation through array elements. This includes an extension to the constant propagation lattice to efficiently record information about array elements and an extension to the worklist algorithm to support *definition-*Φ functions

V. Sarkar (✉)
Georgia Institute of Technology, Atlanta, GA, USA
e-mail: vsarkar@gatech.edu

K. Knobe
Rice University, Houston, TX, USA
e-mail: kath.knobe@comcast.net

S. Fink
Facebook, Yorktown Hieght, NY, USA

F. Rastello, F. Bouchez Tichadou (eds.), *SSA-based Compiler Design*,
https://doi.org/10.1007/978-3-030-80515-9_17

(Sect. 17.2.1) and a further extension to support non-constant (symbolic) array subscripts (Sect. 17.2.2). Section 17.3 shows how Array SSA form can be extended to support elimination of redundant loads of object fields and array elements in strongly typed languages, and Sect. 17.4 contains suggestions for further reading.

17.1   Array SSA Form

To introduce full Array SSA form with runtime evaluation of Φ-functions, we use the concept of an *iteration vector* to differentiate among multiple dynamic instances of a static definition, S_k, that occur in the same dynamic instance of S_k's enclosing procedure, $f()$. Let n be the number of loops that enclose S_k in procedure $f()$. These loops could be for-loops, while-loops, or even loops constructed out of goto statements. For convenience, we treat the outermost region of acyclic control flow in a procedure as a dummy outermost loop with a single iteration, thereby ensuring that $n \geq 1$.

A single point in the iteration space is specified by the iteration vector $\vec{i} = (i_1, \ldots, i_n)$, which is an n-tuple of iteration numbers, one for each enclosing loop. For convenience, this definition of iteration vectors assumes that all loops are single-entry, or equivalently, that the control-flow graph is *reducible*. (This assumption is not necessary for partial Array SSA form.) For single-entry loops, we know that each def executes at most once in a given iteration of its surrounding loops, and hence the iteration vector serves the purpose of a "timestamp." The key extensions in Array SSA form relative to standard SSA form are as follows:

1. **Renamed array variables:** All array variables are renamed so as to satisfy the Static Single Assignment property. Analogous to standard SSA form, control Φ-functions are introduced to generate new names for merging two or more prior definitions at control-flow join points and to ensure that each use refers to precisely one definition.
2. **Array-valued @ variables:** For each static definition A_j, we introduce an @ *variable* (pronounced "at variable") $@A_j$ that identifies the most recent *iteration vector at* which definition A_j was executed. We assume that all @ variables are initialized to the empty vector, ( ), at the start of program execution. Each update of a single array element, $A_j[k] \leftarrow \ldots$, is followed by the statement, $@A_j[k] \leftarrow \vec{i}$, where $\vec{i}$ is the iteration vector for the loops surrounding the definition of A_j.
3. **Definition Φ's:** A *definition* Φ operator Φ_{def} is introduced in Array SSA form to deal with preserving ("non-killing") definitions of arrays. Consider A_0 and A_1, two renamed arrays that originated from the same array variable in the source program such that $A_1[k] \leftarrow \ldots$ is an update of a single array element and A_0 is the prevailing definition at the program point just prior to the definition of A_1. A definition Φ, $A_2 \leftarrow \Phi_{def}(A_1, @A_1, A_0, @A_0)$, is inserted immediately after the definitions for A_1 and $@A_1$. Since definition A_1 only updates one element of A_0,

A_2 represents an element-level merge of arrays A_1 and A_0. Definition Φ's did not need to be inserted in standard SSA form because a scalar definition completely kills the old value of the variable.

4. **Array-valued Φ-functions:** Another consequence of renaming arrays is that a Φ-function for array variables must also return an array value. Consider a (control or definition) Φ-function of the form, $A_2 \leftarrow \Phi(A_1, @A_1, A_0, @A_0)$. Its semantics can be specified precisely by the following conditional expression for each element, $A_2[j]$, in the result array A_2:

$$A_2[j] = \begin{aligned} &\textbf{if} \quad @A_1[j] \succeq @A_0[j] \textbf{ then } A_1[j] \\ &\textbf{else} \quad A_0[j] \\ &\textbf{end if} \end{aligned} \qquad (17.1)$$

The key extension over the scalar case is that the conditional expression specifies an element-level merge of arrays A_1 and A_0.

Figure 17.1 shows an example program with an array variable and the conversion of the program to full Array SSA form as defined above.

We now introduce a *partial Array SSA form* for static analysis, which serves as an approximation of full Array SSA form. Consider a (control or definition) Φ-function, $A_2 \leftarrow \Phi(A_1, @A_1, A_0, @A_0)$. A static analysis will need to approximate the computation of this Φ-function by some data-flow transfer function, $\mathcal{L}_\Phi$. The inputs and output of $\mathcal{L}_\Phi$ will be *lattice elements* for scalar/array variables that are compile-time approximations of their runtime values. We use the notation $\mathcal{L}(V)$ to denote the lattice element for a scalar or array variable V. Therefore, the statement,

---

```
A[*] ← initial value of A
i ← 1
C ← i < 2
if C then
    k ← 2 × i
    A[k] ← i
    print A[k]

print A[2]
```

(a) Example program

```
@A_0[*] ← () ; @A_1[*] ← ()

A_0[*] ← initial value of A
@A_0[*] ← (1)
i ← 1
C ← i < n
if C then
    k ← 2 × i
    A_1[k] ← i
    @A_1[k] ← (1)
    A_2 ← Φ_def(A_1, @A_1, A_0, @A_0)
    @A_2 ← max(@A_1, @A_0)
    print A_2[k]

A_3 ← Φ(A_2, @A_2, A_0, @A_0)
@A_3 ← max(@A_2, @A_0)
print A_3[2]
```

(b) Conversion to Full Array SSA Form

Fig. 17.1 Example program with array variables and its conversion to full array SSA form

$A_2 \leftarrow \Phi(A_1, @A_1, A_0, @A_0)$, will in general be modelled by the data-flow equation, $\mathscr{L}(A_2) = \mathscr{L}_\Phi(\mathscr{L}(A_1), \mathscr{L}(@A_1), \mathscr{L}(A_0), \mathscr{L}(@A_0))$.

While the *runtime* semantics of Φ-functions for array variables critically depends on @ variables (Eq. 17.1), many *compile-time analyses* do not need the full generality of @ variables. For analyses that do not distinguish among iteration instances, it is sufficient to model $A_2 \leftarrow \Phi(A_1, @A_1, A_0, @A_0)$ by a data-flow equation, $\mathscr{L}(A_2) = \mathscr{L}_\phi(\mathscr{L}(A_1), \mathscr{L}(A_0))$, that does not use lattice variables $\mathscr{L}(@A_1)$ and $\mathscr{L}(@A_0)$. For such cases, a *partial* Array SSA form can be obtained by dropping @ variables, and using the ϕ operator, $A_2 \leftarrow \phi(A_1, A_0)$ instead of $A_2 \leftarrow \Phi(A_1, @A_1, A_0, @A_0)$. A consequence of dropping @ variables is that partial Array SSA form does not need to deal with iteration vectors and therefore does not require the control-flow graph to be reducible as in full Array SSA form. For scalar variables, the resulting ϕ-function obtained by dropping @ variables exactly coincides with standard SSA form.

17.2 Sparse Constant Propagation of Array Elements

17.2.1 Array Lattice for Sparse Constant Propagation

In this section, we describe the lattice representation used to model array values for constant propagation. Let $\mathscr{U}_{ind}^A$ and $\mathscr{U}_{elem}^A$ be the universal set of *index values* and the universal set of array *element values*, respectively, for an array variable A. For an array variable, the set denoted by lattice element $\mathscr{L}(A)$ is a subset of index-element pairs in $\mathscr{U}_{ind}^A \times \mathscr{U}_{elem}^A$. There are three kinds of lattice elements for array variables that are of interest in our framework:

1. $\mathscr{L}(A) = \top \quad \Rightarrow \quad \text{SET}(\mathscr{L}(A)) = \{\,\}$
   This "top" case indicates that the set of possible index-element pairs that have been identified thus far for A is the empty set, $\{\,\}$.
2. $\mathscr{L}(A) = \langle (i_1, e_1), (i_2, e_2), \ldots \rangle$
   $\Rightarrow \quad \text{SET}(\mathscr{L}(A)) = \{(i_1, e_1), (i_2, e_2), \ldots\} \cup (\mathscr{U}_{ind}^A - \{i_1, i_2, \ldots\}) \times \mathscr{U}_{elem}^A$
   The lattice element for this "constant" case is represented by a finite list of constant index–element pairs, $\langle (i_1, e_1), (i_2, e_2), \ldots \rangle$. The constant indices, $i_1, i_2, \ldots$, must represent distinct (non-equal) index values. The meaning of this lattice element is that the current stage of analysis has identified some finite number of constant index–element pairs for array variable A, such that $A[i_1] = e_1$, $A[i_2] = e_2$, etc. All other elements of A are assumed to be non-constant. (Extensions to handle non-constant indices are described in Sect. 17.2.2.)
3. $\mathscr{L}(A) = \bot \quad \Rightarrow \quad \text{SET}(\mathscr{L}(A)) = \mathscr{U}_{ind}^A \times \mathscr{U}_{elem}^A$
   This "bottom" case indicates that, according to the approximation in the current stage of analysis, array A may take on any value from the universal set of index–element pairs. Note that $\mathscr{L}(A) = \bot$ is equivalent to an empty list, $\mathscr{L}(A) = \langle \, \rangle$, in case (2) above; they both denote the universal set of index–element pairs.

$\mathcal{L}(A_1[k])$	$\mathcal{L}(k)=\top$	$\mathcal{L}(k)=Constant$	$\mathcal{L}(k)=\bot$
$\mathcal{L}(A_1)=\top$	$\top$	$\top$	$\bot$
$\mathcal{L}(A_1)=\langle(i_1,e_1),\ldots\rangle$	$\top$	e_j, if $\exists (i_j,e_j)\in\mathcal{L}(A_1)$ with $\mathcal{DS}(i_j,\mathcal{L}(k))=\text{true}$ $\bot$, otherwise	$\bot$
$\mathcal{L}(A_1)=\bot$	$\bot$	$\bot$	$\bot$

Fig. 17.2 Lattice computation for $\mathcal{L}(A_1[k]) = \mathcal{L}_{[\,]}(\mathcal{L}(A_1), \mathcal{L}(k))$, where $A_1[k]$ is an array element read operator

$\mathcal{L}(A_1)$	$\mathcal{L}(i)=\top$	$\mathcal{L}(i)=Constant$	$\mathcal{L}(i)=\bot$
$\mathcal{L}(k)=\top$	$\top$	$\top$	$\bot$
$\mathcal{L}(k)=Constant$	$\top$	$\langle(\mathcal{L}(k),\mathcal{L}(i))\rangle$	$\bot$
$\mathcal{L}(k)=\bot$	$\bot$	$\bot$	$\bot$

Fig. 17.3 Lattice computation for $\mathcal{L}(A_1) = \mathcal{L}_{d[\,]}(\mathcal{L}(k), \mathcal{L}(i))$, where $A_1[k] \leftarrow i$ is an array element write operator

We now describe how array lattice elements are computed for various operations that appear in Array SSA form. We start with the simplest operation viz. a read access to an array element. Figure 17.2 shows how $\mathcal{L}(A_1[k])$, the lattice element for array reference $A_1[k]$, is computed as a function of $\mathcal{L}(A_1)$ and $\mathcal{L}(k)$, the lattice elements for A_1 and k. We denote this function by $\mathcal{L}_{[\,]}$, i.e., $\mathcal{L}(A_1[k]) = \mathcal{L}_{[\,]}(\mathcal{L}(A_1), \mathcal{L}(k))$. The interesting case in Fig. 17.2 occurs in the middle cell when neither $\mathcal{L}(A_1)$ nor $\mathcal{L}(k)$ is $\top$ or $\bot$. The notation $\mathcal{DS}$ in the middle cell in Fig. 17.2 represents a "definitely same" binary relation, i.e., $\mathcal{DS}(a, b) = \text{true}$ if and only if a and b are known to have exactly the same value.

Next, consider a write access of an array element, which in general has the form $A_1[k] \leftarrow i$. Figure 17.3 shows how $\mathcal{L}(A_1)$, the lattice element for the array being written into, is computed as a function of $\mathcal{L}(k)$ and $\mathcal{L}(i)$, the lattice elements for k and i. We denote this function by $\mathcal{L}_{d[\,]}$, i.e., $\mathcal{L}(A_1) = \mathcal{L}_{d[\,]}(\mathcal{L}(k), \mathcal{L}(i))$. As before, the interesting case in Fig. 17.3 occurs in the middle cell when both $\mathcal{L}(k)$ and $\mathcal{L}(i)$ are constant. For this case, the value returned for $\mathcal{L}(A_1)$ is simply the singleton list, $\langle (\mathcal{L}(k), \mathcal{L}(i)) \rangle$, which contains exactly one constant index–element pair.

Now, we turn our attention to the ϕ-function. Consider a definition-ϕ-function of the form, $A_2 \leftarrow \phi_{def}(A_1, A_0)$. The lattice computation for $\mathcal{L}(A_2) = \mathcal{L}_{\phi_{def}}(\mathcal{L}(A_1), \mathcal{L}(A_0))$ is shown in Fig. 17.4. Since A_1 corresponds to a definition of a single array element, the list for $\mathcal{L}(A_1)$ can contain at most one pair (see Fig. 17.3). Therefore, the three cases considered for $\mathcal{L}(A_1)$ in Fig. 17.4 are $\mathcal{L}(A_1) = \top$, $\mathcal{L}(A_1) = \langle(i', e')\rangle$, and $\mathcal{L}(A_1) = \bot$.

$\mathscr{L}(A_2)$	$\mathscr{L}(A_0) = \top$	$\mathscr{L}(A_0) = \langle(i_1, e_1), \ldots\rangle$	$\mathscr{L}(A_0) = \bot$
$\mathscr{L}(A_1) = \top$	$\top$	$\top$	$\top$
$\mathscr{L}(A_1) = \langle(i', e')\rangle$	$\top$	$\text{UPDATE}((i', e'), \langle(i_1, e_1), \ldots\rangle)$	$\langle(i', e')\rangle$
$\mathscr{L}(A_1) = \bot$	$\bot$	$\bot$	$\bot$

Fig. 17.4 Lattice computation for $\mathscr{L}(A_2) = \mathscr{L}_{\phi_{def}}(\mathscr{L}(A_1), \mathscr{L}(A_0))$ where $A_2 \leftarrow \phi_{def}(A_1, A_0)$ is a definition ϕ operation

$\mathscr{L}(A_2) = \mathscr{L}(A_1) \sqcap \mathscr{L}(A_0)$	$\mathscr{L}(A_0) = \top$	$\mathscr{L}(A_0) = \langle(i_1, e_1), \ldots\rangle$	$\mathscr{L}(A_0) = \bot$
$\mathscr{L}(A_1) = \top$	$\top$	$\mathscr{L}(A_0)$	$\bot$
$\mathscr{L}(A_1) = \langle(i'_1, e'_1), \ldots\rangle$	$\mathscr{L}(A_1)$	$\mathscr{L}(A_1) \cap \mathscr{L}(A_0)$	$\bot$
$\mathscr{L}(A_1) = \bot$	$\bot$	$\bot$	$\bot$

Fig. 17.5 Lattice computation for $\mathscr{L}(A_2) = \mathscr{L}_\phi(\mathscr{L}(A_1), \mathscr{L}(A_0)) = \mathscr{L}(A_1) \sqcap \mathscr{L}(A_0)$, where $A_2 \leftarrow \phi(A_1, A_0)$ is a control ϕ operation

The notation $\text{UPDATE}((i', e'), \langle(i_1, e_1), \ldots\rangle)$ used in the middle cell in Fig. 17.4 denotes a special update of the list $\mathscr{L}(A_0) = \langle(i_1, e_1), \ldots\rangle$ with respect to the constant index–element pair (i', e'). UPDATE involves four steps:

1. Compute the list $T = \{ (i_j, e_j) \mid (i_j, e_j) \in \mathscr{L}(A_0)$ and $\mathscr{DD}(i', i_j) = \text{true} \}$. Analogous to $\mathscr{DS}$, $\mathscr{DD}$ denotes a "definitely different" binary relation, i.e., $\mathscr{DD}(a, b) = \text{true}$ if and only if a and b are known to have distinct (non-equal) values.
2. Insert the pair (i', e') into T to obtain a new list, I.
3. (Optional) If there is a desire to bound the height of the lattice due to compile-time considerations and the size of list I exceeds a threshold size Z, then one of the pairs in I can be dropped from the output list so as to satisfy the size constraint.
4. Return I as the value of $\text{UPDATE}((i', e'), \langle(i_1, e_1) \ldots\rangle)$.

Finally, consider a control ϕ-function that merges two array values, $A_2 \leftarrow \phi(A_1, A_0)$. The join operator ($\sqcap$) is used to compute $\mathscr{L}(A_2)$, the lattice element for A_2, as a function of $\mathscr{L}(A_1)$ and $\mathscr{L}(A_0)$, the lattice elements for A_1 and A_0, i.e., $\mathscr{L}(A_2) = \mathscr{L}_\phi(\mathscr{L}(A_1), \mathscr{L}(A_0)) = \mathscr{L}(A_1) \sqcap \mathscr{L}(A_0)$. The rules for computing this join operator are shown in Fig. 17.5, depending on different cases for $\mathscr{L}(A_1)$ and $\mathscr{L}(A_0)$. The notation $\mathscr{L}(A_1) \cap \mathscr{L}(A_0)$ used in the middle cell in Fig. 17.5 denotes a simple intersection of lists $\mathscr{L}(A_1)$ and $\mathscr{L}(A_0)$—the result is a list of pairs that appear in both $\mathscr{L}(A_1)$ and $\mathscr{L}(A_0)$.

We conclude this section by discussing the example program in Fig. 17.6a. The partial Array SSA form for this example is shown in Fig. 17.6b, and the data-flow equations for this example are shown in Fig. 17.7a. Each assignment statement in the partial Array SSA form (in Fig. 17.6b) results in one data-flow equation (in Fig. 17.7a); the numbering S1 through S8 indicates the correspondence. Any solver

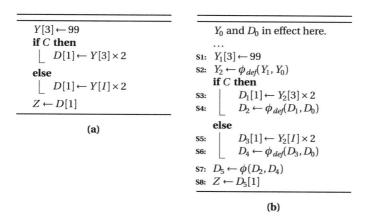

Fig. 17.6 Sparse constant propagation example (**a**) and its Array SSA form (**b**)

S1: $\mathcal{L}(Y_1) = \langle(3,99)\rangle$ $\mathcal{L}(Y_1) = \langle(3,99)\rangle$ $\mathcal{L}(Y_1) = \langle(3,99)\rangle$
S2: $\mathcal{L}(Y_2) = \mathcal{L}_{\phi_{def}}(\mathcal{L}(Y_1),\mathcal{L}(Y_0))$ $\mathcal{L}(Y_2) = \langle(3,99)\rangle$ $\mathcal{L}(Y_2) = \langle(3,99)\rangle$
S3: $\mathcal{L}(D_1) = \mathcal{L}_{d[\ ]}(\mathcal{L}_*(\mathcal{L}_{[\ ]}(\mathcal{L}(Y_2),3)),2))$ $\mathcal{L}(D_1) = \langle(1,198)\rangle$ $\mathcal{L}(D_1) = \langle(1,198)\rangle$
S4: $\mathcal{L}(D_2) = \mathcal{L}_{\phi_{def}}(\mathcal{L}(D_1),\mathcal{L}(D_0))$ $\mathcal{L}(D_2) = \langle(1,198)\rangle$ $\mathcal{L}(D_2) = \langle(1,198)\rangle$
S5: $\mathcal{L}(D_3) = \mathcal{L}_{d[\ ]}(\mathcal{L}_*(\mathcal{L}_{[\ ]}(\mathcal{L}(Y_2),\mathcal{L}(I))),2))$ $\mathcal{L}(D_3) = \bot$ $\mathcal{L}(D_3) = \langle(1,198)\rangle$
S6: $\mathcal{L}(D_4) = \mathcal{L}_{\phi_{def}}(\mathcal{L}(D_3),\mathcal{L}(D_0))$ $\mathcal{L}(D_4) = \bot$ $\mathcal{L}(D_4) = \langle(1,198)\rangle$
S7: $\mathcal{L}(D_5) = \mathcal{L}_{\phi}(\mathcal{L}(D_2),\mathcal{L}(D_4))$ $\mathcal{L}(D_5) = \bot$ $\mathcal{L}(D_5) = \langle(1,198)\rangle$
S8: $\mathcal{L}(Z) = \mathcal{L}_{[\ ]}(\mathcal{L}(D_5),1)$ $\mathcal{L}(Z) = \bot$ $\mathcal{L}(Z) = 198$

          (**a**)   (**b**)   (**c**)

Fig. 17.7 (**a**) Data-flow equations for the sparse constant propagation example of Fig. 17.6 and their solutions assuming (**b**) I is unknown or (**c**) known to be equal to 3

can be used for these data-flow equations, including the standard worklist-based algorithm for constant propagation using scalar SSA form. The fixpoint solution is shown in Fig. 17.7b. This solution was obtained assuming $\mathcal{L}(I) = \bot$. If, instead, variable I is known to equal 3, i.e., $\mathcal{L}(I) = 3$, then the lattice variables that would be obtained after the fixpoint iteration step has completed are shown in Fig. 17.7c. In either case ($\mathcal{L}(I) = \bot$ or $\mathcal{L}(I) = 3$), the resulting array element constants revealed by the algorithm can be used in whatever analyses or transformations the compiler considers to be profitable to perform.

17.2.2 Beyond Constant Indices

In this section, we address constant propagation through *non-constant array subscripts*, as a generalization of the algorithm for constant subscripts described in Sect. 17.2.1. As an example, consider the program fragment in Fig. 17.8. In the loop, we see that the read access of $a[i]$ will have a constant value ($k \times 5 = 10$),

Fig. 17.8 Example of
constant propagation through
non-constant index

```
k ← 2
for i ← ... do
    ...
    a[i] ← k × 5
    ... ← a[i]
```

even though the index/subscript value i is not a constant. We would like to extend
the framework from Sect. 17.2.1 to be able to recognize the read of $a[i]$ as constant
in such programs. There are two key extensions that need to be considered for non-
constant (symbolic) subscript values:

- For constants, C_1 and C_2, $\mathcal{DS}(C_1, C_2) \neq \mathcal{DD}(C_1, C_2)$. However, for two
  symbols, S_1 and S_2, it is possible that both $\mathcal{DS}(S_1, S_2)$ and $\mathcal{DD}(S_1, S_2)$ are
  false, that is, we do not know if they are the same or different.
- For constants, C_1 and C_2, the values for $\mathcal{DS}(C_1, C_2)$ and $\mathcal{DD}(C_1, C_2)$ can be
  computed by inspection. For symbolic indices, however, some program analysis
  is necessary to compute the $\mathcal{DS}$ and $\mathcal{DD}$ relations.

We now discuss the compile-time computation of $\mathcal{DS}$ and $\mathcal{DD}$ for symbolic
indices. Observe that, given index values I_1 and I_2, only one of the following three
cases is possible:

```
Case 1:  𝒟𝒮(I₁, I₂) = false;  𝒟𝒟(I₁, I₂) = false
Case 2:  𝒟𝒮(I₁, I₂) = true;   𝒟𝒟(I₁, I₂) = false
Case 3:  𝒟𝒮(I₁, I₂) = false;  𝒟𝒟(I₁, I₂) = true
```

The first case is the most conservative solution. In the absence of any other
knowledge, it is always correct to state that $\mathcal{DS}(I_1, I_2) = $ false and $\mathcal{DD}(I_1, I_2) = $
false.

The problem of determining if two symbolic index values are the same is
equivalent to the classical problem of *global value numbering*. If two indices i
and j have the same value number, then $\mathcal{DS}(i, j)$ must $= true$. The problem of
computing $\mathcal{DD}$ is more complex. Note that $\mathcal{DD}$, unlike $\mathcal{DS}$, is not an equivalence
relation because $\mathcal{DD}$ is not transitive. If $\mathcal{DD}(A, B) = $ true and $\mathcal{DD}(B, C) = $ true,
it does not imply that $\mathcal{DD}(A, C) = $ true. However, we can leverage past work on
array dependence analysis to identify cases for which $\mathcal{DD}$ can be evaluated to true.
For example, it is clear that $\mathcal{DD}(i, i + 1) = $ true and that $\mathcal{DD}(i, 0) = $ true if i is a
loop index variable that is known to be ≥ 1.

Let us consider how the $\mathcal{DS}$ and $\mathcal{DD}$ relations for symbolic index values are
used by our constant propagation algorithms. Note that the specification of how $\mathcal{DS}$
and $\mathcal{DD}$ are used is a separate issue from the precision of the $\mathcal{DS}$ and $\mathcal{DD}$ values.
We now describe how the lattice and the lattice operations presented in Sect. 17.2.1
can be extended to deal with non-constant subscripts.

First, consider the lattice itself. The $\top$ and $\bot$ lattice elements retain the same
meaning as in Sect. 17.2.1 viz. $\text{SET}(\top) = \{ \}$ and $\text{SET}(\bot) = \mathcal{U}_{ind}^A \times \mathcal{U}_{elem}^A$. Each
element in the lattice is a list of index-value pairs where the value is still required

$\mathscr{L}(A_1[k])$	$\mathscr{L}(k) = \top$	$\mathscr{L}(k) = \mathsf{VN}(k)$	$\mathscr{L}(k) = \bot$
$\mathscr{L}(A_1) = \top$	$\top$	$\top$	$\bot$
$\mathscr{L}(A_1) = \langle (i_1, e_1), \ldots \rangle$	$\top$	e_j, if $\exists (i_j, e_j) \in \mathscr{L}(A_1)$ with $\mathscr{DS}(i_j, \mathsf{VN}(k)) = \mathrm{true}$ $\bot$, otherwise	$\bot$
$\mathscr{L}(A_1) = \bot$	$\bot$	$\bot$	$\bot$

Fig. 17.9 Lattice computation for $\mathscr{L}(A_1[k]) = \mathscr{L}_{[\,]}(\mathscr{L}(A_1), \mathscr{L}(k))$, where $A_1[k]$ is an array element read operator. If $\mathscr{L}(k) = \mathsf{VN}(k)$, the lattice value of index k is a value number that represents a constant or a symbolic value

$\mathscr{L}(A_1)$	$\mathscr{L}(i) = \top$	$\mathscr{L}(i) = Constant$	$\mathscr{L}(i) = \bot$
$\mathscr{L}(k) = \top$	$\top$	$\top$	$\bot$
$\mathscr{L}(k) = \mathsf{VN}(k)$	$\top$	$\langle (\mathsf{VN}(k), \mathscr{L}(i)) \rangle$	$\bot$
$\mathscr{L}(k) = \bot$	$\bot$	$\bot$	$\bot$

Fig. 17.10 Lattice computation for $\mathscr{L}(A_1) = \mathscr{L}_{d[\,]}(\mathscr{L}(k), \mathscr{L}(i))$, where $A_1[k] \leftarrow i$ is an array element write operator. If $\mathscr{L}(k) = \mathsf{VN}(k)$, the lattice value of index k is a value number that represents a constant or a symbolic value

to be constant but the index may be symbolic—the index is represented by its value number ($\mathsf{VN}(i)$ for SSA variable i).

We now revisit the processing of an array element read of $A_1[k]$ and the processing of an array element write of $A_1[k]$. These operations were presented in Sect. 17.2.1 (Figs. 17.2 and 17.3) for constant indices. The versions for non-constant indices appear in Figs. 17.9 and 17.10. For the read operation in Fig. 17.9, if there exists a pair (i_j, e_j) such that $\mathscr{DS}(i_j, k) = \mathrm{true}$ (i.e., i_j and $\mathsf{VN}(k)$ have the same value number), then the result is e_j. Otherwise, the result is $\top$ or $\bot$ as specified in Fig. 17.9. For the write operation in Fig. 17.10, if the value of the right-hand side, i, is a constant, the result is the singleton list $\langle (\mathsf{VN}(k), \mathscr{L}(i)) \rangle$. Otherwise, the result is $\top$ or $\bot$ as specified in Fig. 17.10.

Let us now consider the propagation of lattice values through ϕ_{def} operators. The only extension required relative to Fig. 17.4 is that the $\mathscr{DD}$ relation used in performing the UPDATE operation should be able to determine when $\mathscr{DD}(i', i_j) = \mathrm{true}$ if i' and i_j are symbolic value numbers rather than constants. (If no symbolic information is available for i' and i_j, then it is always safe to return $\mathscr{DD}(i', i_j) = \mathrm{false}$.)

17.3 Extension to Objects: Redundant Load Elimination

In this section, we introduce redundantload elimination as an exemplar of a program optimization that can be extended to heap objects in strongly typed languages by using Array SSA form. This extension models object references (pointers) as indices into hypothetical *heap arrays* (Sect. 17.3.1). We then describe how definitely same and definitely different analyses can be extended to heap array indices (Sect. 17.3.2), followed by a *scalar promotion* transformation that uses the analysis results to perform load elimination (Sect. 17.3.3).

17.3.1 Analysis Framework

We introduce a formalism called *heap arrays* which allows us to model object references as associative arrays. An *extended Array SSA* form is constructed on heap arrays by adding *use-ϕ* functions (denoted ϕ_{use}). For each field x, we introduce a hypothetical one-dimensional heap array, $\mathscr{H}^x$. Heap array $\mathscr{H}^x$ consolidates all instances of field x present in the heap. Heap arrays are indexed by object references. Thus, a GETFIELD of $p.x$ is modelled as a read of element $\mathscr{H}^x[p]$, and a PUTFIELD of $q.x$ is modelled as a write of element $\mathscr{H}^x[q]$. The use of distinct heap arrays for distinct fields leverages the fact that accesses to distinct fields must be directed to distinct memory locations in a strongly typed language. Note that field x is considered to be the same field for objects of types C_1 and C_2, if C_2 is a subtype of C_1. Accesses to one-dimensional array objects with the same element type are modelled as accesses to a single *two-dimensional* heap array for that element type, with one dimension indexed by the object reference as in heap arrays for fields and the second dimension indexed by the integer subscript.

Heap arrays are renamed in accordance with an *extended* Array SSA form that contains three kinds of ϕ-functions:

1. A *control-ϕ-function* (ϕ) from scalar SSA form.
2. A *definition-ϕ-function* (ϕ_{def}) from Array SSA form.
3. A *use-ϕ-function* (ϕ_{use}) function creates a new name whenever a statement reads a heap array element. ϕ_{use}-functions represent the extension in "extended" Array SSA form.

The main purpose of the ϕ_{use}-function is to link together load instructions for the same heap array in control-flow order. While ϕ_{use}-functions are used by the redundant load elimination optimization presented in this chapter, it is not necessary for analysis algorithms (e.g., constant propagation) that do not require the creation of a new name at each use.

17.3.2 Definitely Same and Definitely Different Analyses for Heap Array Indices

In this section, we show how global value numbering and allocation-site information can be used to efficiently compute *definitely same* ($\mathscr{DS}$) and *definitely different* ($\mathscr{DD}$) information for heap array indices, thereby reducing pointer analysis queries

Original program:

$r \leftarrow p$	
$q \leftarrow$ new Type1	
...	
$p.x \leftarrow$ ...	$\triangleright$ $Hx[p] \leftarrow$ ...
$q.x \leftarrow$ ...	$\triangleright$ $Hx[q] \leftarrow$ ...
... $\leftarrow r.x$	$\triangleright$ ... $\leftarrow Hx[r]$

Original program:

$r \leftarrow p$	
$q \leftarrow$ new Type1	
...	
... $\leftarrow p.x$	$\triangleright$ ... $\leftarrow Hx[p]$
$q.x \leftarrow$ ...	$\triangleright$ $Hx[q] \leftarrow$ ...
... $\leftarrow r.x$	$\triangleright$ ... $\leftarrow Hx[r]$

After redundant load elimination:

| $r \leftarrow p$ |
| $q \leftarrow$ new Type1 |
| ... |
| $T1 \leftarrow$ ... |
| $p.x \leftarrow T1$ |
| $q.x \leftarrow$ ... |
| ... $\leftarrow T1$ |

After redundant load elimination:

| $r \leftarrow p$ |
| $q \leftarrow$ new Type1 |
| ... |
| $T2 \leftarrow p.x$ |
| ... $\leftarrow T2$ |
| $q.x \leftarrow$ ... |
| ... $\leftarrow T2$ |

(a)   (b)

Fig. 17.11 Examples of scalar promotion

to array index queries. If more sophisticated pointer analyses are available in a compiler, they can be used to further refine the $\mathcal{DS}$ and $\mathcal{DD}$ information.

As an example, Fig. 17.11 illustrates two different cases of scalar promotion (load elimination) for object fields. The notation Hx[p] refers to a read/write access of heap array element, $\mathcal{H}^x[p]$. For the original program in Fig. 17.11a, introducing a scalar temporary T1 for the store (def) of p.x can enable the load (use) of r.x to be eliminated, i.e., to be replaced by a use of T1. Figure 17.11b contains an example in which a scalar temporary (T2) is introduced for the first load of p.x, thus enabling the second load of r.x to be eliminated, i.e., replaced by T2. In both cases, the goal of our analysis is to determine that the load of r.x is redundant, thereby enabling the compiler to replace it by a use of scalar temporary that captures the value in p.x. We need to establish two facts to perform this transformation: 1. object references p and r are identical (definitely same) in all program executions, and 2. object references q and r are distinct (definitely different) in all program executions.

As before, we use the notation VN(i) to denote the value number of SSA variable i. Therefore, if VN(i) = VN(j), then $\mathcal{DS}(i, j)$ = true. For the code fragment above, the statement, $p \leftarrow r$, ensures that p and r are given the same value number, VN(p) = VN(r), so that $\mathcal{DS}(p, r)$ = true. The problem of computing $\mathcal{DD}$ for object references is more complex than value numbering and relates to pointer alias analysis. We outline a simple and sound approach below, which can be replaced by more sophisticated techniques as needed. It relies on two observations related to allocation-sites:

1. Object references that contain the results of distinct allocation-sites must be different.
2. An object reference containing the result of an allocation-site must be different from any object reference that occurs at a program point that dominates the allocation-site in the control-flow graph. For example, in Fig. 17.11, the presence of the allocation-site in $q \leftarrow$ new Type1 ensures that $\mathcal{DD}(p, q) = \text{true}$.

17.3.3 Scalar Promotion Algorithm

The main program analysis needed to enable redundant load elimination is *index propagation*, which identifies the set of indices that are *available* at a specific def/use A_i of heap array A. Index propagation is a data-flow problem, the goal of which is to compute a lattice value $\mathcal{L}(\mathcal{H})$ for each renamed heap variable $\mathcal{H}$ in the Array SSA form such that a load of $\mathcal{H}[\vec{i}]$ is *available* if $\mathcal{V}(\vec{i}) \in \mathcal{L}(\mathcal{H})$. Note that the lattice element does not include the value of $\mathcal{H}[\vec{i}]$ (as in constant propagation), just the fact that it is available. Figures 17.12, 17.13, and 17.14 give the lattice computations

$\mathcal{L}(A_2)$	$\mathcal{L}(A_0) = \top$	$\mathcal{L}(A_0) = \langle \vec{i_1}, \ldots \rangle$	$\mathcal{L}(A_0) = \bot$
$\mathcal{L}(A_1) = \top$	$\top$	$\top$	$\top$
$\mathcal{L}(A_1) = \langle \vec{i'} \rangle$	$\top$	$\text{UPDATE}(\vec{i'}, \langle (\vec{i_1}), \ldots \rangle)$	$\langle \vec{i'} \rangle$
$\mathcal{L}(A_1) = \bot$	$\bot$	$\bot$	$\bot$

Fig. 17.12 Lattice computation for $\mathcal{L}(A_2) = \mathcal{L}_{\phi_{def}}(\mathcal{L}(A_1), \mathcal{L}(A_0))$ where $A_2 \leftarrow \phi_{def}(A_1, A_0)$ is a definition ϕ operation

$\mathcal{L}(A_2)$	$\mathcal{L}(A_0) = \top$	$\mathcal{L}(A_0) = \langle (\vec{i_1}), \ldots \rangle$	$\mathcal{L}(A_0) = \bot$
$\mathcal{L}(A_1) = \top$	$\top$	$\top$	$\top$
$\mathcal{L}(A_1) = \langle \vec{i'} \rangle$	$\top$	$\mathcal{L}(A_1) \cup \mathcal{L}(A_0)$	$\mathcal{L}(A_1)$
$\mathcal{L}(A_1) = \bot$	$\bot$	$\bot$	$\bot$

Fig. 17.13 Lattice computation for $\mathcal{L}(A_2) = \mathcal{L}_{\phi_{use}}(\mathcal{L}(A_1), \mathcal{L}(A_0))$ where $A_2 \leftarrow \phi_{use}(A_1, A_0)$ is a use ϕ operation

$\mathcal{L}(A_2) = \mathcal{L}(A_1) \sqcap \mathcal{L}(A_0)$	$\mathcal{L}(A_0) = \top$	$\mathcal{L}(A_0) = \langle (\vec{i_1}), \ldots \rangle$	$\mathcal{L}(A_0) = \bot$
$\mathcal{L}(A_1) = \top$	$\top$	$\mathcal{L}(A_0)$	$\bot$
$\mathcal{L}(A_1) = \langle \vec{i_1}, \ldots \rangle$	$\mathcal{L}(A_1)$	$\mathcal{L}(A_1) \sqcap \mathcal{L}(A_0)$	$\bot$
$\mathcal{L}(A_1) = \bot$	$\bot$	$\bot$	$\bot$

Fig. 17.14 Lattice computation for $\mathcal{L}(A_2) = \mathcal{L}_\phi(\mathcal{L}(A_1), \mathcal{L}(A_0)) = \mathcal{L}(A_1) \sqcap \mathcal{L}(A_0)$, where $A_2 \leftarrow \phi(A_1, A_0)$ is a control ϕ operation

(a)	(b)	(c)
$r \leftarrow p$ $q \leftarrow \text{new Type1}$ ... $\mathcal{H}_1^x[p] \leftarrow \ldots$ $\mathcal{H}_2^x \leftarrow \phi_{def}(\mathcal{H}_1^x, \mathcal{H}_0^x)$ $\mathcal{H}_3^x[q] \leftarrow \ldots$ $\mathcal{H}_4^x \leftarrow \phi_{def}(\mathcal{H}_3^x, \mathcal{H}_2^x)$ $\ldots \leftarrow \mathcal{H}_4^x[r]$	$\mathcal{L}(\mathcal{H}_0^x) = \{\}$ $\mathcal{L}(\mathcal{H}_1^x) = \{VN(p) = VN(r)\}$ $\mathcal{L}(\mathcal{H}_2^x) = \{VN(p) = VN(r)\}$ $\mathcal{L}(\mathcal{H}_3^x) = \{VN(q)\}$ $\mathcal{L}(\mathcal{H}_4^x) = \{VN(p) = VN(r), VN(q)\}$	$r \leftarrow p$ $q \leftarrow \text{new Type1}$ ... $T1 \leftarrow \ldots$ $p.x \leftarrow T1$ $q.x \leftarrow \ldots$ $\ldots \leftarrow T1$

(a) Extended Array SSA form   **(b)** After index propagation   **(c)** After transformation

Fig. 17.15 Trace of index propagation and load elimination transformation for program in Fig. 17.11a

which define the index propagation solution. The notation UPDATE($\vec{i}'$, $\langle \vec{i}_1, \ldots \rangle$) used in the middle cell in Fig. 17.12 denotes a special update of the list $\mathcal{L}(A_0) = \langle \vec{i}_1, \ldots \rangle$ with respect to index $\vec{i}'$. UPDATE involves four steps:

1. Compute the list $T = \{ \vec{i}_j \mid \vec{i}_j \in \mathcal{L}(A_0) \text{ and } \mathcal{DD}(\vec{i}', \vec{i}_j) = \text{true} \}$. List T contains only those indices from $\mathcal{L}(A_0)$ that are *definitely different* from $\vec{i}'$.
2. Insert $\vec{i}'$ into T to obtain a new list, I.
3. (Optional) As before, if there is a desire to bound the height of the lattice due to compile-time considerations and the size of list I exceeds a threshold size Z, then any one of the indices in I can be dropped from the output list.
4. Return I as the value of UPDATE($\vec{i}'$, $\langle \vec{i}_1, \ldots \rangle$).

After index propagation, the algorithm selects a load, $A_j[\vec{x}]$, for scalar promotion if and only if index propagation determines that an index with value number $VN(()\vec{x})$ is available at the def of A_j. Figure 17.15 illustrates a trace of this load elimination algorithm for the example program in Fig. 17.11a. Figure 17.15a shows the extended Array SSA form computed for this example program. The results of index propagation are shown in Fig. 17.15b. These results depend on definitely different analysis establishing that $VN(p) \neq VN(q)$ and definitely same analysis establishing that $VN(p) = VN(r)$. Figure 17.15c shows the transformed code after performing the scalar promotion actions. The load of p.x has thus been eliminated in the transformed code and replaced by a use of the scalar temporary, T1.

17.4 Further Reading

In this chapter, we introduced an Array SSA form that captures element-level data-flow information for array variables, illustrated how it can be used to extend program analyses for scalars to array variables using constant propagation as an exemplar, and illustrated how it can be used to extend optimizations for scalars to array variables using load elimination in heap objects as an exemplar. In addition to reading the other chapters in this book for related topics, the interested reader can consult [168] for details on full Array SSA form, [169] for details on constant propagation using Array SSA form, [120] for details on load and store elimination for pointer-based objects using Array SSA form, [253] for efficient dependence analysis of pointer-based array objects, and [18] for extensions of this load elimination algorithm to parallel programs.

There are many possible directions for future research based on this work. The definitely same and definitely different analyses outlined in this chapter are sound but conservative. In restricted cases, they can be made more precise using array subscript analysis from polyhedral compilation frameworks. Achieving a robust integration of Array SSA and polyhedral approaches is an interesting goal for future research. Past work on Fuzzy Array Data-flow Analysis (FADA) [19] may provide a useful foundation for exploring such an integration. Another interesting direction is to extend the value numbering and definitely different analyses mentioned in Sect. 17.2.2 so that they can be combined with constant propagation rather than performed as a pre-pass. An ultimate goal is to combine conditional constant, type propagation, value numbering, partial redundancy elimination, and otion analyses within a single framework that can be used to analyse scalar variables, array variables, and pointer objects with a unified approach.

Part IV
Machine Code Generation and Optimization

Chapter 18
SSA Form and Code Generation

Benoît Dupont de Dinechin

In a compiler for imperative languages such as C, C++, or FORTRAN, the code generator covers the set of code transformations and optimizations that are performed on a program representation close to the target machine's Instruction Set Architecture (ISA). The code generator produces an assembly source or relocatable file with debugging information.

The main duties of code generation are: lowering the program Intermediate Representation (IR) [270] to machine instruction level with appropriate calling conventions; laying out data objects in sections; composing the stack frames; allocating variable live ranges to architectural registers; scheduling instructions in order to exploit micro-architecture features; and producing assembly source or object code.

Historically, the 1986 edition of the "Compilers Principles, Techniques, and Tools" Dragon Book by Aho et al. [2] lists the tasks of code generation as

- Instruction selection and lowering of the calling conventions;
- Control-flow (dominators, loops) and data-flow (variable liveness) analyses;
- Register allocation and stack frame building;
- Peephole optimizations.

Ten years later, the 1997 textbook "Advanced Compiler Design & Implementation" by Muchnick [205] extends code generation with the following tasks:

- Loop unrolling and basic block replication
- Instruction scheduling and software pipelining
- Branch optimizations and basic block alignment

B. Dupont de Dinechin (✉)
Kalray, Montbonnot-Saint-Martin, France
e-mail: benoit.dinechin@kalray.eu

© The Author(s), under exclusive license to Springer Nature Switzerland AG 2022
F. Rastello, F. Bouchez Tichadou (eds.), *SSA-based Compiler Design*,
https://doi.org/10.1007/978-3-030-80515-9_18

In high-end compilers such as Open64 [64, 65], GCC [268], or LLVM [176], code generation techniques have significantly evolved, as they are mainly responsible for exploiting the performance-oriented features of architectures and microarchitectures. In those compilers, code generator optimizations include:

- If-conversion using `select`, conditional move, or predicated instructions (Chap. 20)
- Use of specialized addressing modes such as auto-modified addressing [180] and modulo addressing [66]
- Exploitation of hardware looping [183] or static branch prediction hints
- Matching fixed-point arithmetic and SIMD idioms to special instructions
- Optimization with regard to memory hierarchy, including cache prefetching and register preloading [99]
- VLIW instruction bundling, where parallel instruction groups constructed by postpass instruction scheduling are encoded into instruction bundles [159]

This sophistication of modern compiler code generation is one of the reasons behind the introduction of the SSA form for machine code, in order to simplify certain analyses and optimizations. In particular, liveness analysis (Chap. 9), if-conversion (Chap. 20), unrolling-based loop optimizations (Chap. 10), and exploitation of special instructions or addressing modes benefit significantly from the SSA form. Chapter 19 presents an advanced technique for instruction selection on the SSA form by solving a specialized quadratic assignment problem (PBQP). Although there is a debate as to whether or not SSA form should be used in a register allocator, Chap. 22 makes a convincing case for it. The challenge of correct and efficient SSA form destruction under the constraints of machine code is addressed in Chap. 21. Finally, Chap. 23 illustrates how the SSA form has been successfully applied to hardware compilation.

Basing ourselves on our experience on a family of production code generators and linear assembly optimizers for the ST120 DSP core [102, 103, 240, 275] and the Lx/ST200 VLIW family [34–36, 100, 101, 113], this chapter reviews some of the issues of using the SSA form in a code generator. Section 18.1 presents the challenges of maintaining the SSA form on a program representation based on machine instructions. Section 18.2 discusses two code generator optimizations that seem at odds with the SSA form, yet must occur before register allocation. One is if-conversion, whose modern formulations require an extension of the SSA form. The other is prepass instruction scheduling, for which the benefit of using the SSA form has not been assessed by any implementation yet. Using the SSA form at machine-code level requires the ability to construct and destruct SSA form at that level. Section 18.3 characterizes various SSA form destruction algorithms in terms of satisfying the constraints of machine code.

18.1 SSA Form Engineering Issues

18.1.1 Instructions, Operands, Operations, and Operators

An *instruction* is a member of the machine instruction set architecture (ISA). Instructions access values and modify the machine state through *operand*s. We distinguish *explicit operands*, which are associated with a specific bit-field in the instruction encoding, from *implicit operands*, without any encoding bits. Explicit operands correspond to allocatable architectural registers, immediate values, or operand modifiers (such as shift, extend, saturate). Implicit operands correspond to dedicated architectural registers, and to registers implicitly used by some instructions, for instance, the status register, the procedure link register, or the stack pointer.

An *operation* is an instance of an instruction that composes a program. It is seen by the compiler as an *operator* applied to a list of operands (explicit and implicit), along with operand naming constraints, and has a set of clobbered registers (i.e., that can be trashed or modified in unpredictable way). The compiler view of operations also involves *indirect operands*, which are not apparent in the instruction behaviour, but are required to connect the flow of values between operations. *Implicit operands* correspond to the registers used for passing arguments and returning results at function call sites, and may also be used for the registers encoded in register mask immediates.

18.1.2 Representation of Instruction Semantics

Unlike IR operators, there is no straightforward mapping between machine instructions and their semantics. For instance, a subtract instruction with operands (a, b, c) may either compute $c \leftarrow a - b$ or $c \leftarrow b - a$ or any such expression with permuted operands. Yet basic SSA form code cleanups, such as constant propagation or sign extension removal, need to know what is actually computed by machine instructions. Machine instructions may also have multiple target operands, such as memory accesses with auto-modified addressing, combined division-modulus instructions, or side-effects on status registers.

There are two ways to associate machine instructions with semantics:

- Add properties to the instruction operator and to its operands, a technique used by the Open64 compiler. Operator properties include *isAdd*, *isLoad*, etc. Typical operand properties include *isLeft*, *isRight*, *isBase*, *isOffset*, *isPredicated*, etc. Extended properties that involve the operator and some of its operands include *isAssociative*, *isCommutative*, etc.

- Associate a *semantic combinator* [300], that is, a tree of IR-like operators, to each target operand of a machine instruction. This alternative has been implemented in the SML/NJ [182] compiler and the LAO compiler [102].

An issue related to the representation of instruction semantics is how to encode it. Most information can be statically tabulated by the instruction operator, yet properties such as safety for control speculation, or equivalence to a simple IR instruction, can be refined by the context where the instruction appears. For instance, range propagation may ensure that an addition cannot overflow, that a division by zero is impossible, or that a memory access is safe for control speculation. Context-dependent semantics, which needs to be associated with specific machine instructions in the code generator's internal representation, can be provided as annotations that override the statically tabulated information.

Finally, code generation for some instruction set architectures requires that pseudo-instructions with standard semantics be available, as well as variants of ϕ-functions and parallel copy operations.

- Machine instructions that operate on register pairs, such as the long multiplies on the ARM, or more generally on register tuples, must be handled. In such cases there is a need for pseudo-instructions to compose wide operands in register tuples, and to independently extract register allocatable operands from wide operands.
- Embedded processor architectures such as the Tensilica Xtensa [130] provide zero-overhead loops (hardware loops), where an implicit conditional branch back to the loop header is taken whenever the program counter matches some addresses. The implied loop-back branch is also conveniently materialized by a pseudo-instruction.
- Register allocation for predicated architectures requires that the live ranges of temporary variables with predicated definitions be contained by pseudo-instructions [128] that provide backward kill points for liveness analysis.

18.1.3   Operand Naming Constraints

Implicit operands and indirect operands are constrained to specific architectural registers, either by the instruction set architecture (ISA constraints) or by the application binary interface (ABI constraints). An effective way to deal with such *dedicated register* naming constraints in the SSA form is to insert parallel copy operations that write to the constrained source operands, or read from the constrained target operands of instructions. The new SSA variables thus created are pre-coloured with the required architectural register. With modern SSA form destruction [35, 267] (see Chap. 21), copy operations are aggressively coalesced, and the remaining ones are sequentialized into machine operations.

Explicit instruction operands may be constrained to use the same resource (an unspecified architectural register) between a source and a target operand, as illustrated by most x86 instructions [257] and by DSP-style auto-modified addressing modes [180]. A related naming constraint is to require different resources between source and destination operands, as with the `mul` instructions on the ARMv5 processors. The *same resource* naming constraints are represented under the SSA form by inserting a copy operation between the constrained source operand and a new variable, then using this new variable as the constrained source operand. In case of multiple constrained source operands, a parallel copy operation is used. Again, these copy operations are processed by the SSA form destruction.

A more widespread case of an operand naming constraint is when a variable must be bound to a specific architectural register at all points in the program. This is the case with the stack pointer, as interrupt handling may reuse the runtime stack at any program point. One possibility is to inhibit the promotion of the stack pointer to a SSA variable. Stack pointer definitions, including memory allocations through `alloca()`, activation frame creation/destruction, are then encapsulated as instances of a specific pseudo-instruction. Instructions that use the stack pointer must be treated as special cases for SSA form analyses and optimizations.

18.1.4 Non-kill Target Operands

The SSA form requires variable definitions to be killing definitions (see e.g. [208] or kill points in data-flow analysis). This is not the case for target operands such as a status register, which contains several independent bit-fields. Moreover, some instruction effects on bit-fields may be *sticky*, that is, with an implied disjunction with the previous value. Typical sticky bits include exception flags of the IEEE 754 arithmetic, or the integer overflow flag on DSPs with fixed-point arithmetic. When mapping a status register to a SSA variable, any operation that partially reads or modifies the register bit-fields should appear as reading and writing the corresponding variable.

Predicated execution and conditional execution are other sources of definitions that do not kill their target register. The execution of predicated instructions is guarded by the evaluation of a single bit operand. The execution of conditional instructions is guarded by the evaluation of a condition on a multi-bit operand. We extend the ISA classification of [193] to include four classes:

Partial predicated execution support: `select` instructions, first introduced by the Multiflow TRACE architecture [81], are provided. These instructions write to a destination register the value of one among two source operands, depending on the condition tested on a third source operand. The Multiflow TRACE 500 architecture was planned to include predicated store and floating-point instructions [189].

Full predicated execution support: Most instructions accept a boolean predicate operand, which nullifies the instruction effects if the predicate evaluates to false. EPIC-style architectures also provide predicate define instructions (pdi) to efficiently evaluate predicates corresponding to nested conditions: Unconditional, Conditional, parallel or, parallel and [128].

Partial conditional execution support: Conditional move (cmov) instructions, first introduced by the Alpha AXP architecture [31], are provided. cmov instructions have been available in the ia32 ISA since the Pentium Pro.

Full conditional execution support: Most instructions are conditionally executed depending on the evaluation of a condition of a source operand. On the ARM architecture, the implicit source operand is a bit-field in the status register and the condition is encoded on 4 bits. On the VelociTI™ TMS230C6x architecture [256], the source operand is a general register encoded on 3 bits and the condition is encoded on 1 bit.

18.1.5 *Program Representation Invariants*

Engineering a code generator requires decisions about which information is transient and which belongs to the invariants of the program representation. An *invariant* is a property that is ensured before and after each phase. Transient information is recomputed as needed by some phases from the program representation invariants. The applicability of the SSA form only spans the early phases of the code generation process: from instruction selection down to register allocation. After register allocation, program variables are mapped to architectural registers or to memory locations, so the SSA form analyses and optimizations no longer apply. In addition, a program may be only partially converted to the SSA form. This is the reason for the engineering of the SSA form as extensions to a baseline code generator program representation.

Some extensions to the program representation required by the SSA form are better engineered as invariants, in particular for operands, operations, basic blocks, and control-flow graphs. Operands that are SSA variables need to record the unique operation that defines them as a target operand, and possibly to maintain the list of the places where they appear as source operands. Operations such as ϕ-functions , σ-functions of the SSI form [36] (see Chap. 13), and parallel copies may appear as regular operations constrained to specific places in the basic blocks. The incoming (resp. outgoing) edges of basic blocks also need be kept in the same order as the operands of each of their ϕ-functions (resp. σ-functions).

A program representation invariant that impacts the engineering of SSA form is the structure of loops. The modern way of identifying loops in a CFG is the construction of a loop nesting forest as defined by Ramalingam [236]. High-level information such as loop-carried memory dependencies, or user-level loop annotations, should be provided to the code generator. Such information is attached to a loop structure, which thus becomes an invariant. The impact on the SSA form is that some loop nesting forests, such as the Havlak [142] loop structure, are more suitable than others as they can be used to attach to basic blocks the results of key analyses such as SSA variable liveness [36] (see Chap. 9).

Live-in and live-out sets at basic block boundaries are also candidates for program representation invariants. However, when using and updating liveness information under the SSA form, it appears convenient to distinguish the ϕ-function contributions from the results of data-flow fixed-point computation. In particular, Sreedhar et al. [267] introduced the ϕ-function semantics that became later known as *multiplexing mode* (see Chap. 21), where a ϕ-function $B_0 : a_0 = \phi(B_1 : a_1, \ldots, B_n : a_n)$ makes a_0 live-in of basic block B_0, and $a_1, \ldots a_n$ live-out of basic blocks $B_1, \ldots B_n$. The classical basic block invariants $\mathrm{LiveIn}(B)$ and $\mathrm{LiveOut}(B)$ are then complemented with $\mathrm{PhiDefs}(B)$ and $\mathrm{PhiUses}(B)$.

Finally, some compilers adopt the invariant that the SSA form be *conventional* across the code generation phases. This approach is motivated by the fact that classical optimizations such as SSAPRE [164] (see Chap. 11) require that "the live ranges of different versions of the same original program variable do not overlap," implying that the SSA form should be conventional. Other compilers that use SSA numbers and omit the ϕ-functions from the program representation [175] are similarly constrained. Work by Sreedhar et al. [267] and by Boissinot et al. [35] have clarified how to convert the transformed SSA form into conventional form wherever required, so there now is no reason for this property to be an invariant.

18.2 Code Generation Phases and the SSA Form

18.2.1 If-Conversion

If-conversion refers to optimizations that convert a program region to straight-line code. It is primarily motivated by instruction scheduling on instruction-level parallel cores [193], as removing conditional branches serves to:

- Eliminate branch resolution stalls in the instruction pipeline
- Reduce uses of the branch unit, which is often single-issue
- Increase the size of the instruction scheduling regions

In case of inner loop bodies, if-conversion further enables vectorization [5] and software pipelining [220] (modulo scheduling). Consequently, control-flow regions selected for if-conversion are acyclic, even though seminal techniques [5, 220] consider more general control-flow.

The scope and effectiveness of if-conversion depend on the ISA support. In principle, any if-conversion technique targeted to full predicated or conditional execution support may be adapted to partial predicated or conditional execution support. For instance, non-predicated instructions with side-effects such as memory accesses can be used in combination with `select` to provide a harmless effective address in the event that the operation must be nullified [193].

Besides predicated or conditional execution, architectural support for if-conversion is improved by supporting speculative execution. Speculative execution (control speculation) refers to executing an operation before knowing that its execution is required, which happens when moving code above a branch [189] or promoting operation predicates [193]. Speculative execution assumes that instructions have reversible side effects, so speculating potentially excepting instructions requires architectural support. On the Multiflow TRACE 300 architecture and later on the Lx VLIW architecture [113], non-trapping memory loads known as *dismissible* loads are provided. The IMPACT EPIC architecture speculative execution [13] is generalized from the *sentinel* model [192].

Classical Contributions to If-Conversion Without SSA Form
The initial classical contributions to if-conversion did not consider the SSA form. The goal of this paragraph is to provide a short overview of those contributions.

Allen et al. [5]   Conversion of control dependencies to data dependencies, motivated by inner loop vectorization. The authors distinguish forward branches, exit branches, and backward branches and compute boolean guards accordingly. As this work pre-dates the Program Dependence Graph [118] (see Chap. 14), the complexity of the resulting boolean expressions is an issue. When compared to later if-conversion techniques, only the conversion of forward branches is relevant.

Park and Schlansker [220]   Formulation of the "*RK algorithm*" based on control dependencies. The authors assume a fully predicated architecture with only conditional pdi instructions. The R function assigns a minimal set of boolean predicates to basic blocks, and the K function expresses the way those predicates are computed (Fig. 18.1). The algorithm is general enough to process cyclic and irreducible rooted flow graphs, but in practice it is applied to single-entry acyclic regions.

Blickstein et al. [31]   Pioneering use of cmov instructions to replace conditional branches in the GEM compilers for the Alpha AXP architecture.

Lowney et al. [189]   Matching of the innermost if-then constructs in the Multiflow Trace Scheduling compiler, in order to generate the select and the predicated memory store operations.

Fang [112]   The proposed algorithm assumes a fully predicated architecture with conditional pdi instructions. It is tailored to acyclic regions with single entry and multiple exits, and as such is able to compute both R and K functions without relying on explicit control dependencies. The main improvement of this algorithm over [220] is that it also speculates instructions up the dominance tree through *predicate promotion*,[1] except for stores and pdi instructions. This work further proposes a pre-optimization pass performed before predication and speculation that hoist or sink common sub-expressions.

Leupers [184]   The technique focuses on if-conversion of nested if-then-else statements for architectures with full conditional execution support. A dynamic programming technique appropriately selects either a

---

[1] The predicate used to guard an operation is promoted to a weaker condition.

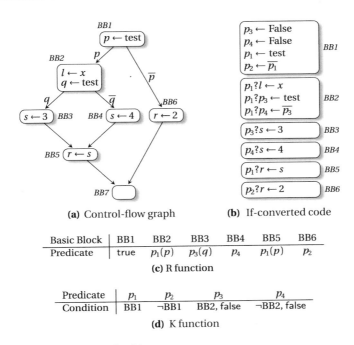

(a) Control-flow graph   **(b)** If-converted code

Basic Block	BB1	BB2	BB3	BB4	BB5	BB6
Predicate	true	$p_1(p)$	$p_3(q)$	p_4	$p_1(p)$	p_2

(c) R function

Predicate	p_1	p_2	p_3	p_4
Condition	BB1	¬BB1	BB2, false	¬BB2, false

(d) K function

Fig. 18.1 Illustration of the RK algorithm

conditional `jump` or a conditional instruction-based implementation scheme for each `if-then-else` statement. The objective is the reduction of worst-case execution time (WCET).

Classical Contributions to If-Conversion with Internal Use of SSA Form

We now describe two contributions to if-conversion that did use the SSA form, but only internally.

Jacome et al. [148]   Proposition of the Static Single Assignment—Predicated Switching (SSA-PS) transformation. This assumes a clustered VLIW architecture fitted with conditional move instructions that operate inside clusters (internal moves) or between clusters (external moves). The first idea of the SSA-PS transformation is to perform the conditional assignments corresponding to ϕ-functions via predicated switching operations, in particular conditional move operations. The second idea is that the conditional external moves leverage the penalties associated with inter-cluster data transfers. The SSA-PS transformation predicates non-move operations and is apparently restricted to innermost if-then-else statements.

Chuang et al. [74]   A predicated execution support aimed at removing non-kill register writes from the micro-architecture. They propose `select` instructions called *PHI-ops* (similar to the ϕ_{if}-function described in Sect. 14.4), predicated memory accesses, unconditional `pdi` instructions, and `orp` instructions for or-ing multiple predicates. The RK algorithm is simplified for the case of single-entry single-exit regions, and adapted to the proposed architectural support. The other contribution is the generation of PHI-ops, whose insertion points are computed like the SSA form placement of ϕ-functions. The ϕ-functions' source operands are replaced by *phi-lists*, where each operand is associated with the predicate of its source basic block. The phi-lists are processed by topological order of the predicates to generate the PHI-ops.

If-Conversion Under SSA Form

The ability to perform if-conversion on the SSA form of a program representation requires the handling of operations that do not kill the target operand because of predicated or conditional execution. The following papers address this issue:

Stoutchinin and Ferrière [275]   Introduction of ψ-functions in order to represent fully predicated code under the SSA form, which is then called the ψ-SSA form. The ψ-functions' arguments are paired with predicates and are ordered in dominance order in the ψ-function argument list. This ordering is a correctness condition re-discovered by Chuang et al. [74] for their PHI-ops. The ψ-SSA form is presented in Chap. 15.

Stoutchinin and Gao [276]   Proposition of an if-conversion technique based on the predication of Fang [112] and the replacement of ϕ-functions by ψ-functions. The authors prove that the conversion is correct provided that the SSA form is conventional. The technique is implemented in Open64 for the IA-64 architecture.

Bruel [52]   The technique targets VLIW architectures with select and dismissible load instructions. The proposed framework reduces acyclic control-flow constructs from innermost to outermost. A benefit criterion is used to stop the reduction process. The core technique performs control speculation in addition to tail duplication, and reduces the height of predicate computations. It can also generate ψ-functions instead of select operations. A generalization of this framework, which also accepts ψ-SSA form as input, is described in Chap. 20.

Ferrière [104]   Extension of the ψ-SSA form algorithms of [275] to architectures with partial predicated execution support, by formulating simple correctness conditions for the predicate promotion of operations that do not have side-effects. This work also details how to transform the ψ-SSA form to conventional ψ-SSA form by generating cmov operations. A self-contained explanation of these techniques appears in Chap. 15.

Thanks to these contributions, virtually all if-conversion techniques formulated without the SSA form can be adapted to the ψ-SSA form, with the added benefit that already predicated code may be part of the input. In practice, these contributions follow the generic steps of if-conversion proposed by Fang [112]:

* If-conversion region selection
* Code hoisting and sinking of common sub-expressions
* Assignment of predicates to the basic blocks

- Insertion of operations to compute the basic block predicates
- Predication or speculation of operations
- Conditional branch removal

The result of an if-converted region is a hyper-block, that is, a sequence of basic blocks with predicated or conditional operations, where control may only enter from the top, but may exit from one or more locations [191].

Although if-conversion based on the ψ-SSA form appears effective for the different classes of architectural support, the downstream phases of the code generator require some adaptations of the plain SSA form algorithms to handle the ψ-functions. The largest impact of handling ψ-functions is apparent in the ψ-SSA form destruction [104], whose original description [275] was incomplete.

In order to avoid such complexities, a code generator may adopt a simpler solution than the ψ-functions to represent the non-kill effects of conditional operations on target operands. The key observation is that under the SSA form, a `cmov` operation is equivalent to a `select` operation with the *same resource* naming constraint between one source and the target operand. Unlike other predicated or conditional instructions, a `select` instruction kills its target register. Generalizing this observation provides a simple way to handle predicated or conditional operations in plain SSA form:

- For each target operand of the predicated or conditional instruction, add a corresponding source operand in the instruction signature.
- For each added source operand, add the same resource naming constraint with the corresponding target operand.

This simple transformation enables the SSA form analyses and optimizations to remain oblivious to predicated or conditional execution. The drawback of this solution is that non-kill definitions of a given variable (before SSA variable renaming) remain in dominance order across program transformations, as opposed to ψ-SSA where predicate value analysis may enable this order to be relaxed.

18.2.2 Prepass Instruction Scheduling

Further down the code generator, the last major phase before register allocation is prepass instruction scheduling. Innermost loops with a single basic block, super-block, or hyper-block body are candidates for software pipelining techniques such as modulo scheduling [241]. For innermost loops that are not software pipelined, and for other program regions, acyclic instruction scheduling techniques apply: basic block scheduling [131]; super-block scheduling [146]; hyper-block scheduling [191]; tree region scheduling [140]; or trace scheduling [189].

By definition, prepass instruction scheduling operates before register allocation. At this stage, instruction operands are mostly virtual registers, except for instructions with ISA or ABI constraints that bind them to specific architectural registers.

Moreover, preparation for prepass instruction scheduling includes virtual register renaming, also known as register web construction, whose goal is to reduce the number of anti-dependencies and output dependencies in the instruction scheduling problem. Such code preconditioning would not be required under SSA form, but the design of an SSA based prepass instruction scheduling algorithm would face several difficulties including:

- Ordering constraints related to control dependencies and true data dependencies have to be fulfiled so as to avoid violating the semantic of the original code. Anti-dependencies, on the other hand, are artifacts of the storage allocation. Temporary variable renaming (with the most aggressive form being obtained by SSA construction) removes such dependencies, thus giving more freedom to the scheduling heuristic. The more freedom, the more potential effect on interferences between variables, and thus on register pressure. Without adding any bias on the scheduling strategy to account for register pressure and renaming constraints, important negative impacts might be observed with the insertion of shuffle code such as spill and register-to-register copies.
- Some machine instructions have partial effects on special resources such as the status register. Representing special resources as SSA variables even though they are accessed at the bit-field level requires coarsening the instruction effects to the whole resource, as discussed in Sect. 18.1.4. Doing so naively would lead to the creation of spurious data-flow dependencies associated with the def-use chain of the special resource. Moreover, def-use ordering implied by SSA form is not adapted to resources composed of sticky bits, as the definitions of these resources can be reordered. Scheduling OR-type predicate define operations [255] raises the same issues. An instruction scheduler is also expected to precisely track accesses to unrelated or partially overlapping bit-fields in a status register. More generally, some sets of operations might be considered as commutative in the IR. A reordering with relaxed constraints may, under SSA, create inconsistencies (that have to be fixed) among the implicit def-use relationships between impacted instructions.

To summarize, trying to keep the SSA form inside the prepass instruction scheduling seems reasonable but requires non-trivial adaptations.

18.3 SSA Form Destruction Algorithms

The destruction of the SSA form in a code generator is required at some point. A weaker form of SSA destruction is the conversion of transformed SSA form to conventional SSA form, which is required by a few classical SSA form optimizations such as SSAPRE (see Chap. 11). For all such cases, the main objective is the lowering to machine-code representation (getting rid of pseudo-instructions and satisfying naming constraints) by inserting the necessary copy/spill instructions.

The contributions to SSA form destruction techniques can be characterized as an evolution towards correctness, the ability to manage operand naming constraints, the handling of critical edges, and the reduction of algorithmic time and memory requirements. One of the main contributions is also the coupling with register allocation and the generation of high-quality code in terms of edge splitting and the amount of inserted copies.

Cytron et al. [90]   First technique for *translating out of SSA*, by "naive replace-ment preceded by dead code elimination and followed by colouring." The authors replace each ϕ-function $B_0 : a_0 = \phi(B_1 : a_1, \ldots, B_n : a_n)$ by n copies $a_0 = a_i$, one per basic block B_i, before applying Chaitin-style coalescing.

Briggs et al. [50]   The correctness issues of Cytron et al. [90] out of (transformed) SSA form translation are identified and illustrated by the *lost-copy problem* and the *swap problem*. These problems appear in relation to the critical edges and when the parallel assignment semantics of a sequence of ϕ-functions at the start of a basic block is not accounted for [35]. Two SSA form destruction algorithms are proposed, depending on the presence of critical edges in the control-flow graph. However, the need for parallel copy operations to represent code after ϕ-function removal is not recognized.

Sreedhar et al. [267]   This work is based on the definition of ϕ-congruence classes as the sets of SSA variables that are transitively connected by a ϕ-function. When none of the ϕ-congruence classes has members that interfere, the SSA form is called *conventional* and its destruction is trivial: replace all the SSA variables of a ϕ-congruence class by a temporary variable, and remove the ϕ-functions. In general, the SSA form is *transformed* after program optimizations, that is, some ϕ-congruence classes contain interferences. In Method I, the SSA form is made conventional by isolating ϕ-functions using copies both at the end of direct predecessor blocks and at the start of the current block. The latter is the key for not depending on critical edge splitting [35]. The code is then improved by running a new SSA variable coalescer that grows the ϕ-congruence classes with copy-related variables, while keeping the SSA form conventional. In Method II and Method III, the ϕ-congruence classes are initialized as singletons, then merged as the ϕ-functions are processed. In Method II, two variables of the current ϕ-function that interfere directly or through their ϕ-congruence classes are isolated by inserting copy operations for both. This ensures that the ϕ-congruence class that is grown from the classes of the variables related by the current ϕ-function is interference-free. In Method III, if possible only one copy operation is inserted to remove the interference, and more involved choices about which variables to isolate from the ϕ-function congruence class are resolved by a maximum independent set heuristic. Both methods are correct except for a detail about the live-out sets to consider when testing for interferences [35].

Leung and George [182]   This work is the first to address the problem of satisfy-ing the *same resource* and the *dedicated register* operand naming constraints of the SSA form on machine code. They identify that Chaitin-style coalescing after

SSA form destruction is not sufficient, and that adapting the SSA optimizations to enforce operand naming constraints is not practical. They work in three steps: collect the renaming constraints; mark the renaming conflicts; and reconstruct code, which adapts the SSA destruction of Briggs et al. [50]. This work is also the first to make explicit use of parallel copy operations. A few correctness issues were later identified and corrected by Rastello et al. [240].

Budimlić et al. [53]   Contribution of a lightweight SSA form destruction motivated by JIT compilation. It uses the (strict) SSA form property of dominance of variable definitions over uses to avoid the maintenance of an explicit interference graph. Unlike previous approaches to SSA form destruction that coalesce increasingly larger sets of non-interfering ϕ-related (and copy-related) variables, they first construct SSA webs with early pruning of obviously interfering variables, then de-coalesce the SSA webs into non-interfering classes. They propose the *dominance forest* explicit data structure to speed up these interference tests. This SSA form destruction technique does not handle the operand naming constraints, and also requires critical edge splitting.

Rastello et al. [240]   The problem of satisfying the *same resource* and *dedicated register* operand naming constraints of the SSA form on machine code is revisited, motivated by erroneous code produced by the technique of Leung and George [182]. Inspired by the work of Sreedhar et al. [267], they include the ϕ-related variables as candidates in the coalescing that optimizes the operand naming constraints. This work avoids the patent of Sreedhar et al. (US patent 6182284).

Boissinot et al. [35]   Formulation of a generic approach to SSA form destruction that is proved correct handles operand naming constraints and can be optimized for speed (see Chap. 21 for details of this generic approach). The foundation of this approach is to transform the program to conventional SSA form by isolating the ϕ-functions like in Method I of Sreedhar et al. [267]. However, the copy operations inserted are parallel, so a parallel copy sequentialization algorithm is provided. The task of improving the conventional SSA form is then seen as a classical aggressive variable coalescing problem, but thanks to the SSA form the interference relation between SSA variables is made precise and frugal to compute. Interference is obtained by combining the intersection of SSA live ranges, and the equality of values, which is easily tracked under the SSA form across copy operations. Moreover, the use of the dominance forest data structure of Budimlić et al. [53] to speed up interference tests between congruence classes is obviated by a linear traversal of these classes in pre-order of the dominance tree. Finally, the same resource operand constraints are managed by pre-coalescing, and the dedicated register operand constraints are represented by pre-colouring the congruence classes. Congruence classes with a different pre-colouring always interfere.

Chapter 19
Instruction Code Selection

Dietmar Ebner, Andreas Krall, and Bernhard Scholz

Instruction code selection is a transformation step in a compiler that translates a machine-independent intermediate code representation into a low-level intermediate representation or into machine code for a specific target architecture. Instead of hand-crafting an instruction selector for each target architecture, generator tools have been developed that generate the instruction code selector based on a specification. The specification describes the machine of the target and how the intermediate instructions are mapped to the machine code. This approach is used in large compiler infrastructures such as GCC or LLVM that target a range of architectures. A possible scenario of a code generator in a compiler is depicted in Fig. 19.1. The Intermediate Representation (IR) of an input program is passed on to an optional lowering phase that breaks down instructions and performs other machine-dependent transformations. Thereafter, the instruction selection performs the mapping to machine code or lowered IR based on the machine description of the target architecture.

One of the widely used techniques in code generation is tree-pattern matching. The unit of translation for tree-pattern matching is expressed as a tree structure that is called a data-flow tree (DFT). The basic idea is to describe the target instruction set using an *ambiguous* cost-annotated graph grammar. The instruction code selector seeks a cost-minimal *cover* of the DFT. Each of the selected rules

D. Ebner
Waymo, San Jose, CA, USA
e-mail: ebner@waymo.com

A. Krall
TU Wien, Vienna, Austria
e-mail: andi@complang.tuwien.ac.at; andreas.krall@tuwien.ac.at

B. Scholz (✉)
University of Sydney, Sydney, NSW, Australia
e-mail: Bernhard.Scholz@sydney.edu.au

© The Author(s), under exclusive license to Springer Nature Switzerland AG 2022
F. Rastello, F. Bouchez Tichadou (eds.), *SSA-based Compiler Design*,
https://doi.org/10.1007/978-3-030-80515-9_19

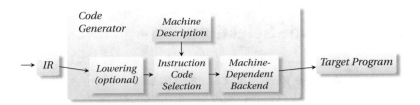

Fig. 19.1 Scenario: an instruction code selector translates a compiler's IR to a low-level machine-dependent representation

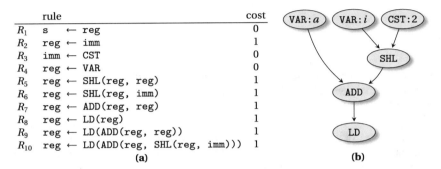

	rule		cost
R_1	s	← reg	0
R_2	reg	← imm	1
R_3	imm	← CST	0
R_4	reg	← VAR	0
R_5	reg	← SHL(reg, reg)	1
R_6	reg	← SHL(reg, imm)	1
R_7	reg	← ADD(reg, reg)	1
R_8	reg	← LD(reg)	1
R_9	reg	← LD(ADD(reg, reg))	1
R_{10}	reg	← LD(ADD(reg, SHL(reg, imm)))	1

(a)   (b)

Fig. 19.2 Example of a data-flow tree (**b**) and a rule fragment with associated costs (**a**)

has an associated semantic action that is used to emit the corresponding machine instructions, either by constructing a new intermediate representation or by rewriting the DFT bottom-up.

An example of a DFT along with a set of rules representing valid ARM instructions are shown in Fig. 19.2. Each rule consists of non-terminals (shown in lower case) and terminal symbols (shown in upper case). Non-terminals are used to chain individual rules together. Non-terminal s denotes a distinguished start symbol for the root node of the DFT. Terminal symbols match the corresponding labels of nodes in the data-flow trees. The terminals of the grammar are VAR, CST, SHL, ADD, and LD. Rules that translate from one non-terminal to another are called *chain rules*, e.g., reg ← imm, which translates an immediate value to a register. Note that there are multiple possibilities to obtain a cover of the data-flow tree for the example shown in Fig. 19.2b. Each rule has associated costs. The cost of a tree cover is the sum of the costs of the selected rules. For example, the DFT could be covered by rules R_3, R_4, and R_{10}, which would give a total cost for the cover of one cost unit. Alternatively, the DFT could be covered by rules R_2, R_3, R_5, R_7, and R_8 yielding four cost units for the cover for issuing four assembly instructions. A dynamic programming algorithm selects a cost-optimal cover for the DFT.

Tree-pattern matching on a DFT is limited to the scope of tree structures. To overcome this limitation, we can extend the scope of the matching algorithm to the computational flow of a whole procedure. The use of the SSA form as an intermediate representation improves the code generation by making def-use relationships explicit. Hence, SSA exposes the data flow of a translation unit and utilizes the code generation process. Instead of using a textual SSA representation, we employ a graph representation of SSA called the *SSA graph*[1], which is an extension of DFTs and represents the data flow for scalar variables of a procedure in SSA form. SSA graphs are a suitable representation for code generation: First, SSA graphs capture acyclic and cyclic information flow beyond basic block boundaries. Second, SSA graphs often arise naturally in modern compilers, as the intermediate code representation is usually already in SSA form. Third, output dependencies or anti-dependencies do not exist in SSA graphs.

As even acyclic SSA graphs are generally not restricted to a tree, no dynamic programming approach can be employed for instruction code selection. To get a handle on instruction code selection for SSA graphs, below we will discuss an approach based on a problem reduction to a quadratic mathematical programming problem (PBQP). Consider the code fragment of a dot-product routine and the corresponding SSA graph shown in Fig. 19.3. The code implements a simple vector dot-product using fixed-point arithmetic. Nodes in the SSA graph represent a single operation while edges describe the flow of data that is produced at the source node and consumed at the target node. Incoming edges have an order which reflects the argument order of the operation.

The example in Fig. 19.3 has fixed-point computations that have to be expressed in the grammar. For fixed-point values most arithmetic and bit-wise operations are identical to their integer equivalents. However, some operations have different semantics, e.g., multiplying two fixed-point values in format $m.i$ results in a value with $2i$ fractional digits. The result of the multiplication has to be adjusted by a shift operation to the right (LSR). To accommodate for fixed-point values, we add the following rules to the grammar introduced in Fig. 19.2a:

rule	cost	instruction
reg ← VAR	is_fixed_point ? ∞ otherwise 0	
fp ← VAR	is_fixed_point ? 0 otherwise ∞	
fp2 ← MUL(fp, fp)	1	MUL Rd, Rm, Rs
fp ← fp2	1	LSR Rd, Rm, i
fp ← ADD(fp, fp)	1	ADD Rd, Rm, Rs
fp2 ← ADD(fp2, fp2)	1	ADD Rd, Rm, Rs
fp ← PHI(fp, . . .)	0	
fp2 ← PHI(fp2, . . .)	0	

In the example, the accumulation for double-precision fixed-point values (fp2) can be performed at the same cost as for the single-precision format (fp). Thus, it would be beneficial to move the necessary shift from the inner loop to the return

---

[1] We consider its data-based representation here, see Chap. 14.

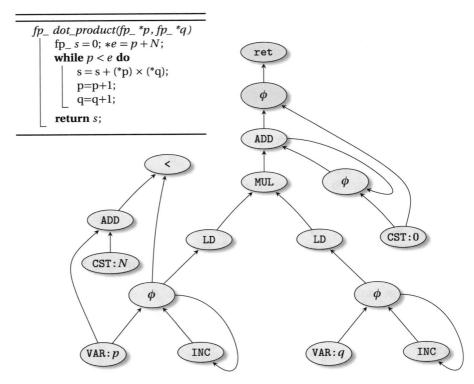

Fig. 19.3 Instruction code selection SSA Graph for a vector dot-product in fixed-point arithmetic. "fp_" stands for unsigned short fixed-point type and "*" for pointer manipulations (like in C). The colours of the nodes indicate which basic block the operations belong to

block, performing the intermediate calculations in the extended format. Note that a tree-pattern matcher generates code at statement level, and hence the information of having values as double-precision cannot be hoisted across basic block boundaries. However, an instruction code selector that is operating on the SSA graph is able to propagate non-terminal fp2 across the ϕ node prior to the return and emits the code for the shift to the right in the return block.

In the following, we will explain how to perform instruction code selection on SSA graphs by means of a specialized quadratic assignment problem (PBQP). First, we discuss the instruction code selection problem by employing a discrete optimization problem called a partitioned boolean quadratic problem. An extension of *patterns* to arbitrary acyclic graph structures, which we refer to as DAG grammars, is discussed in Sect. 19.2.1.

19.1 Instruction Code Selection for Tree Patterns on SSA Graphs

The matching problem for SSA graphs is reduced to a discrete optimization problem called a Partitioned Boolean Quadratic Problem (PBQP). First, we will introduce the PBQP problem and then we will describe the mapping of the instruction code selection problem to PBQP.

19.1.1 Partitioned Boolean Quadratic Problem

Partitioned boolean quadratic programming (PBQP) is a generalized quadratic assignment problem that has proven to be effective for a wide range of applications in embedded code generation, e.g., register assignment, address mode selection, or bank selection for architectures with partitioned memory. Instead of problem-specific algorithms, these problems can be modelled in terms of generic PBQPs that are solved using a common solver library. PBQP is flexible enough to model the irregularities of embedded architectures that are hard to cope with using traditional heuristic approaches.

Consider a set of discrete variables $X = \{x_1, \ldots, x_n\}$ and their finite domains $\{\mathbb{D}_1, \ldots, \mathbb{D}_n\}$. A solution of PBQP is a mapping h of each variable to an element in its domain, i.e., an element of $\mathbb{D}_i$ needs to be chosen for variable x_i. The chosen element imposes *local costs* and *related costs* with neighbouring variables. Hence, the quality of a solution is based on the contribution of two sets of terms.

1. For assigning variable x_i to the element d_i in $\mathbb{D}_i$. The quality of the assignment is measured by a *local cost function* $c(x_i, d_i)$.
2. For assigning two related variables x_i and x_j to the elements $d_i \in \mathbb{D}_i$ and $d_j \in \mathbb{D}_j$. We measure the quality of the assignment with a *related cost function* $C(x_i, x_j, d_i, d_j)$.

The total cost of a solution h is given as

$$f = \sum_{1 \le i \le n} c(x_i, h(x_i)) + \sum_{1 \le i < j \le n} C\left(x_i, x_j, h(x_i), h(x_j)\right). \tag{19.1}$$

The PBQP problem seeks an assignment of variables x_i with minimum total costs.

In the following we represent both the local cost function and the related cost function in matrix form, i.e., the related cost function $C(x_i, x_j, d_i, d_j)$ is decomposed for each pair (x_i, x_j). The costs for the pair are represented as $|\mathbb{D}_i|$-by-$|\mathbb{D}_j|$ matrix/table $\mathscr{C}_{ij}$. A matrix element corresponds to an assignment (d_i, d_j). Similarly, the local cost function $c(x_i, d_i)$ is represented by a cost vector $\overrightarrow{c_i}$ enumerating the costs of the elements. A PBQP problem has an underlying graph structure, expressed as graph $G = (V, E, C, c)$, which we refer to as a PBQP graph.

For each decision variable x_i we have a corresponding node $v_i \in V$ in the graph, and for each cost matrix $\mathscr{C}_{i,j}$ that is not the zero matrix, we introduce an edge $e = (v_i, v_j)$. The cost functions c and C map nodes and edges to the original cost vectors and matrices, respectively. We will present an example later in this chapter in the context of instruction code selection.

In general, finding a solution to this minimization problem is NP-hard. However, in many practical cases, the PBQP instances are sparse, i.e., many of the cost matrices $\mathscr{C}_{i,j}$ are zero matrices and do not contribute to the overall solution. Thus, optimal or near-optimal solutions can often be found within reasonable time limits. Currently, there are two algorithmic approaches for PBQP that have been proven to be efficient in practice for instruction code selection problems, i.e., a polynomial-time heuristic algorithm and a branch-&-bound based algorithm with exponential worst-case complexity. For a certain subclass of PBQP, the algorithm produces provably optimal solutions in time $\mathcal{O}(nm^3)$, where n is the number of discrete variables and m is the maximal number of elements in their domains, i.e., $m = \max(|\mathbb{D}_1|, \ldots, |\mathbb{D}_n|)$. For general PBQPs, however, the solution may not be optimal. To still obtain an optimal solution outside the subclass, branch-&-bound techniques can be applied.

19.1.2 Instruction Code Selection with PBQP

In this section, we describe the modelling of instruction code selection for SSA graphs as a PBQP problem. In the basic modelling, SSA and PBQP graphs coincide. The variables x_i of the PBQP are decision variables reflecting the choices of applicable rules (represented by $\mathbb{D}_i$) for the corresponding node of x_i. The local costs reflect the costs of the rules, and the related costs reflect the costs of chain rules making rules compatible with each other. This means that the number of decision vectors and the number of cost matrices in the PBQP are determined by the number of nodes and edges in the SSA graph, respectively. The sizes of $\mathbb{D}_i$ depend on the number of rules in the grammar. A solution for the PBQP instance induces a complete cost-minimal cover of the SSA graph.

As in traditional tree-pattern matching, an ambiguous graph grammar consisting of tree patterns with associated costs and semantic actions is used. Input grammars have to be *normalized*. This means that each rule is either a so-called *base rule* or a *chain rule*. A base rule is a production of the form $\mathrm{nt}_0 \leftarrow OP(\mathrm{nt}_1, \ldots, \mathrm{nt}_{k_p})$ where nt_i are non-terminals and OP is a terminal symbol, i.e., an operation represented by a node in the SSA graph. A chain rule is a production of the form $\mathrm{nt}_0 \leftarrow \mathrm{nt}_1$, where nt_0 and nt_1 are non-terminals. A production rule $\mathrm{nt} \leftarrow OP_1(\alpha, OP_2(\beta), \gamma))$ can be normalized by rewriting the rule into two production rules $\mathrm{nt} \leftarrow OP_1(\alpha, \mathrm{nt}', \gamma)$ and $\mathrm{nt}' \leftarrow OP_2(\beta)$ where nt' is a new non-terminal symbol and α, β, and γ denote arbitrary pattern fragments. This transformation can be iteratively applied until all production rules are either chain rules or base rules. To illustrate this transformation, consider the grammar in Fig. 19.4, which is

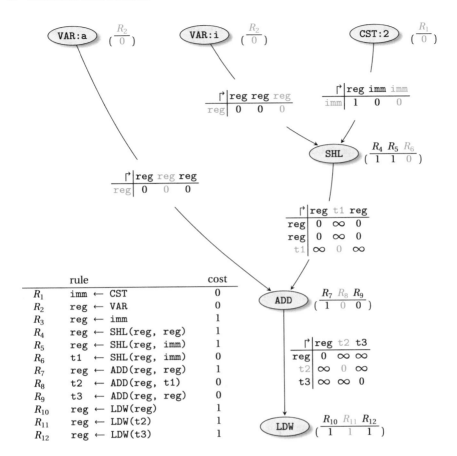

Fig. 19.4 PBQP instance derived from the example shown in Fig. 19.2. The grammar has been normalized by introducing additional non-terminals. Highlighted elements show a cost-minimal solution

a normalized version of the tree grammar introduced in Fig. 19.2a. Temporary non-terminal symbols t1, t2, and t3 are used to decompose larger tree patterns into simple base rules. Each base rule spans across a single node in the SSA graph.

The instruction code selection problem for SSA graphs is modelled in PBQP as follows. For each node u in the SSA graph, a PBQP variable x_u is introduced. The domain of variable x_u is determined by the subset of base rules whose terminal symbol matches the operation of the SSA node, e.g., there are three rules (R_4, R_5, R_6) that can be used to cover the shift operation SHL in our example. The last rule is the result of automatic normalization of a more complex tree pattern. The cost vector $\vec{c_u} = w_u \cdot \langle c(R_1), \ldots, c(R_{k_u}) \rangle$ of variable x_u encodes the local costs for a particular assignment where $c(R_i)$ denotes the associated cost of base rule R_i. Weight w_u is used as a parameter to optimize for various objectives including speed (e.g., w_u is the expected execution frequency of the operation at node u) and space (e.g., the w_u is set to one). In our example, both R_4 and R_5 have associated costs of one. Rule R_6 contributes no local costs, as we account for the full costs of a complex tree pattern at the root node. All nodes have the same weight of one, thus the cost vector for the SHL node is $\langle 1, 1, 0 \rangle$.

An edge in the SSA graph represents data transfer between the result of an operation u, which is the source of the edge, and the operand v, which is the tail of the edge. To ensure consistency among base rules and to account for the costs of chain rules, we impose costs that are dependent on the selection of variable x_u and variable x_v in the form of a cost matrix $\mathscr{C}_{uv}$. An element in the matrix corresponds to the costs of selecting a specific base rule $r_u \in R_u$ of the result and a specific base rule $r_v \in R_v$ of the operand node. Assume that r_u is nt $\leftarrow OP(\ldots)$ and r_v is $\cdots \leftarrow OP(\alpha, \text{nt}', \beta)$ where nt' is the non-terminal of operand v whose value is obtained from the result of node u. There are three possible cases:

1. If the non-terminals nt and nt' are identical, the corresponding element in matrix $\mathscr{C}_{uv}$ is zero, since the result of u is compatible with the operand of node v.
2. If the non-terminals nt and nt' differ and there exists a rule $r : \text{nt}' \leftarrow \text{nt}$ in the transitive closure of all chain rules, the corresponding element in $\mathscr{C}_{uv}$ has the costs of the chain rule, i.e., $w_v \cdot c(r)$.
3. Otherwise, the corresponding element in $\mathscr{C}_{uv}$ has infinite costs prohibiting the selection of incompatible base rules.

As an example, consider the edge from CST:2 to node SHL in Fig. 19.4. There is a single base rule R_1 with local cost 0 and result non-terminal imm for the constant. Base rules R_4, R_5, and R_6 are applicable for the shift, the first of which expects non-terminal reg as its second argument, and rules R_5 and R_6 both expect imm. Consequently, the corresponding cost matrix accounts for the cost of converting from reg to imm at index $(1, 1)$ and is zero otherwise.

Highlighted elements in Fig. 19.4 show a cost-minimal solution of the PBQP with cost one. A solution of the PBQP directly induces a selection of base and chain rules for the SSA graph. The execution of the semantic action rules inside a basic block follows the order of basic blocks. Special care is necessary for chain rules that correspond to data flow across basic blocks. Such chain rules may be placed inefficiently, and a placement algorithm is required for some grammars.

19.2 Extensions and Generalizations

19.2.1 Instruction Code Selection for DAG Patterns

In the previous section we have introduced an approach based on code patterns that resemble simple tree fragments. This restriction often complicates code generators

for modern CPUs with specialized instructions and SIMD extensions, e.g., there is
no support for machine instructions with multiple results.

Consider the introductory example shown in Fig. 19.3. Many architectures have
some form of auto-increment addressing modes. On such a machine, the load and
the increment of both p and q can be done in a single instruction benefiting both
code size and performance. However, post-increment loads cannot be modelled
using a single tree-shaped pattern. Instead, it produces multiple results and spans
across two non-adjacent nodes in the SSA graph, with the only restriction that their
arguments have to be the same.

Similar examples can be found in most architectures, e.g., the DIVU instruction
in the Motorola 68K architecture performs the division and the modulo operation for
the same pair of inputs. Other examples are the RMS (read-modify-store) instructions
on the IA32/AMD64 architecture, autoincrement- and decrement-addressing modes
of several embedded systems architectures, the IRC instruction of the HPPA
architecture, or fsincos instructions of various math libraries. Compiler writers
are forced to pre- or post-process these patterns heuristically, often missing much
of the optimization potential. These architecture-specific tweaks also complicate
re-targeting, especially in situations where patterns are automatically derived from
generic architecture descriptions.

We will now outline, through the example in Fig. 19.5, a possible problem
formulation for these generalized patterns in the PBQP framework discussed so
far. The code fragment contains three feasible instances of a post-increment store
pattern. Assuming that p, q, and r point to mutually distinct memory locations,
there are no further dependencies apart from the edges shown in the SSA graph. If
we select *all* three instances of the post-increment store pattern concurrently, the
cover induced by SSA edges becomes cyclic, and the code cannot be emitted. To
overcome this difficulty, the idea is to express in the PBQP model a numbering of
chosen nodes that reflects the existence of a topological order in the cover-avoiding
cycles. PBQP has no constraints as such, but they can be simulated by imposing

Fig. 19.5 DAG patterns may
introduce cyclic data
dependencies

$*p = r + 1;$
$*q = p + 1;$
$*r = q + 1;$

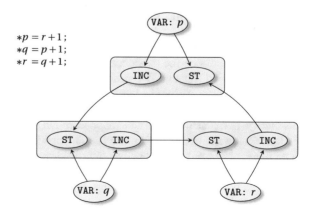

arbitrary high costs denoted by ∞ for certain combinations that do not satisfy the topological constraint.

Modelling

The first step is to explicitly enumerate *instances* of complex patterns, i.e., concrete tuples of nodes that match the terminal symbols specified in a particular production. There are three instances of the post-increment store pattern (surrounded by boxes) in the example shown in Fig. 19.5. As for tree patterns, DAG patterns are decomposed into simple base rules for the purpose of modelling, e.g., the post-increment store pattern

P_1: tmt $\leftarrow$ ST(x: reg, reg), reg $\leftarrow$ INC(x) : 3

is decomposed into the individual pattern fragments

$P_{1,1}$: stmt $\leftarrow$ ST(reg, reg)
$P_{1,2}$: reg $\leftarrow$ INC(reg)

For our modelling, new variables are created for each enumerated instance of a complex production. They encode whether a particular instance is chosen or not, i.e., the domain basically consists of the elements on and off. The local costs are set to the combined costs for the particular pattern for the on state and to 0 for the off state. Furthermore, the domain of existing nodes is augmented with the base rule fragments obtained from the decomposition of complex patterns. We can safely squash all identical base rules obtained from this process into a single state. Thus, each of these new states can be seen as a proxy for the whole set of instances of (possibly different) complex productions including the node. The local costs for these proxy states are set to 0.

Continuing our example, the PBQP for the SSA graph introduced in Fig. 19.5 is shown in Fig. 19.6. In addition to the post-increment store pattern with costs three, we assume regular tree patterns for the store and the increment nodes with costs two denoted by P_2 and P_3, respectively. Rules for the VAR nodes are omitted for simplicity.

Nodes 1–6 correspond to the nodes in the SSA graph. Their domain is defined by the simple base rule with costs two and the proxy state obtained from the decomposition of the post-increment store pattern. Nodes 7, 8, and 9 correspond to the three instances identified for the post-increment store pattern. As noted before, we have to guarantee the existence of a topological order in the cover among the chosen nodes. To this end, we refine the state on such that it reflects a particular index in a concrete topological order. Matrices among these nodes account for data dependencies, e.g., consider the matrix established among nodes 7 and 8. Assuming instance 7 is on at index 2 (i.e., mapped to on_2), the only remaining choices for instance 8 are not to use the pattern (i.e., mapped to off) or to enable it at index 3 (i.e., mapped to on_3), as node 7 has to precede node 8.

Additional cost matrices are required to ensure that the corresponding proxy state is selected on all the variables forming a particular pattern instance (which can be modelled with combined costs of 0 or ∞, respectively). However, this formulation allows for the trivial solution where all of the related variables encoding the selection

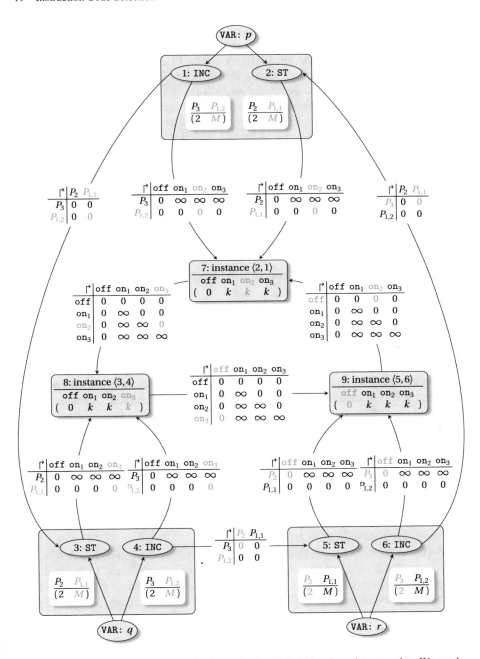

Fig. 19.6 PBQP Graph for the example shown in Fig. 19.5. M is a large integer value. We use k as a shorthand for the term $3 - 2M$

of a complex pattern are set to off (accounting for 0 costs) even though the artificial proxy state has been selected. We can overcome this problem by adding a large integer value M to the costs for all proxy states. In exchange, we subtract these costs from the cost vector of instances. Thus, the penalties for the proxy states are effectively eliminated unless an invalid solution is selected.

Cost matrices among nodes 1–6 do not differ from the basic approach discussed before and reflect the costs of converting the non-terminal symbols involved. It should be noted that for general grammars and irreducible graphs, the heuristic solver of PBQP cannot guarantee delivering a solution that satisfies all constraints modelled in terms of ∞ costs. This would be an NP-complete problem. One way to work around this limitation is to include a small set of rules that cover each node individually and that can be used as a fallback rule in situations where no feasible solution has been obtained, which is similar to macro substitution techniques and ensures a correct but possibly non-optimal matching. These limitations do not apply to exact PBQP solvers such as the branch-&-bound algorithm. It is also straightforward to extend the heuristic algorithm with a backtracking scheme on RN reductions, which would of course also be exponential in the worst case.

19.3 Concluding Remarks and Further Reading

Aggressive optimizations for the instruction code selection problem are enabled by the use of SSA graphs. The whole flow of a function is taken into account for instruction selection rather than a local scope of statements. The move from basic tree-pattern matching [1] to SSA-based DAG matching is a relatively small step as long as a PBQP library and some basic infrastructure (graph grammar translator, etc.) are provided. The complexity of the approach is hidden in the discrete optimization problem called PBQP. Free PBQP libraries are available from the web-pages of the authors and a library is implemented as part of the LLVM [186] framework.

Many aspects of the PBQP formulation presented in this chapter could not be covered in detail. The interested reader is referred to the relevant literature [108, 109] for an in-depth discussion.

As we move from acyclic linear code regions to whole-functions, it becomes less clear in which basic block the selected machine instructions should be emitted. For chain rules, the obvious choices are often non-optimal. In [254], a polynomial-time algorithm based on generic network flows is introduced that allows a more efficient placement of chain rules across basic block boundaries. This technique is orthogonal to the generalization to complex patterns.

Chapter 20
If-Conversion

Christian Bruel

Very long instruction word (VLIW) or explicitly parallel instruction computing (EPIC) architectures make instruction-level Parallelism (ILP) visible within the Instruction Set Architecture (ISA), relying on static schedulers to organize the compiler output such that multiple instructions can be issued in each cycle.

If-conversion is the process of transforming a control-flow region with conditional branches into an equivalent predicated or speculated sequence of instructions (into a region of basic blocks, possibly single) referred to as a Hyperblock. If-converted code replaces control dependencies by data dependencies and thus exposes instruction-level Parallelism very naturally within the new region at the software level.

Removing control hazards improves performance in several ways. When the misprediction penalty is removed, the instruction fetch throughput is increased and the instruction cache locality is improved. Enlarging the size of basic blocks allows earlier execution of long latency operations and the merging of multiple control-flow paths into a single flow of execution that can later be exploited by scheduling frameworks such as VLIW scheduling, hyperblock scheduling, or modulo scheduling.

Consider the simple example given in Fig. 20.1, which represents the execution of an `if-then-else-end` statement on a 4-issue processor with non-biased branches. In this figure, $r = q \ ? \ r_1 : r_2$ stands for a `select` instruction where r is assigned r_1 if q is true, and r_2 otherwise. With standard basic block ordering, assuming that all instructions have a one-cycle latency, the schedule height rises from five cycles in the most optimistic case to six cycles. After if-conversion, the execution path is reduced to four cycles with no branches, regardless of the test

C. Bruel (✉)
STMicroelectronics, Grenoble, France
e-mail: christian.bruel@st.com

© The Author(s), under exclusive license to Springer Nature Switzerland AG 2022   269
F. Rastello, F. Bouchez Tichadou (eds.), *SSA-based Compiler Design*,
https://doi.org/10.1007/978-3-030-80515-9_20

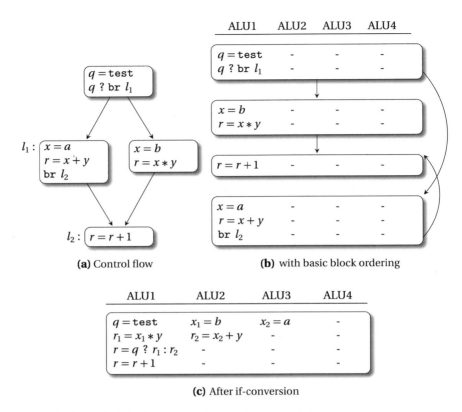

(a) Control flow   **(b)** with basic block ordering

(c) After if-conversion

Fig. 20.1 Static schedule of a simple region on a 4-issue processor

outcome, and assuming a very optimistic one-cycle branch penalty. The main benefit here is that it can be executed without branch disruption.

From this introductory example, we can observe that:

- The two possible execution paths have been merged into a single execution path, implying a better exploitation of the available resources.
- The schedule height has been reduced, because instructions can be control-speculated before the branch.
- The variables have been renamed, and a *merge* pseudo-instruction has been introduced.

Thanks to SSA, the merging point is already materialized in the original control flow as a ϕ pseudo-instruction, and register renaming has been performed by SSA construction. Given this, the transformation to generate if-converted code seems natural locally. Exploiting these properties on larger-scale control-flow regions requires a framework that we will develop further.

20.1 Architectural Requirements

The *merge* pseudo-operations need to be mapped to a conditional form of execution in the target's architecture. As illustrated in Fig. 20.2, we differentiate the following three models of conditional execution:

- *Fully predicated execution*: Any instruction can be executed conditionally on the value of a predicate operand.
- *(Control) speculative execution*: The instruction is executed unconditionally and then committed using conditional move (cmov or select) instructions.
- *Partially predicated execution*: Only a subset of the ISA is predicated, usually memory operations that are not easily speculated; other instructions are speculated.

In this figure, we use the notation $r = c ? r_1 : r_2$ to represent a select -like operation. Its semantic is identical to the gating ϕ_{if}-function presented in Chap. 14: r takes the value of r_1 if c is true, and r_2 otherwise. Similarly, we also use the notation $c ? r = op$ to represent the predicated execution of op if the predicate c is true; $\bar{c} ? r = op$ if the predicate c is false.

To be speculated, an instruction must not have any side effects or hazards. For instance, a memory load must not trap because of an invalid address. Memory operations are a major impediment to if-conversion. This is regrettable, because like any other long latency instructions, speculative loads can be very effective in fetching data earlier in the instruction stream, thereby reducing stalls. Modern architectures provide architectural support to dismiss invalid address exceptions. Examples are the ldw.d dismissible load operation in the Multiflow Trace series of computers, or in the STMicroelectronics ST231 processor, but also the speculative load of the Intel IA64. The main difference is that with a dismissible model, invalid memory access exceptions are not delivered, which can be problematic in an embedded or kernel environment that relies on memory exception for correct behaviour. A speculative model serves to catch the exception thanks to the token bit check instruction. Some architectures, such as the IA64, offer both speculative and predicated memory operations. Stores can also be executed conditionally by speculating part of their address value, with additional constraints on the ordering of the memory operations due to possible aliases between the two paths. Figure 20.3 shows examples of various forms of speculative memory operations.

$p ? x = a + b$	$t_1 = a + b$	$x = a + b$
$\bar{p} ? x = a * b$	$t_2 = a * b$	$t = a * b$
	$x = p ? t_1 : t_2$	$x = \text{cmov } p, t$
(a) fully predicated	**(b)** speculative using *select*	**(c)** speculative using *cmov*

Fig. 20.2 Conditional execution using different models

$t = \texttt{ld.s}(addr)$
$\texttt{chk.s (t)}$
$p \, ? \, x = t$

(a) IA64 speculative load

$t = \texttt{ldw.d}(addr)$
$x = \texttt{select} \, p \, ? \, t : x$

(b) Multiflow/ST231 dismissible load

$x \; = \; \texttt{select} \, p \; ? \; addr \; :$
$dummy$
$\texttt{stw}(x, value)$

(c) base store hoisting

$index = \texttt{select} \, p \, ? \, i : j$
$\texttt{stw}(x[index], value)$

(d) index store hoisting

Fig. 20.3 Examples of speculated memory operations

Note that the `select` instruction is an architecture instruction that does not need to be replaced during the SSA destruction phase. If the target architecture does not provide such a gating instruction, it can be emulated using two conditional moves. This translation can be done afterwards, and the `select` instruction can still be used as an intermediate form. It allows the program to stay in full SSA form where all the data dependencies are made explicit, and can thus be fed to all SSA optimizers.

This chapter is organized as follows. We begin by describing the SSA techniques to convert a CFG region into SSA form to produce an if-converted SSA representation using speculation. We then describe how this framework is extended to use predicated instructions, using the ψ-SSA form presented in Chap. 15. Finally, we outline a global framework to pull together these techniques, incrementally enlarging the scope of the if-converted region to its maximum beneficial size.

20.2   Basic Transformations

Unlike global approaches that identify a control-flow region and if-convert it in one shot, the technique described in this chapter is based on incremental reductions. To this end, we consider basic SSA transformations whose goal is to isolate a simple diamond-DAG structure (informally an `if-then-else-end`) that can be easily if-converted. The complete framework, which identifies and incrementally performs the transformation, is described in Sect. 20.3.

20.2.1   SSA Operations on Basic Blocks

The basic transformation that actually if-converts the code is the ϕ *removal*, which takes a simple diamond-DAG as an input, i.e., a single-entry node/single-exit node (SESE) DAG with only two distinct forward paths from its entry node to its

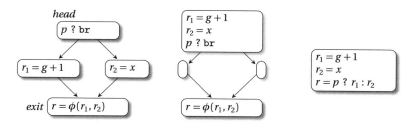

Fig. 20.4 ϕ removal

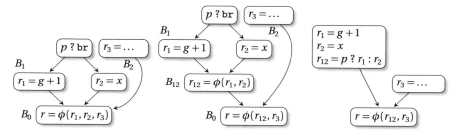

Fig. 20.5 ϕ reduction

exit node. The ϕ removal consists in (1) speculating the code of both branches in the entry basic block (denoted *head*); then (2) replacing the ϕ-function by a `select`; and finally (3) simplifying the control flow to a single basic block. This transformation is illustrated in Fig. 20.4.

The goal of the ϕ *reduction* transformation is to isolate a diamond-DAG from a structure that resembles a diamond-DAG but has side entries to its exit block. This diamond-DAG can then be reduced using the ϕ removal transformation. Nested `if-then-else-end` in the original code can create such a control-flow region. Of note is the similarity with the nested arity-two ϕ_{if}-functions used for gated SSA (see Chap. 14). In the most general case, the joint node of the considered region has n direct predecessors with ϕ-functions of the form $B_0 : r = \phi(B_1 : r_1, B_2 : r_2, \ldots, B_n : r_n)$ and is such that removing edges from $B_3, \ldots, B_n$ would give a diamond-DAG. After the transformation, B_1 and B_2 point to a freshly created basic block, say B_{12}, that itself points to B_0; a new variable $B_{12} : r_{12} = \phi(B_1 : r_1, B_2 : r_2)$ is created in this new basic block; the ϕ-function in B_0 is replaced by $B_0 : r = \phi(B_{12} : r_{12}, \ldots, B_n : r_n)$. This is illustrated in Fig. 20.5.

The goal of *path duplication* is to isolate a diamond-DAG from a structure that resembles a diamond-DAG but has side exit edges. Through path duplication, all edges that point to a node different from the exit node or to the willing entry node are "redirected" to the exit node. ϕ reduction can then be applied to the region obtained. More formally, consider two distinguished nodes, the first named *head* and the second a single-exit node of the region named *exit*, such that there are exactly two different control-flow paths from *head* to *exit*; consider (if it exists) the first node

$side_i$ on one of the forward paths $head \rightarrow side_0 \rightarrow \ldots side_p \rightarrow exit$, which has at least two direct predecessors. The transformation duplicates the path $P = side_i \rightarrow \cdots \rightarrow side_p \rightarrow exit$ into $P' = side'_i \rightarrow \cdots \rightarrow side'_p \rightarrow exit$ and redirects $side_{i-1}$ (or $head$ if $i = 0$) to $side'_i$. All the ϕ-functions that are along P and P' for which the number of direct predecessors has changed have to be updated accordingly. Hence, a $r = \phi(side_p : r_1, B_2 : r_2, \ldots, B_n : r_n)$ in $exit$ will be updated into $r = \phi(side'_p : r_1, B_2 : r_2, \ldots, B_n : r_n, side_p : r_1)$; a $r = \phi(side_{i-1} : r_0, r_1, \ldots, r_m)$ originally in $side_i$ will be updated into $r = \phi(r_1, \ldots, r_m)$ in $side_i$ and into $r = \phi(r_0)$, i.e., $r = r_0$ in $side'_i$. Variable renaming (see Chap. 5) along with copy folding can then be performed on P and P'. All steps are illustrated in Fig. 20.6.

The last transformation, namely the *conjunctive predicate merge*, concerns the if-conversion of a control-flow pattern that sometimes appears on codes to represent logical `and` or `or` conditional operations. As illustrated in Fig. 20.7, the goal is to isolate a diamond-DAG from a structure that resembles a diamond-DAG but has side exit edges. As opposed to path duplication, the transformation is actually restricted

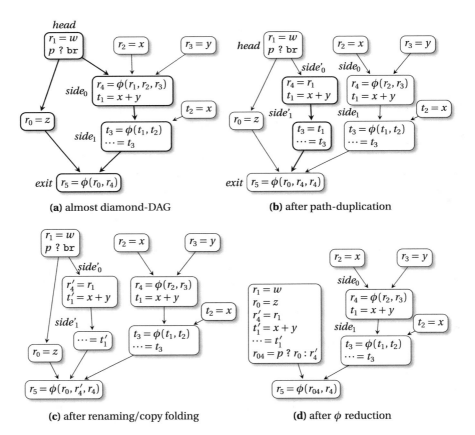

Fig. 20.6 Path duplication

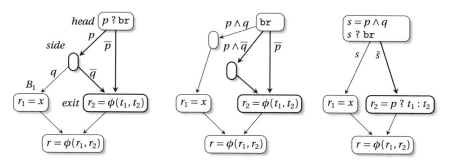

Fig. 20.7 Convergent conjunctive merge

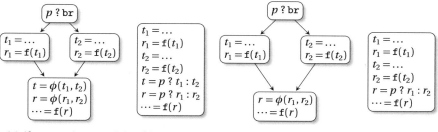

(**a**) if-conversion on minimal SSA   (**b**) if-conversion on pruned SSA

Fig. 20.8 SSA predicate minimality

to a very simple pattern, highlighted in Fig. 20.7, made up of three distinct basic blocks: *head*, which branches with predicate p to *side*, or *exit*. *side*, which is empty and branches itself with predicate q to another basic block outside of the region, or *exit*. Conceptually, the transformation can be understood as first isolating the outgoing path $p \to q$ and then if-converting the obtained diamond-DAG.

Implementing the same framework on a non-SSA-form program would require more effort: The ϕ reduction would do the renaming, involving either a global data-flow analysis or the insertion of copies at the *exit* node of the diamond-DAG; inferring the minimum amount of select operations would also require the upkeep of liveness information. SSA form solves the renaming issue without additional effort, and, as illustrated in Fig. 20.8, the minimality and the pruned type of the SSA form avoid inserting useless select operations.

20.2.2 Handling of Predicated Execution Model

The ϕ removal transformation described above involved a speculative execution model. As we will illustrate below, in the context of a predicated execution model, the choice of speculation versus predication is an optimization decision that should not be imposed by the intermediate representation. Also, transforming speculated code into predicated code can be viewed as a coalescing problem. The use of ψ-SSA (see Chap. 15) as the intermediate form of if-conversion serves to postpone the decision to speculate some code, while the coalescing problem is naturally handled by the ψ-SSA destruction phase.

Just as (control) speculating an operation on a control-flow graph means ignoring the control dependence with the conditional branch, speculating an operation on an if-converted code is the same as removing the data dependence with the corresponding predicate. On the other hand, after register allocation, speculation adds anti-dependencies. This trade-off is illustrated in the example of Fig. 20.9: For the fully predicated version of the code, the computation of p has to be done before the computations of x_1 and x_2; speculating the computation of x_1 removes the dependence with p and allows it to be executed in parallel with the test $(a \nleq b)$; if the computation of both x_1 and x_2 is speculated, they cannot be coalesced, and when destruction of ψ-SSA, the ψ-function will give rise to some select instructions; if only the computation of x_1 is speculated, then x_1 and x_2 can be coalesced to x, but then an anti-dependence from $x = a + b$ and $p\,?\,x = c$ appears, forbidding its execution in parallel.

In practice, speculation is performed during the ϕ removal transformation, whenever it is possible (operations with side effect cannot be speculated) and considered as beneficial. As illustrated in Fig. 20.10b, only the operations part of

$$p = (a \nleq b)$$
$$p\,?\,x_1 = a + b$$
$$\overline{p}\,?\,x_2 = c$$
$$x = \psi(p\,?\,x_1, \overline{p}\,?\,x_2)$$

(a) predicated code

$$p = (a \nleq b)$$
$$x_1 = a + b$$
$$x_2 = c$$
$$x = \psi(p\,?\,x_1, \overline{p}\,?\,x_2)$$

(b) fully speculated

$$p = (a \nleq b)$$
$$x_1 = a + b$$
$$\overline{p}\,?\,x_2 = c$$
$$x = \psi(p\,?\,x_1, \overline{p}\,?\,x_2)$$

(c) partially speculated

$$p = (a \nleq b)$$
$$x = a + b$$
$$\overline{p}\,?\,x = c$$

(d) after coalescing

Fig. 20.9 Speculation removes the dependency with the predicate but adds anti-dependencies between concurrent computations

if q **then**
$$p = (a \nleq b)$$
$$x_1 = a + b$$
$$\overline{p}\,?\,x_2 = c$$
$$x = \psi(x_1, \overline{p}\,?\,x_2)$$
$$d_1 = \mathtt{f}(x)$$
else
$$d_2 = 3$$
$$d = \phi(d_1, d_2)$$

(a) nested if

$$p = (a \nleq b)$$
$$x_1 = a + b$$
$$\overline{p}\,?\,x_2 = c$$
$$x = \psi(x_1, \overline{p}\,?\,x_2)$$
$$d_1 = \mathtt{f}(x)$$

$$\overline{q}\,?\,d_2 = 3$$

$$d = \psi(q\,?\,d_1, \overline{q}\,?\,d_2)$$

(b) speculating the **then** branch

$$p = (a \nleq b)$$
$$s = q \wedge \overline{p}$$
$$q\,?\,x_1 = a + b$$
$$s\,?\,x_2 = c$$
$$x = \psi(q\,?\,x_1, s\,?\,x_2)$$
$$q\,?\,d_1 = \mathtt{f}(x)$$

$$\overline{q}\,?\,d_2 = 3$$

$$d = \psi(q\,?\,d_1, \overline{q}\,?\,d_2)$$

(c) predicating both branches

Fig. 20.10 Inner region ψ-functions

one of the diamond-DAG branches are actually speculated. This partial speculation leads to the manipulation of predicated code.

Speculating code is the easiest part, as it could be done prior to the actual if-conversion by simply hoisting the code above the conditional branch. Still, it is worth pointing out that since ψ-functions are part of the intermediate representation, they can be considered for inclusion in a candidate region for if-conversion, and in particular for speculation. However, the strength of ψ-SSA allows ψ-functions to be treated just like any other operation. Consider the code in Fig. 20.10a containing a subregion that has already been processed. To speculate the operation $d_1 = \mathtt{f}(x)$, the operation defining x, i.e., the ψ-function, also has to be speculated. Similarly, all the operations defining the operands x_1 and x_2 should also be speculated. If one of them can produce hazardous execution, then the ψ-function cannot be speculated, which in turn forbids the operation $d_1 = \mathtt{f}(x)$ from being speculated. Marking operations that cannot be speculated can be done easily using a forward propagation along def-use chains.

All operations that cannot be speculated, possibly including some ψ-functions, should be predicated. Suppose we are considering a non-speculated operation that we aim to if-convert and that is part of the `then` branch on predicate q. Just as for $x_2 = c$ in Fig. 20.10a, this operation might already be predicated (on $\overline{p}$ here) prior to the if-conversion. In that case, a *projection* on q is performed, meaning that instead of predicating $x_2 = c$ by $\overline{p}$ it gets predicated by $q \wedge \overline{p}$. A ψ-function can also be projected on a predicate q, as described in Chap. 15: All gates of each operand are individually projected on q. As an example, originally non-gating operand x_1 gets gated by q, while the $\overline{p}$-gated operand x_2 gets gated by $s = q \wedge \overline{p}$. Note that as opposed to speculating it, predicating a ψ-function does not impose predicating the operations that defined its operands. The only subtlety related to projection is the generation of the new predicate as the logical conjunction of the original guard (e.g., $\overline{p}$) and the current branch predicate (e.g., q). Here, s needs to be computed at some point. The heuristic consists in first listing the set of all necessary predicates and then emitting the corresponding code at the earlier place. Here, the used predicates are $q, \overline{q}$, and $q \wedge \overline{p}$, and q and $\overline{q}$ are already available. The earlier place where $q \wedge \overline{p}$ can be computed is just after calculating p.

Once operations have been speculated or projected (on q for the `then` branch, on $\overline{q}$ for the `else`), each ϕ-function at the merge point is replaced by a ψ-function: operands of speculated operations are placed first and guarded by true; operands of projected operations follow, guarded by the predicate of the corresponding branch.

20.3 Global Analysis and Transformations

Frequently executed regions are rarely just composed of simple `if-then-else` control-flow regions, but processors have limited resources: the number of registers will determine the acceptable level of data dependencies to minimize register pressure; the number of predicate registers will determine the depth of the if-conversion so that the number of conditions does not exceed the number of available predicates; and the number of processing units will determine the number of instructions that can be executed simultaneously. The inner–outer incremental process advocated in this chapter serves to evaluate precisely the profitability of if-conversion.

20.3.1 SSA Incremental If-Conversion Algorithm

The algorithm takes as input a CFG in SSA form and applies incremental reductions using the list of candidate-conditional basic blocks sorted in post-order. Each basic block in the list designates the head of a sub-graph that can be if-converted using the transformations described in Sect. 20.2. Post-order traversal serves to process each region from the inner to the outer. When the if-converted region cannot grow anymore because of resources, or because a basic block cannot be if-converted, then the next sub-graph candidate is considered until the entire CFG is explored. Note that as the reduction proceeds, maintaining SSA can be done using the general technique described in Chap. 5. Basic local ad hoc updates can also be implemented instead.

Consider for example the CFG reported in Fig. 20.11a. The exit node B_6 and basic block B_3 (which contains a function call) cannot be if-converted. The post-order list of conditional blocks (represented in bold) is [B_9, B_{14}, B_{13}, B_{11}, B_8, B_7, B_5, B_2]. (1) The first candidate region is composed of $\{B_9, B_2, B_{10}\}$; ϕ reduction can be applied, promoting the instructions of B_{10} in B_9; B_2 becomes the single direct successor of B_9. (2) The region headed by B_{14} is then considered; B_{15} cannot yet be promoted because of the side entries coming both from B_{12} and B_{13}; B_{15} is duplicated into a $B_{15'}$ with B_2 as direct successor; $B_{15'}$ can then be promoted into B_{14}, which now has a single direct successor B_2. (3) The region headed by B_{13}, which has B_{14} and B_{15} as direct successors, is now considered; B_{15} is again duplicated into $B_{15'}$', so as to promote B_{14} and $B_{15'}$ into B_{13} through ϕ reduction; $B_{15'}$ already contains predicated operations from the previous transformation, so a new merging predicate is computed and inserted. After the completion of ϕ removal, B_{13} has a unique direct successor, B_2. (4) B_{11} is the head of the new candidate region; here, B_{12} and B_{13} can be promoted. Again, since B_{13} contains predicated and predicate setting operations, a fresh predicate must be created to hold the merged conditions. (5) B_8 is then considered; B_{11} needs to be duplicated to $B_{11'}$. The process finishes with the region head B_7.

20.3.2 Tail Duplication

As shown in Fig. 20.11, some basic blocks (such as B_3) may have to be excluded from the region to if-convert. *Tail duplication* can be used for this purpose. Similar

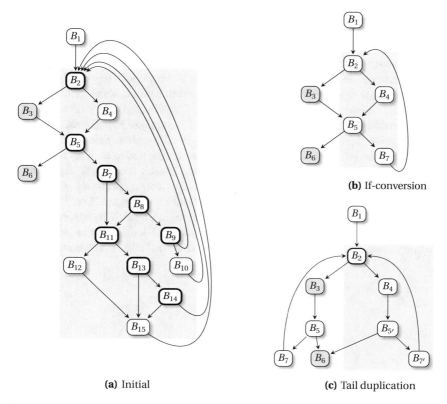

(b) If-conversion

(a) Initial   **(c)** Tail duplication

Fig. 20.11 If-conversion of wc (*word count* program). Basic blocks in the highlighted region cannot be if-converted. Tail duplication can be used to exclude B_3 from the to-be-if-converted region

to path duplication described in Sect. 20.1, the goal of tail duplication is to get rid of the incoming edges of a region to if-convert. This is usually done in the context of hyperblock formation, a technique that, unlike the inner-outer incremental technique described in this chapter, consists in if-converting a region in "one shot." Consider again the example of Fig. 20.11a, and suppose that the set of selected basic blocks defining the region to if-convert consists of all basic blocks from B_2 to B_{15} excluding B_3 and B_6. Getting rid of the incoming edge from B_3 to B_5 is possible by duplicating all basic blocks of the region reachable from B_5, as shown in Fig. 20.11c.

Consider a region $\mathscr{R}$ made up of a set of basic blocks, a distinguished one *entry* and the others denoted $(B_i)_{2 \leq i \leq n}$, such that any B_i is reachable from *entry* in $\mathscr{R}$. Suppose a basic block B_s has some direct predecessors $out_1, \ldots, out_m$ that are not in $\mathscr{R}$. Tail duplication involves the following steps: (1) for all B_j (including B_s) reachable from B_s in $\mathscr{R}$, create a basic block B_j' as a copy of B_j; (2) any branch from B_j' that points to a basic block B_k of the region is rerouted to its duplicate B_k';

(3) any branch from a basic block out_k to B_s is rerouted to B'_s. In our example, we would have $entry = B_2$, $B_s = B_6$, and $out = B_4$.

A global approach would follow the steps in Fig. 20.11c: First, select the region; second, get rid of the incoming edges using tail duplication; and finally, perform if-conversion of the whole region in one shot. It is worth pointing out that there is no phasing issue with tail duplication. To illustrate this point, consider the example of Fig. 20.12 where B_2 cannot be if-converted. The selected region is made up of all other basic blocks. Using a global approach as in standard hyperblock formation, tail duplication would be performed prior to any if-conversion. This would result in the CFG on the left part of the figure. Note that a new node, B_7, has been added here after the tail duplication by a process called branch coalescing. Applying if-conversion on the two disjoint regions, whose heads are, respectively, B_4 and $B_{4'}$, would result in the final code shown at the bottom of the figure. Our incremental scheme would first perform if-conversion of the region headed by B_4, resulting in the code depicted in the CFG on the right. Applying tail duplication to get rid of the side entry from B_2 would result in exactly the same final code at the bottom.

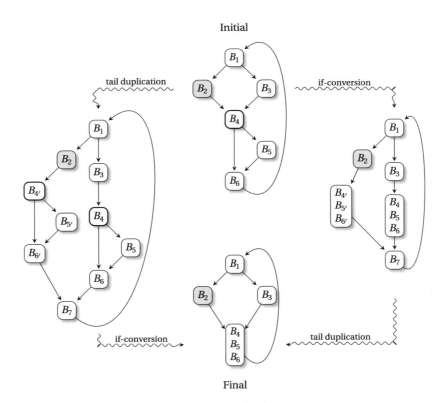

Fig. 20.12 The absence of phasing issue for tail duplication

20.3.3 *Profitability*

Fusing execution paths can overcommit the architectural ability to execute the multiple instructions in parallel: Data dependencies and register renaming introduce new register constraints. Moving operations earlier in the instruction stream increases live ranges. Aggressive if-conversion can easily exceed processor resources, leading to excessive register pressure or moving infrequently used long latency instructions into the critical path. The prevalent idea is that a region can be if-converted if the cost of the resulting if-converted basic block is smaller than the cost of each individual basic block of the original region weighted by their respective execution probability. To evaluate these costs, we consider all possible paths impacted by the if-conversion.

For all transformations except the conjunctive predicate merge, there are two such paths starting at the basic block *head*. For the code in Fig. 20.13, we would have $path_p = [head, B_1, exit[$ and $path_{\overline{p}} = [head, B'_1, B'_2, exit[$ of respective probabilities $\text{prob}(p)$ and $\text{prob}(\overline{p})$. For a path $P_q = [B_0, B_1, \ldots, B_n[$ of probability $\text{prob}(q)$, its cost is given by $\widehat{P_q} = \text{prob}(q) \times \sum_{i=0}^{n-1} [\widehat{B_i, B_{i+1}}[$, where $[\widehat{B_i, B_{i+1}}[$ represents the cost of basic block $[\widehat{B_i}]$ estimated using its schedule height plus the branch latency br_lat, if the edge (B_i, B_{i+1}) corresponds to a conditional branch, 0 otherwise. Note that if B_i branches to S_q on predicate q and falls through to $S_{\overline{q}}$, we have

$$\widehat{B_i} = \text{prob}(q) \times \left([\widehat{B_i, S_q}[\right) + \text{prob}(\overline{q}) \times \left([\widehat{B_i, S_{\overline{q}}}[\right) = [\widehat{B_i}] + \text{prob}(q) \times \text{br\_lat}.$$

Let $path_p = [head, B_1, \ldots, B_n, exit[$ and $path_{\overline{p}} = [head, B'_1, \ldots, B'_m, exit[$ be the two paths starting at the basic block head, where p is the branch predicate. Then, the overall cost before if-conversion simplifies to the following:

$$\text{cost}_{\text{control}} = \widehat{path}_p + \widehat{path}_{\overline{p}}$$

$$= [\widehat{head}] + \text{prob}(p) \times \left(\text{br\_lat} + \sum_{i=0}^{n} [\widehat{B_i}] \right) + \text{prob}(\overline{p}) \times \sum_{i=0}^{m} [\widehat{B'_i}].$$

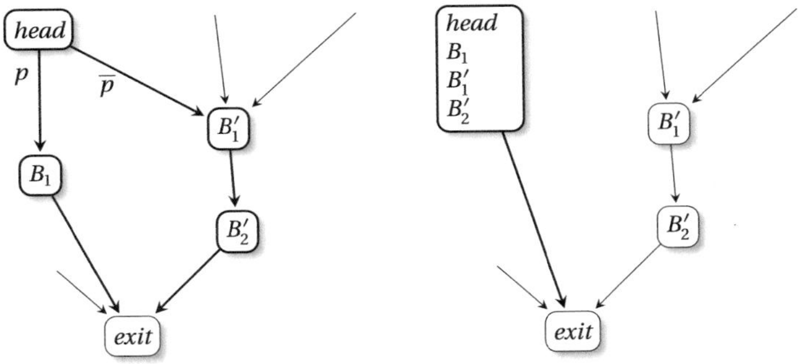

Fig. 20.13 ϕ reduction

Fig. 20.14 Conjunctive predicate merge

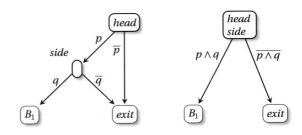

This is to be compared to the cost after if-conversion:

$$\text{cost}_{\text{predicated}} = \left[head \circ \left(\bigcirc_{i=1}^{n} \widehat{B_i} \right) \circ \left(\bigcirc_{i=1}^{m} \widehat{B_i'} \right) \right],$$

where $\circ$ is the composition function that merges basic blocks together, removes associated branches, and creates the predicate operations.

The profitability of the logical conjunctive merge in Fig. 20.14 can be evaluated similarly. There are three paths impacted by the transformation: $path_{p \wedge q} = [head, side, B_1[$, $path_{p \wedge \overline{q}} = [head, side, exit[$, and $path_{\overline{p}} = [head, exit[$ of respective probabilities $\text{prob}(p \wedge q)$, $\text{prob}(p \wedge \overline{q})$, and $\text{prob}(\overline{p})$. The overall cost before the transformation (if branches are on p and q) $\widehat{path_{p \wedge q}} + \widehat{path_{p \wedge \overline{q}}} + \widehat{path_{\overline{p}}}$ simplifies to

$$\text{cost}_{\text{control}} = \widehat{head} + \widehat{side} = \widehat{[head]} + \text{prob}(p) \times (1 + \text{prob}(q)) \times \text{br\_lat},$$

which should be compared to (if the branch on the new *head* block is on $p \wedge q$)

$$\text{cost}_{\text{predicated}} = \widehat{head \circ side} = \widehat{[head \circ side]} + \text{prob}(p) \times \text{prob}(q) \times \text{br\_lat}.$$

Note that if $\text{prob}(p) \ll 1$, emitting a conjunctive merge might not be beneficial. In that case, another strategy such as path duplication from the *exit* block will be evaluated. Profitability for any conjunctive predicate merge (disjunctive or conjunctive; convergent or not) is evaluated similarly.

A speed-oriented objective function needs the target machine description to derive the instruction latencies, resource usage, and scheduling constraints. The local dependencies computed between instructions are used to compute the dependence height. The branch probability is obtained from static branch prediction heuristics, profile information, or user-inserted directives. Naturally, this heuristic can be either pessimistic, because it does not take into account new optimization opportunities introduced by the branch removal or explicit new dependencies, or optimistic because of inaccurate register pressure estimation leading to register spilling on the critical path, or uncertainty in the branch prediction. But since the SSA incremental if-conversion framework reduces the scope for the decision function to a localized part of the CFG, the size and complexity of the inner region under consideration make the profitability a comprehensive process. This cost function is fast enough to be reapplied to each region during the incremental processing, with the advantage that all the instructions introduced by the if-conversion process in the inner regions, such as new predicate merging instructions or new temporary pseudo-registers, can be taken into account.

20.4 Concluding Remarks and Further Reading

In this chapter, we presented how an if-conversion algorithm can take advantage of the SSA properties to efficiently assign predicates and lay out the new control flow in an incremental, inner–outer process. As opposed to the alternative top-down approach, the region selection can be reevaluated at each nested transformation, using local analysis. Hyperblocks [191] were proposed as the primary if-converted scheduling framework, excluding basic blocks that do not justify their inclusion in the if-converted flow of control. Region selection and if-conversion can be performed as a single pass, hyperblocks being created lazily, with well-known techniques such as tail duplication or branch coalescing used only when the benefit is established. Predication and speculation are often presented as two different alternatives for if-conversion. They can coexist in an efficient if-conversion process such that every model of conditional execution is accepted in the same framework, thanks to the conditional moves and ψ [275] transformations.

Chapter 21
SSA Destruction for Machine Code

Fabrice Rastello

Chapter 3 provides a basic algorithm for destructing SSA that suffers from several limitations and drawbacks: first, it works under implicit assumptions that are not necessarily fulfiled at machine level; second, it must rely on subsequent phases to remove the numerous copy operations it inserts; finally, it subsequently increases the size of the intermediate representation, thus making it unsuitable for just-in-time compilation.

Correctness

SSA at machine level complicates the process of destruction that can potentially lead to bugs if not performed carefully. The algorithm described in Sect. 3.2 involves the splitting of every critical edge. Unfortunately, because of specific architectural constraints, region boundaries, or exception handling code, edge splitting is not always possible. As we will see further on, this obstacle could easily be overcome by appending copy operations at the very beginning and very end of basic blocks. Unfortunately, appending a copy operation at the very end of a basic block might not be possible either (it has to be before the jump operation). Also, care must be taken with duplicated edges, i.e., when the same basic block appears twice in the list of direct predecessors. This can occur after control-flow graph structural optimizations such as dead code elimination or empty block elimination, etc.

SSA imposes a strict discipline on variable naming: every "name" must be associated with only one definition which, most of the time, is obviously not compatible with the instruction set of the target architecture. As an example, a two-address mode instruction such as auto-increment ($x = x + 1$) would force its definition to use the same resource as one of its arguments (defined elsewhere), thus imposing two different definitions for the same temporary variable. This is

F. Rastello (✉)
Inria, Grenoble, France
e-mail: fabrice.rastello@inria.fr

why some compiler designers prefer using, for SSA construction, the notion of versioning in place of renaming. Implicitly, two versions of the same original variable should not interfere, while two names can. Such a flavour corresponds to the C-SSA form described in Chap. 2. The former simplifies the SSA destruction phase, while the latter simplifies and allows more transformations to be performed under SSA (updating C-SSA is very difficult) . Apart from dedicated registers for which optimizations are usually very careful in managing their live range, register constraints related to calling conventions or instruction set architecture might be handled by the register allocation phase. However, as we will see, enforcement of register constraints impacts the register pressure as well as the number of copy operations. For those reasons we may want those constraints to be expressed earlier (such as for the pre-pass scheduler), in which case the SSA destruction phase might have to cope with them.

Code Quality

The natural way of lowering ϕ-functions and expressing register constraints is through the insertion of copies (when edge splitting is not mandatory as discussed above). If done carelessly, the resulting code will contain many temporary-to-temporary copy operations. In theory, reducing the number of these copies is the role of the coalescing during the register allocation phase. A few memory and time-consuming existing coalescing heuristics mentioned in Chap. 22 are quite effective in practice. The difficulty comes both from the size of the interference graph (the information of colourability is spread out) and from the presence of many overlapping live ranges that carry the same value (so are non-interfering). With less effort, coalescing can also be performed prior to the register allocation phase. As opposed to a (so-called conservative) coalescing during register allocation, this aggressive coalescing would not cope with the interference graph colourability. As we will see, strict SSA form is really helpful for both computing and representing equivalent variables. This makes the SSA destruction phase the right candidate for eliminating (or not inserting) those copies.

Speed and Memory Footprint

The cleanest and simplest way to perform SSA destruction with good code quality is to first insert copy instructions to make the SSA form conventional, then take advantage of the SSA form to efficiently run aggressive coalescing (without breaking the conventional property), before eventually renaming ϕ-webs and getting rid of ϕ-functions. Unfortunately, in a transitional stage this approach will lead to an intermediate representation with a substantial number of variables: The liveness sets and the interference graph classically used to perform coalescing become prohibitively large for dynamic compilation. To overcome this difficulty liveness and interference can be computed on demand, which, as we already mentioned, is made simpler by the use of SSA form (see Chap. 9). There remains the process of copy insertion itself that might still take a substantial amount of time. To fulfil memory and time constraints imposed by just-in-time compilation, one idea is to *virtually* insert those copies, and only *effectively* insert the non-coalesced ones.

This chapter addresses these three issues: handling of machine-level constraints, code quality (elimination of copies), and algorithm efficiency (speed and memory footprint). The layout falls into three corresponding sections.

21.1   Correctness

Isolating ϕ-Node Using Copies

In most cases, edge splitting can be avoided by treating ϕ-uses and ϕ-definition operands symmetrically: Instead of just inserting copies on the incoming control-flow edges of the ϕ-*node* (one for each use operand), a copy is also inserted on the outgoing edge (one for its defining operand). This has the effect of isolating the value associated with the ϕ-node, thus avoiding (as discussed further on) SSA destruction issues such as the well-known lost-copy problem. The process of ϕ-node isolation is illustrated by Fig. 21.1. The corresponding pseudo-code is given in Algorithm 21.1. If, because of different ϕ-functions, several copies are introduced at the same place, they should be viewed as parallel copies. For that reason, an empty parallel copy is initially inserted both at the beginning (i.e., right after ϕ-functions, if any) and at the end of each basic block (i.e., just before the branching operation, if any). Note that, as far as correctness is concerned, these copies can be sequentialized in any order, as they concern different variables (this is a consequence of ϕ-node isolation—see below).

When incoming edges are not split, it is important to insert a copy not only for each argument of the ϕ-function but also for its result: Without the copy $a'_0 \leftarrow a_0$, the ϕ-function defines directly a_0 whose live range can be long enough to intersect the live range of some a'_i, $i > 0$. Prior SSA destruction algorithms that have not performed the copy $a'_0 \leftarrow a_0$ have identified two problems. (1) In the "lost-copy problem", a_0 is used in a direct successor of $B_i \neq B_0$, and the edge from B_i to B_0 is *critical*. (2) In the "swap problem," a_0 is used in B_0 as a ϕ-function argument. In this latter case, if parallel copies are used, a_0 is dead before a'_i is defined. But, if copies are sequentialized blindly, the live range of a_0 can go beyond the definition

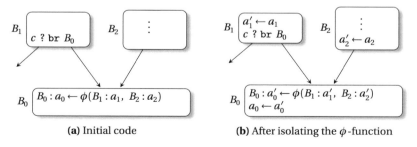

(a) Initial code   (b) After isolating the ϕ-function

Fig. 21.1 Isolation of a ϕ-node

point of a'_i and lead to an incorrect code after renaming a_0 and a'_i with the same name. ϕ-node isolation can be used to solve most of the issues that may be faced at machine level. However, the subtleties listed below remain.

Limitations

A tricky case is where the basic block contains variables *defined after* the point of copy insertion. This, for example, is the case of the PowerPC `bclr` branch instructions with a behaviour similar to hardware loop. In addition to the condition, a counter u is decremented by the instruction itself. If u is used in a ϕ-function in a direct successor block, no copy insertion can split its live range. It must then be given the same name as the variable defined by the ϕ-function. If both variables interfere, this is simply impossible! For example, suppose that for the code in Fig. 21.2a, the instruction selection chooses a branch with decrement (denoted `br_dec`) for Block B_1 (Fig. 21.2b). Then, the ϕ-function of Block B_2, which uses u, cannot be translated out of SSA by standard copy insertion because u interferes with t_1 and its live range cannot be split. To destruct SSA, one could add $t_1 \leftarrow u - 1$ in Block B_1 to anticipate the branch. Or one could split the critical edge between B_1 and B_2 as in Fig. 21.2c. In other words, simple copy insertions are not enough in this case. We see several alternatives to solve the problem: (1) The SSA optimization could be designed with more care; (2) the counter variable must not be promoted to SSA; (3) some instructions must be changed; (4) the control-flow edge must be split somehow.

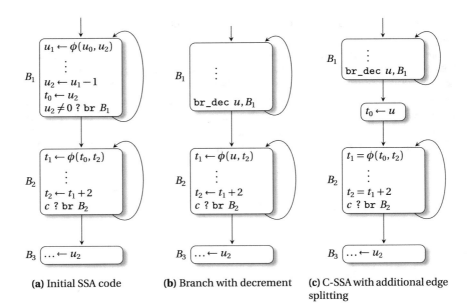

(a) Initial SSA code   (b) Branch with decrement   (c) C-SSA with additional edge splitting

Fig. 21.2 Copy insertion may not be sufficient. `br_dec` u, B_1 decrements u, then branches to B_1 if $u \neq 0$

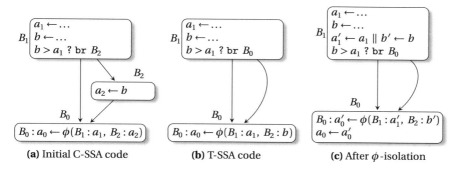

Fig. 21.3 Copy folding followed by empty block elimination can lead to SSA code for which destruction is not possible through simple copy insertion

Another tricky case is when a basic block has the same direct predecessor block twice. This can result from consecutively applying copy folding and control-flow graph structural optimizations such as dead code elimination or empty block elimination. This is the case for the example of Fig. 21.3 where copy folding would remove the copy $a_2 \leftarrow b$ in Block B_2. If B_2 is eliminated, there is no way to implement the control dependence of the value to be assigned to a_3 other than through predicated code (see Chaps. 15 and 14) or through the reinsertion of a basic block between B_1 and B_0 by splitting one of the edges.

The last difficulty SSA destruction faces when performed at machine level is related to register constraints such as instruction set architecture (ISA) or application binary interface (ABI) constraints. For the sake of the discussion we differentiate two kinds of resource constraints that we will refer as *operand pinning* and *live range pinning*. The live range pinning of a variable v to resource R will be represented by R_v, just as if v were a version of temporary R. An operand pinning to a resource R will be represented using the exponent $^{\uparrow R}$ on the corresponding operand. Live range pinning expresses the fact that the *entire* live range of a variable must reside in a given resource (usually a dedicated register). Examples of live range pinning are versions of the stack-pointer temporary that must be assigned back to register *SP*. On the other hand, the pinning of an operation's operand to a given resource does not impose anything on the live range of the corresponding variable. The scope of the constraint is restricted to the operation. Examples of operand pinning are operand constraints such as *two-address mode* where two operands of one instruction must use the same resource, or where an operand must use a given register. This last case encapsulates ABI constraints.

Note that looser constraints where the live range or the operand can reside in more than one resource are not handled here. We assume that the handling of this latter constraint is the responsibility of the register allocation. We first simplify the problem by transforming any operand pinning into a live range pinning, as sketched

$$p_2^{\uparrow T} \leftarrow p_1^{\uparrow T} + 1$$

$$T_{p_1} \leftarrow p_1$$
$$T_{p_2} \leftarrow T_{p_1} + 1$$
$$p_2 \leftarrow T_{p_2}$$

(a) Operand pinning of an auto-increment

(b) Corresponding live range pinning

$$a^{\uparrow R0} \leftarrow f(b^{\uparrow R0}, c^{\uparrow R1})$$

$$R0_{b'} = b \parallel R1_{c'} = c$$
$$R0_{a'} = f(R0_{b'}, R1_{c'})$$
$$a = R0_{a'}$$

(c) Operand pinning of a function call

(d) Corresponding live range pinning

Fig. 21.4 Operand pinning and corresponding live range pinning

in Fig. 21.4: Parallel copies with new variables pinned to the corresponding resource are inserted just before (for use operand pinning) and just after (for definition-operand pinning) the operation.

Detection of Strong Interferences
The scheme we propose in this section to perform SSA destruction that deals with machine level constraints does not address compilation cost (in terms of speed and memory footprint). It is designed to be simple. It first inserts parallel copies to isolate ϕ-functions and operand pinning. Then it checks for interferences that may persist. We will denote such interferences as *strong*, as they cannot be tackled through the simple insertion of temporary-to-temporary copies in the code. We consider that fixing strong interferences should be done on a case-by-case basis and restrict the discussion here to their detection.

As far as correctness is concerned, Algorithm 21.1 splits the data flow between variables and ϕ-nodes through the insertion of copies. For a given ϕ-function $a_0 \leftarrow \phi(a_1, \ldots, a_n)$, this transformation is correct as long as the copies can be inserted close enough to the ϕ-function. This might not be the case if the insertion point (for a use operand) of copy $a_i' \leftarrow a_i$ is not dominated by the definition point of a_i (such as for argument u of the ϕ-function $t_1 \leftarrow \phi(u, t_2)$ for the code in Fig. 21.2b); symmetrically, it will not be correct if the insertion point (for the definition-operand) of copy $a_0 \leftarrow a_0'$ does not dominate all the uses of a_0. More precisely, this leads to the insertion of the following tests in Algorithm 21.1:

- Line 9: "**if** the definition of a_i does not dominate PC_i **then** continue."
- Line 16: "**if** one use of a_0 is not dominated by PC_0 **then** continue."

For the discussion, we will denote as *split operands* the newly created local variables to differentiate them from the ones concerned by the two previous cases (designated as *non-split operands*). We suppose that a similar process has been performed for operand pinning to express them in terms of live range pinning with (when possible) very short live ranges around the concerned operations.

At this point, the code is still under SSA, and the goal of the next step is to check that it is conventional: This will obviously be the case only if all the variables of a

ϕ-web can be coalesced together. But this is not the only constraint: The set of all variables pinned to a common resource must also be interference free. We say that x and y are *pinned-ϕ-related* to one another if they are ϕ-related or if they are pinned to a common resource. The transitive closure of this relation defines an equivalence relation that partitions the variables defined locally in the procedure into equivalence classes, the pinned-ϕ-webs. Intuitively, the pinned-ϕ-equivalence class of a resource represents a set of resources "connected" via ϕ-functions and resource pinning. The computation of ϕ-webs given by Algorithm 3.4 can be generalized easily to compute pinned-ϕ-webs. The resulting pseudo-code is given by Algorithm 21.2.

Algorithm 21.1: Algorithm making non-conventional SSA form conventional by isolating ϕ-nodes

---

 1 **foreach** B: basic block of the CFG **do**
 2 $\qquad$ insert an empty parallel copy at the beginning of B
 3 $\qquad$ insert an empty parallel copy at the end of B

 4 **foreach** B_0: basic block of the CFG **do**
 5 $\qquad$ **foreach** ϕ-function at the entry of B_0 of the form $a_0 = \phi(B_1 : a_1, \ldots, B_n : a_n)$ **do**
 6 $\qquad\qquad$ **foreach** a_i (argument of the ϕ-function corresponding to B_i) **do**
 7 $\qquad\qquad\qquad$ **let** PC_i be the parallel copy at the end of B_i
 9
10 $\qquad\qquad\qquad$ **let** a_i' be a freshly created variable
11 $\qquad\qquad\qquad$ add copy $a_i' \leftarrow a_i$ to PC_i
12 $\qquad\qquad\qquad$ replace a_i by a_i' in the ϕ-function;

13 $\qquad\qquad$ **begin**
14 $\qquad\qquad\qquad$ **let** PC_0 be the parallel copy at the beginning of B_0
16
17 $\qquad\qquad\qquad$ **let** a_0' be a freshly created variable
18 $\qquad\qquad\qquad$ add copy $a_0 \leftarrow a_0'$ to PC_0
19 $\qquad\qquad\qquad$ replace a_0 by a_0' in the ϕ-function
 $\qquad\qquad\quad$ ▷ *all a_i' can be coalesced and the ϕ-function removed*

Now we need to check that each web is interference free. A web contains variables and resources. The notion of interferences between two variables is the one discussed in Sect. 2.6 for which we will propose an efficient implementation later in this chapter. A variable and a physical resource do not interfere while two distinct physical resources interfere with one another.

If any interference has been discovered, it has to be fixed on a case-by-case basis. Note that some interferences such as the one depicted in Fig. 21.3 can be detected and handled initially (through edge splitting if possible) during the copy insertion phase.

Algorithm 21.2: The pinned-ϕ-webs discovery algorithm, based on the union-find pattern

---

1 **for** each resource R **do**
2 $\quad$ web$(R) \leftarrow \{R\}$

3 **for** each variable v **do**
4 $\quad$ web$(v) \leftarrow \{v\}$
5 $\quad$ **if** v pinned to a resource R **then**
6 $\quad\quad$ union(web(R), web(v))

7 **for** each instruction of the form $a_{\text{dest}} = \phi(a_1, \ldots, a_n)$ **do**
8 $\quad$ **for** each source operand a_i in instruction **do**
9 $\quad\quad$ union(web(a_{dest}), web(a_i))

21.2 Code Quality

Once the code is in conventional SSA, the correctness problem is solved: Destructing it is by definition straightforward, as it consists in renaming all variables in each ϕ-web into a unique representative name and then removing all ϕ-functions. To improve the code, however, it is important to remove as many copies as possible.

Aggressive Coalescing

Aggressive coalescing can be treated with standard non-SSA coalescing technique. Indeed, conventional SSA allows us to coalesce the set of all variables in each ϕ-web together. Coalesced variables are no longer SSA variables, but ϕ-functions can be removed. Liveness and interferences can then be defined as for a regular code (with parallel copies). An interference graph (as depicted in Fig. 21.5e) can be used. A solid edge between two nodes (e.g., between x_2 and x_3) materializes the presence of an interference between the two corresponding variables, i.e., expressing the fact that they cannot be coalesced and share the same resource. A dashed edge between two nodes materializes an affinity between the two corresponding variables, i.e., the presence of a copy (e.g., between x_2 and x_2') that could be removed by their coalescing.

This process is illustrated by Fig. 21.5: the isolation of the ϕ-function leads to the insertion of the three copies that respectively define x_1', define x_3', and use x_2'; the corresponding ϕ-web $\{x_1', x_2', x_3'\}$ is coalesced into a representative variable x; according to the interference graph in Fig. 21.5e, x_1, x_3 can then be coalesced with x leading to the code in Fig. 21.5c.

Liveness Under SSA

If the goal is not to destruct SSA completely but remove as many copies as possible while maintaining the conventional property, liveness of ϕ-function operands should reproduce the behaviour of the corresponding non-SSA code as if the variables of the ϕ-web were coalesced all together. The semantic of the ϕ-operator in the

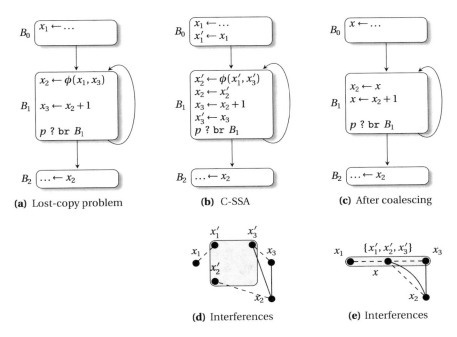

Fig. 21.5 SSA destruction for the lost-copy problem

so-called *multiplexing* mode fits the requirements. The corresponding interference graph on our example is depicted in Fig. 21.5d.

Definition 21.1 (Multiplexing Mode) Let a ϕ-function $B_0 : a_0 = \phi(B_1 : a_1, \ldots, B_n : a_n)$ be in *multiplexing* mode, then its liveness follows the following semantic: Its definition-operand is considered to be at the entry of B_0, in other words variable a_0 is live-in of B_0; its use operands are at the exit of the corresponding direct predecessor basic blocks, in other words, variable a_i for $i > 0$ is live-out of basic block B_i.

Value-Based Interference

As outlined earlier, after the ϕ-isolation phase and the treatment of operand pinning constraints, the code contains many overlapping live ranges that carry the same value. Because of this, to be efficient coalescing must use an accurate notion of interference. As already mentioned in Chap. 2, the ultimate notion of interference contains two dynamic (i.e., execution-related) notions: the notion of liveness and the notion of value. Analysing statically if a variable is live at a given execution point or if two variables carry identical values is a difficult problem. The scope of variable coalescing is usually not so large, and graph colouring-based register allocation commonly takes the following conservative test: *Two variables interfere if one is live at a definition point of the other and this definition is not a copy between the two variables.*

It can be noted that, with this conservative interference definition, when a and b are coalesced, the set of interferences of the new variable may be strictly smaller than the union of interferences of a and b. Thus, simply merging the two corresponding nodes in the interference graph is an over-approximation with respect to the interference definition. For example, in a block with two successive copies $b = a$ and $c = a$ where a is defined before, and b and c (and possibly a) are used after, it is considered that b and c interfere but that neither of them interfere with a. However, after coalescing a and b, c should no longer interfere with the coalesced variable. Hence the interference graph would have to be updated or rebuilt.

However, in SSA, each variable has, statically, a *unique* value, given by its unique definition. Furthermore, the "has-the-same-value" binary relation defined on variables is, if the SSA form fulfils the dominance property , an equivalence relation. The *value* of an equivalence class[1] is the variable whose definition dominates the definitions of all other variables in the class. Hence, using the same scheme as in SSA copy folding, finding the value of a variable can be done by a simple topological traversal of the dominance tree: When reaching an assignment of a variable b, if the instruction is a copy $b = a$, $V(b)$ is set to $V(a)$, otherwise $V(b)$ is set to b. The interference test is now both simple and accurate (no need to rebuild/update after a coalescing): if live(x) denotes the set of program points where x is live,

a *interferes with* b *if* live(a) *intersects* live(b) *and* $V(a) \neq V(b)$.

The first part reduces to def(a) $\in$ live(b) or def(b) $\in$ live(a) thanks to the dominance property. In the previous example, a, b, and c have the same value $V(c) = V(b) = V(a) = a$, thus they do not interfere.

Note that our notion of values is limited to the live ranges of SSA variables, as we consider that each ϕ-function defines a new variable. We could propagate information through a ϕ-function when its arguments are equivalent (same value). But we would face the complexity of global value numbering (see Chap. 11). By comparison, our equality test in SSA comes for free.

21.3   Speed and Memory Footprint

Implementing the technique of the previous section may be considered too costly. First, it inserts many instructions before realizing that most are useless. Also, copy insertion is already in itself time-consuming. It introduces many new variables, too: The size of the variable universe has an impact on the liveness analysis and the interference graph construction. Finally, if a general coalescing algorithm is used, a graph representation with adjacency lists (in addition to the bit matrix) and a working graph to explicitly merge nodes when coalescing variables, would be

---

[1] Dominance property is required here, e.g., consider the following loop body if($i \neq 0$) {$b \leftarrow a$; } $c \leftarrow \ldots$; $\cdots \leftarrow b$; $a \leftarrow c$; the interference between b and c is actual.

required. All these constructions, updates, and manipulations are time-consuming and memory-consuming. We may improve the whole process by: (a) avoiding the use of any interference graph and liveness sets; (b) avoiding the quadratic complexity of interference checks between two sets of variables by adopting an optimistic approach that first coalesces all copy-related variables (even interfering ones), then traverses each set of coalesced variables and un-coalesces all the interfering ones one by one; (c) emulating ("virtualizing") the introduction of the ϕ-related copies.

Interference Check

Liveness sets and the interference graph are the main source of memory usage. In the context of JIT compilation, this is a good reason not to build an interference graph at all, and rely on the liveness check described in Chap. 9 to test if two live ranges intersect or not. Let us suppose for this purpose that a "has-the-same-value" equivalence relation is available thanks to a mapping V of variables to symbolic values:

$$\text{variables } a \text{ and } b \text{ have the same value} \Leftrightarrow V(a) = V(b)$$

As explained in Paragraph 21.2, this can be done linearly (without requiring a hash map-table) on a single traversal of the program if under strict SSA form. We also suppose that the liveness check is available, meaning that for a given variable a and program point p, one can answer if a is live at this point through the boolean value of $a.islive(p)$. This can directly be used, under strict SSA form, to check if two variables live ranges intersect:

$$\text{intersect}(a, b) \Leftrightarrow liverange(a) \cap liverange(b) \neq \emptyset$$
$$\Leftrightarrow \begin{cases} a.def.op = b.def.op \\ a.def.op \text{ dominates } b.def.op \wedge a.islive\,(out(b.def.op)) \\ b.def.op \text{ dominates } a.def.op \wedge b.islive\,(out(a.def.op)) \end{cases}$$

Which leads to our refined notion of interference:

$$\text{interfere}(a, b) \Leftrightarrow \text{intersect}(a, b) \bigwedge V(a) \neq V(b)$$

De-coalescing in Linear Time

The interference check outlined in the previous paragraph serves to avoid building an interference graph of the SSA form program. However, coalescing has the effect of merging vertices, and interference queries are actually to be done between sets of vertices. To overcome this complexity issue, the technique proposed here is based on a de-coalescing scheme. The idea is to first merge all copy- and ϕ-function-related variables together. A merged set might contain interfering variables at this point. The principle is to identify some variables that interfere with some other variables within the merged set and remove them (along with the one they are pinned with) from the merged set. As we will see, thanks to the dominance property, this can be done linearly using a single traversal of the set.

In reference to register allocation, and graph colouring, we will associate the notion of colours with merged sets: All the variables of the same set are assigned the same colour, and different sets are assigned different colours. The process of *de-coalescing* a variable is to extract it from its set; it is not put in another set, just isolated. We will say *uncoloured*. Actually, variables pinned together have to stay together. We denote the (interference free) set of variables pinned to a common resource that contains variable v, atomic-merged-set(v). So the process of uncolouring a variable might have the effect of uncolouring some others. In other words, a coloured variable is to be coalesced with variables of the same colour, and any uncoloured variable v is to be coalesced only with the variables it is pinned with, i.e., atomic-merged-set(v).

We suppose that variables have already been coloured, and the goal is to uncolour some of them (preferably not all of them) so that each merged set becomes interference free. We suppose that if two variables are pinned together they have been assigned the same colour, and that a merged set cannot contain variables pinned to different physical resources. Here we focus on a single merged set and the goal is to make it interference free within a single traversal. The idea exploits the tree shape of variables' live ranges under strict SSA. To this end, variables are identified by their definition point and ordered accordingly using dominance.

Algorithm 21.3 performs a traversal of this set along the dominance order, enforcing at each step the subset of already considered variables to be interference free. From now, we will abusively design as the dominators of a variable v, the set of variables of *colour identical to v* whose definition dominates the definition of v. Variables defined at the same program point are arbitrarily ordered, so as to use the standard definition of immediate dominator (denoted v.idom, set to $\perp$ if they do not exist, updated lines 6–8). To illustrate the role of v.eanc in Algorithm 21.3, let us consider the example of Fig. 21.6 where all variables are assumed to be originally in the same merged set: v.eanc (updated line 16) represents the immediate intersecting dominator with the same value as v; so we have b.eanc $= \perp$ and d.eanc $= a$. When line 14 is reached, cur_anc (if not $\perp$) represents a dominating variable interfering with v and with the same value than v.idom: when v is set to c (c.idom $= b$), as b does not intersect c and as b.eanc $= \perp$, $cur\_anc = \perp$, which allows us to conclude that there is no dominating variable that interferes with c; when v is set to e, d does not intersect e but as a intersects and has the same value as d (otherwise a or d would have been uncoloured), we have d.eanc $= a$ and thus $cur\_anc = a$. This allows us to detect on line 18 the interference of e with a.

Virtualizing ϕ-Related Copies

The last step towards a memory-friendly and fast SSA destruction algorithm consists in emulating the initial introduction of copies and only actually inserting them on the fly when they appear to be required. We use *exactly the same algorithms as for the solution without virtualization*, and use a special location in the code, identified as a "virtual" parallel copy, where the real copies, if any, will be placed.

Algorithm 21.3: De-coalescing of a merged set

---

1 cur_idom = ⊥
2 **foreach** variable v of the merged set in DFS pre-order of the dominance tree **do**
3 | DeCoalesce(v, cur_idom)
4 |_   cur_idom ← v

5 **Function** *DeCoalesce(v, u)*
6 | **while** $(u \neq \bot) \bigwedge (\neg(u\text{dominates } v) \vee \text{uncolored}(u))$ **do** $u \leftarrow u.\text{idom}$
7 |
8 | $v.\text{idom} \leftarrow u$
9 | $v.\text{eanc} \leftarrow \bot$
10 | $cur\_anc \leftarrow v.\text{idom}$
11 | **while** $cur\_anc \neq \bot$ **do**
12 | | **while** $cur\_anc \neq \bot \bigwedge \neg (\text{colored}(cur\_anc) \wedge \text{intersect}(cur\_anc, v))$ **do**
13 | | |_ $cur\_anc \leftarrow cur\_anc.\text{eanc}$
14 | | **if** $cur\_anc \neq \bot$ **then**
15 | | | **if** $V(cur\_anc) = V(v)$ **then**
16 | | | | $v.\text{eanc} \leftarrow cur\_anc$
17 | | | |_   break
18 | | | **else**   ▷ *cur_anc and v interfere*
19 | | | | **if** preferable to uncolor v **then**
20 | | | | | uncolor atomic-merged-set(v)
21 | | | | |_   break
22 | | | | **else**
23 | | | | | uncolor atomic-merged-set(cur_anc)
24 | | | | |_ $cur\_anc \leftarrow cur\_anc.\text{eanc}$

Fig. 21.6 Variables' live
ranges are sub-trees of the
dominator tree

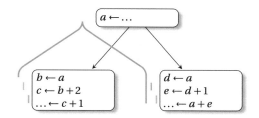

Because of this we consider a different semantic for ϕ-functions from the
multiplexing mode previously defined. To this end we differentiate ϕ-operands for
which a copy cannot be inserted (such as for the `br_dec` of Fig. 21.2b) from the
others. We use the terms non-split and split operands introduced in Sect. 21.1. For
a ϕ-function $B_0 : a_0 = \phi(B_1 : a_1, \ldots, B_n : a_n)$ and a split operand $B_i : a_i$, we
denote the program point where the corresponding copy would be inserted as the
early point of B_0 (early(B_0)—right after the ϕ-functions of B_0) for the definition-
operand, and as the *late point* of B_i (late(B_i)—just before the branching instruction)
for a use operand.

Definition 21.2 (Copy Mode) Let a ϕ-function $B_0 : a_0 = \phi(B_1 : a_1, \ldots, B_n : a_n)$ be in *copy mode*, then the liveness for any split operand follows the following semantic: Its definition-operand is considered to be at the early point of B_0, in other words, variable a_0 is *not* live-in of B_0; its use operands are at the late point of the corresponding direct predecessor basic blocks, in other words variable a_i for $i > 0$ is (unless used further) *not* live-out of basic block B_i. The liveness for non-split operands follows the multiplexing mode semantic.

When the algorithm decides that a virtual copy $a_i' \leftarrow a_i$ (resp. $a_0 \leftarrow a_0'$) cannot be coalesced, it is *materialized* in the parallel copy and a_i' (resp. a_0') becomes explicit in its merged set. The corresponding ϕ-operator is replaced and the use of a_i' (resp. def of a_0') is now assumed, as in the multiplexing mode, to be on the corresponding control-flow edge. This way, only copies that the first approach would finally leave un-coalesced are introduced. We choose to postpone the materialization of all copies along a single traversal of the program at the very end of the de-coalescing process. Because of the virtualization of ϕ-related copies, the de-coalescing scheme given by Algorithm 21.3 has to be adapted to emulate live ranges of split operands. The pseudo-code for processing a local virtual variable is given by Algorithm 21.5. The trick is to use the ϕ-function itself as a placeholder for its set of local virtual variables. As the live range of a local virtual variable is very small, the cases to consider are quite limited: A local virtual variable can interfere with another virtual variable (lines 2–4) or with a "real" variable (lines 5–18).

The overall scheme works as follows: (1) All copy-related (even virtual) variables are first coalesced (unless pinning to physical resources forbids this); (2) then merged sets (identified by their associated colour) are traversed and interfering variables are de-coalesced; (3) finally, materialization of the remaining virtual copies is performed through a single traversal of all ϕ-functions of the program: Whenever one of the two operands of a virtual copy is uncoloured, or whenever the colours are different, a copy is inserted.

A key implementation aspect is related to the handling of pinning. In particular, for the purpose of correctness, coalescing is performed in two separated steps. First pinned-ϕ-webs have to be coalesced. Detection and fixing of strong interferences are handled at this point. The merged sets thus obtained (that contain local virtual variables) have to be identified as atomic, i.e., they cannot be separated. After the second step of coalescing, atomic merged sets will compose larger merged sets. A variable cannot be de-coalesced from a set without also de-coalescing its atomic merged-set from the set. Non-singleton atomic merged sets have to be represented somehow. For ϕ-functions, the trick is again to use the ϕ-function itself as a placeholder for its set of local virtual variables: The pinning of the virtual variables is represented through the pinning of the corresponding ϕ-function. As a consequence, any ϕ-function will be pinned to all its non-split operands.

Without any virtualization, the process of transforming operand pinning into live range pinning also introduces copies and new local variables pinned together. This systematic copy insertion can also be avoided and managed lazily, just as for ϕ-nodes isolation. We do not address this aspect of the virtualization here: To

---

Algorithm 21.4: De-coalescing with virtualization of ϕ-related copies

---

```
 1  foreach c ∈ COLORS do c.cur_idom = ⊥
 2  foreach basic block B in CFG in DFS pre-order of the dominance tree do
 3        foreach program point l of B in topological order do
 4              if l = late(B) then
 5                    foreach c ∈ COLORS do c.curphi = ⊥
 6                    foreach basic-block B' direct successor of B do
 7                          foreach operation Phi: "B' : a₀ = φ(..., B : v, ...)" in B' do
 8                                if ¬colored(Phi) then continue
 9                                else c ← color(Phi)
10
11                                DeCoalesce_virtual(Phi, B : v, c.curphi, c.cur_idom)
12                                if colored(Phi) then c.curphi ← Phi
13
14              else
15                    foreach operation OP at l (including φ-functions) do
16                          foreach variable v defined by OP do
17                                if ¬colored(v) then continue
18                                else c ← color(v)
19
20                                DeCoalesce(v, c.cur_idom)
21                                if colored(v) then c.cur_idom ← v
22
```

simplify, we consider any operand pinning to be either ignored (handled by register allocation) or expressed as live range pinning.

Sequentialization of Parallel Copies

Throughout the algorithm, we treat the copies placed at a given program point as *parallel copies*, which are indeed the semantics of ϕ-functions. This gives several benefits: a simpler implementation, in particular for defining and updating liveness sets, a more symmetric implementation, and fewer constraints for the coalescer. However, at the end of the process, we need to go back to standard code, i.e., write the final copies in sequential order.

As explained in Sect. 3.2, (Algorithm 3.6) the sequentialization of a parallel copy can be done using a simple traversal of a windmill farm-shaped graph from the tips of the blades to the wheels. Algorithm 21.6 emulates a traversal of this graph (without building it), allowing a variable to be overwritten as soon as it is saved in some other variable.

When a variable a is copied in a variable b, the algorithm remembers b as the last location where the initial value of a is available. This information is stored into `loc(a)`. The initial value that must be copied into b is stored in `directpred(b)`. The initialization consists in identifying the variables whose values are not needed (tree leaves), which are stored in the list `ready`. The list `to_do` contains the

---

Algorithm 21.5: Process (de-coalescing) a virtual variable

---

1 **Function** *DeCoalesce_virtual(Phi, B : v, Phi', u)*
2 | **if** *Phi'* $\neq \bot \wedge V$ (operand_from_B(*Phi'*)) $\neq V(a')$ **then**   ▷ *Interference*
3 | | uncolor atomic-merged-set(*Phi*)
4 | | return
5 | **while** $(u \neq \bot) \bigwedge (\neg(u\text{dominates } B) \vee \neg\text{colored}(u))$ **do**
6 | | $u \leftarrow u$.idom
7 | v.idom $\leftarrow u$;
8 | v.eanc $\leftarrow \bot$; *cur_anc* $\leftarrow u$
9 | **while** *cur_anc* $\neq \bot$ **do**
10 | | **while** *cur_anc* $\neq \bot \bigwedge \neg$ (colored(*cur_anc*) $\wedge$ *cur_anc*.islive(out(*B*))) **do**
11 | | | *cur_anc* $\leftarrow$ *cur_anc*.eanc
12 | | **if** *cur_anc* $\neq \bot$ **then**   ▷ *interference*
13 | | | **if** preferable to uncolor *Phi* **then**
14 | | | | uncolor atomic-merged-set(*Phi*)
15 | | | | break
16 | | | **else**
17 | | | | uncolor atomic-merged-set(*cur_anc*)
18 | | | | *cur_anc* $\leftarrow$ *cur_anc*.eanc

---

destination of all copies to be treated. Copies are first treated by considering leaves (while loop on the list `ready`). Then, the `to_do` list is considered, ignoring copies that have already been treated, possibly breaking a circuit with no duplication, thanks to an extra copy into the fresh variable n.

21.4 Further Reading

SSA destruction was first addressed by Cytron et al. [90] who propose to simply replace each ϕ-function by copies in the direct predecessor basic block. Although this naive translation seems, at first sight, correct, Briggs et al. [50] pointed out subtle errors due to parallel copies and/or critical edges in the control-flow graph. Two typical situations are identified, namely the "lost-copy problem" and the "swap problem." The first solution, both simple and correct, was proposed by Sreedhar et al. [267]. They address the associated problem of coalescing and describe three solutions. The first one consists of three steps: (a) translate SSA into CSSA, by isolating ϕ-functions; (b) eliminate redundant copies; (c) eliminate ϕ-functions and leave CSSA. The third solution, which turns out to be nothing else than the first solution except that it virtualizes the isolation of ϕ-functions, has the advantage of introducing fewer copies. The reason for that, identified by Boissinot et al., is

---

Algorithm 21.6: Parallel copy sequentialization algorithm

Data: Set P of parallel copies of the form $a \mapsto b$, $a \neq b$, one extra fresh variable n
Output: List of copies in sequential order

```
1  ready ← [] ; to_do ← [] ; directpred(n) ← ⊥
2  forall (a ↦ b) ∈ P do
3  |   loc(b) ← ⊥ ; directpred(a) ← ⊥                                    ▷ initialization

4  forall (a ↦ b) ∈ P do
5  |   loc(a) ← a                                          ▷ needed and not copied yet
6  |   directpred(b) ← a                                   ▷ (unique) direct predecessor
7  |   to_do.push(b)                                              ▷ copy into b to be done

8  forall (a ↦ b) ∈ P do
9  |   if loc(b) = ⊥ then ready.push(b)            ▷ b is not used and can be overwritten

10 while to_do ≠ [] do
11 |   while ready ≠ [] do
12 |   |   b ← ready.pop()                                         ▷ pick a free location
13 |   |   a ← directpred(b) ; c ← loc(a)                               ▷ available in c
14 |   |   emit_copy(c ↦ b)                                         ▷ generate the copy
15 |   |   loc(a) ← b                                            ▷ now, available in b
16 |   |   if a = c and directpred(a) ≠⊥ then ready.push(a)  ▷ just copied, can be overwritten

17 |   b ← to_do.pop()                                          ▷ look for remaining copy
18 |   if b = loc(directpred(b)) then
19 |   |   emit_copy(b ↦ n)                                    ▷ break circuit with copy
20 |   |   loc(b) ← n                                              ▷ now, available in n
21 |   |   ready.push(b)                                      ▷ b can be overwritten
```

---

the fact that in the presence of many copies the code contains many intersecting variables that do not actually interfere. Boissinot et al. [35] revisited Sreedhar et al.'s approach in the light of this remark and proposed the value-based interference described in this chapter.

The ultimate notion of interference was discussed by Chaitin et al. [60] in the context of register allocation. They proposed a simple conservative test: *Two variables interfere if one is live at a definition point of the other and this definition is not a copy between the two variables.* This interference notion is the most commonly used, see for example how the interference graph is computed in [10]. Still they noticed that, with this conservative interference definition, after coalescing some variables the interference graph has to be updated or rebuilt. A counting mechanism to update the interference graph was proposed, but it was considered to be too space-consuming. Recomputing it from time to time was preferred [59, 60].

The value-based technique described here can also obviously be used in the context of register allocation even if the code is not under SSA form. The notion of value may be approximated using data-flow analysis on specific lattices [6], and under SSA form simple global value numbering [249] can be used.

Leung and George [182] addressed SSA destruction for machine code. Register renaming constraints, such as calling conventions or dedicated registers, are treated with pinned variables. A simple data-flow analysis scheme is used to place repairing copies. By revisiting this approach to address the coalescing of copies, Rastello et al. [240] pointed out and fixed a few errors present in the original algorithm. While being very efficient in minimizing the introduced copies, this algorithm is quite complicated to implement and not suited to just-in-time compilation.

The first technique to address speed and memory footprint was proposed by Budimlić et al. [53]. It proposes the de-coalescing technique, revisited in this chapter, that exploits the underlying tree structure of the dominance relation between variables of the same merged set.

Last, this chapter describes a fast sequentialization algorithm that requires the minimum number of copies. A similar algorithm has already been proposed by C. May [196].

Chapter 22
Register Allocation

Florent Bouchez Tichadou and Fabrice Rastello

Register allocation maps the variables of a program to physical memory locations: usually either CPU registers or the main memory. The compiler determines the location for each variable and each program point. Ideally, as many operations as possible should draw their operands from processor registers without loading them from memory beforehand. As there is only a small number of registers available in a CPU (with usual values ranging from 8 to 128), it is usually not possible to only use registers, and the task of register allocation is also to decide which variables should be evicted from registers and at which program points to store and load them from memory (spilling).

Furthermore, register allocation has to remove spurious copy instructions (copy coalescing) inserted by previous phases in the compilation process and to deal with allocation restrictions that the instruction set architecture and the run time system impose (register targeting). Classical register allocation algorithms address those different issues with either complex and sometimes expensive schemes (usually graph-based) or simpler and faster (but less efficient) algorithms such as linear scan.

The goal of this chapter is to illustrate how SSA form can help in designing both simpler and faster schemes with similar or even better quality than the most complex existing ones.

F. Bouchez Tichadou
University of Grenoble Alpes, Grenoble, France
e-mail: florent.bouchez-tichadou@imag.fr

F. Rastello (✉)
Inria, Grenoble, France
e-mail: fabrice.rastello@inria.fr

© The Author(s), under exclusive license to Springer Nature Switzerland AG 2022
F. Rastello, F. Bouchez Tichadou (eds.), *SSA-based Compiler Design*,
https://doi.org/10.1007/978-3-030-80515-9_22

303

22.1   Introduction

Let us first review the basics of register allocation, to help us understand the choices made by graph-based and linear scan style allocations.

Register allocation is usually performed per procedure. In each procedure, a liveness analysis (see Chap. 9) determines for each variable the program points where the variable is live. The set of all program points where a variable is live is called the *live range* of the variable, and all along this live range, storage needs to be allocated for that variable, ideally a register. When two variables "exist" at the same time, they are conflicting for resources, i.e., they cannot reside in the same location.

This resource conflict of two variables is called *interference* and is usually defined via liveness: two variables interfere if (and only if) there exists a program point where they are simultaneously live, i.e., their live ranges intersect.[1] It represents the fact that those two variables cannot share the same register. For instance, in Fig. 22.1, variables a and b interfere as a is live at the definition of b.

22.1.1   Questions for Register Allocators

There are multiple questions that arise at that point that a register allocator has to answer to:

- Are there enough registers for all my variables? (*spill test*)
- If yes, how do I choose which register to assign to which variable? (*assignment*)
- If no, how do I choose which variables to store in memory? (*spilling*)

Without going into the details, let us see how linear scan and graph-based allocators handle these questions. Figure 22.1 will be used in the next paragraphs to illustrate how these allocators work.

Linear Scan
The linear scan principle is to consider that a procedure is a long basic block and, hence, live ranges are approximated as intervals. For instance, the procedure shown in Fig. 22.1b is viewed as the straight-line code of Fig. 22.1a. The algorithm then proceeds in scanning the block from top to bottom. When encountering the definition of a variable (i.e., the beginning of a live range), we check if some registers are free (*spill test*). If yes, we pick one to assign the variable to (*assignment*). If no, we choose from the set of currently live variables the one that

---

[1] This definition of interference by liveness is an over-approximation (see Sect. 2.6 of Chap. 2), and there are refined definitions that create less interferences (see Chap. 21). However, in this chapter, we will restrict ourselves to this definition and assume that two variables whose live ranges intersect cannot be assigned the same register.

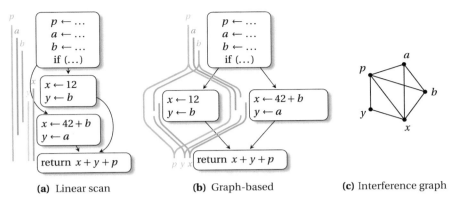

(a) Linear scan   **(b)** Graph-based   **(c)** Interference graph

Fig. 22.1 Linear scan makes an over-approximation of live ranges as intervals, while graph-based allocator creates an interference graph capturing the exact interferences. Linear scan requires 5 registers in this case, while colouring the interference graph can be done with 4 registers

has the farthest use and spill it (*spilling*). When we encounter the end of its live range (e.g., a last use), we free the register it was assigned to. On our example, we would thus greedily colour along the following order: p, a, b, x, and finally y. When the scan encounters the first definition of y in the second basic block, four other variables are live and a fifth colour is required to avoid spilling.

Graph-Based

Graph-based allocators, such as the "Iterated Register Coalescing" allocator (IRC), represent interferences of variables as an undirected *interference graph*: the nodes are the variables of the program, and two nodes are connected by an edge if they interfere, i.e., if their live ranges intersect. For instance, Fig. 22.1c shows the interference graph of the code example presented in Fig. 22.1b. In this model, two neighbouring nodes must be assigned different registers, so the assignment of variables to registers amounts to *colouring* the graph—two neighbouring nodes must have a different colour—using at most R colours, the number of registers.[2]

Here, the allocator will try to colour the graph. If it succeeds (*spill test*), then the colouration represents a valid assignment of registers to variables (*assignment*). If not, the allocator will choose some nodes (usually the ones with the highest number of neighbours) and remove them from the graph by storing the corresponding variables in memory (*spilling*).

On our example, one would need to use four different colours for a, b, p, and x, but y could use the same colour as a or b.

---

[2] Hence, the terms "register" and "colour" will be used interchangeably in this chapter.

Comparison

Linear scan is a very fast allocator where a procedure is modelled as a straight-line code. In this model, the colouring scheme is considered to be optimal. However, the model itself is very imprecise: Procedures generally are not just straight-line codes but involve complex flow structures such as if-conditions and loops. The live ranges are artificially longer and produce more interferences than there actually are. If we look again at the example Fig. 22.1, we have a simple code with an if-condition. Linear scan would decide that, because four variables are live at the definition $y \leftarrow b$, it needs five registers (spill test). But one can observe that a is actually not live at that program point: Modelling the procedure as a straight-line code artificially increases the live ranges of variables.

On the other hand, a graph-based allocator has a much more precise notion of interference. Unfortunately, graph k-colouring is known to be an NP-complete problem. Control-flow structures create cycles in the interference graph that can get arbitrarily complex. The allocator uses a heuristic for colouring the graph and will base its spill decisions on this heuristic.

In our example, IRC would create the graph depicted in Fig. 22.1c, which includes a 4-clique (i.e., a complete sub-graph of size 4, here with variables a, b, p, and x), and, hence, would require at least 4 colours. This simple graph would actually be easily 4-colourable with a heuristic; hence, the spill test would succeed with four registers.

Still, one could observe that at each point of the procedure, no more than three variables are simultaneously live. However, since x interferes with b on the left branch and with a on the right branch, with the model used by IRC, it is indeed impossible to use only three registers.

The question we raise here is: can't we do better? The answer is yes, as depicted in Fig. 22.2a. If it was possible for x to temporarily use the same register as b in the right branch as long as a lives (short live range denoted x' in the figure), then x could use the same colour as a (freed after a's last use). In this context, three registers are enough. In fact, one should expect that, for a code where only three variables are live at any program point, it should be possible to register allocate without spilling with only three registers and proper reshuffling of variables in registers from time to time.

So what is wrong with the graph colouring based scheme we just described? We will develop below that its limitation is mostly due to the fact that it arbitrarily enforces all variables to be assigned to only one register for their entire live range.

In conclusion, linear scan allocators are faster, and graph colouring ones have better results in practice, but both approaches have an inexact spill test: linear scan has artificial interferences, and graph colouring uses a colouring heuristic. Moreover, both require variables to be assigned to *exactly one register* for all their live ranges. This means both allocators will potentially spill more variables than strictly necessary. We will see how SSA can help with this problem.

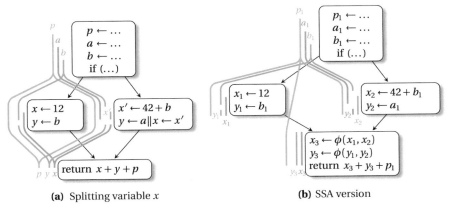

(a) Splitting variable x **(b)** SSA version

Fig. 22.2 Splitting variable x in the previous example breaks the interference between x and a. By using copies between x/x' and y/a in parallel, now only 3 registers are required on the left. SSA introduces splitting that guarantees *Maxlive* registers are enough

22.1.2   Maxlive and Variable Splitting

The number of simultaneously live variables at a program point is called the *register pressure* at that program point.[3] The maximum register pressure over all program points in a procedure is called the register pressure of that procedure, or "*Maxlive*." One can observe that *Maxlive* expresses the *minimum* number of required registers for a spill-free register assignment, i.e., an allocation that does not require memory. For instance, in Fig. 22.1b, *Maxlive* $= 3$, so a minimum number of 3 registers is required. If a procedure is restricted to a single basic block (straight-line code), then *Maxlive* also constitutes a *sufficient number of registers* for a spill-free assignment. But in general, a procedure is made of several basic blocks, and under the standard model described so far, an allocation might require *more than Maxlive registers* to be spill-free.

This situation changes if we permit *live range splitting*. This means inserting a variable-to-variable copy instruction at a program point that creates a new version of a variable. Thus, the value of a variable is allowed to reside in different registers at different times. For instance, in Fig. 22.1b, we can split x in the right branch by changing the definition to x' and using a copy $x \leftarrow x'$ at the end of the block, producing the code shown in Fig. 22.2a. It means the node x in the graph is split into two nodes, x and x'. Those nodes do not interfere, and also interfere differently with the other variables: Now, x can use the same register as a because only x' interferes with a. Conversely, x' can use the same register as b, because only x interferes with b. In this version, we only need *Maxlive* $= 3$ registers, which means that, if the

---

[3] See Fig. 13.1 of Chap. 13 for a visualization of program points.

number of registers was tight (i.e., $R = 3$), we have traded a spill (here one store and one load) for one copy, which is an excellent bargain.

This interplay between live range splitting and colourability is one of the key issues in register allocation, and we will see in the remainder of this chapter how SSA, which creates live range splitting at particular locations, can play a role in register allocation.

22.1.3 The Spill Test Under SSA

As said above, a register allocation scheme needs to address the following three sub-problems: spill test, assignment, and spilling. We focus here on the spill test, that is, verify whether there are enough registers for all variables without having to store any of them in memory.

As already mentioned in Sect. 2.3 of Chap. 2, the live ranges in an SSA-form program with dominance property have interesting structural properties: In that flavour, SSA requires that all uses of a variable are dominated by its definition. Hence, the whole live range is dominated by the definition of the variable. Dominance, however, induces a tree on the control-flow graph (see for instance the dominance edges of Fig. 4.1 in Chap. 4). Thus, the live ranges of SSA variables are all tree-shaped. They can branch downward on the dominance tree but have a single root: the program point where the variable is defined. Hence, a situation like in Fig. 22.1 can no longer occur: x and y had two "roots" because they were defined twice. Under SSA form, the live ranges of those variables are split by ϕ-functions, which creates the code shown in Fig. 22.2b, where we can see that live ranges form a "tree." The argument and result variables of the ϕ-functions constitute new live ranges, giving more freedom to the register allocator since they can be assigned to different registers.

This structural property is interesting as we can now perform exact polynomial colouring schemes that work both for graph-based and linear-style allocators.

Graph-Based

Graph-based allocators such as the IRC mentioned above use a *simplification scheme* that works quite well in practice but is a heuristic for colouring general graphs. We will explain it in more detail in Sect. 22.3.1, but the general idea is to remove from the interference graph nodes that have strictly less than R neighbours, as there will always be a colour available for them. If the whole graph can be simplified, nodes are given a colour in the reverse order of their simplification. We also call this method the *greedy colouring scheme*. On our running example, the interference graph of Fig. 22.1c, candidates for simplification with $R = 4$ would be nodes with strictly less than 4 neighbours, that is, a, b, or y. As soon as one of them is simplified (removed from the graph), p and x now have only 3 neighbours and can also be simplified. So a possible order would be to simplify a, then y, p, b, and finally x. Colours can be greedily assigned during the reinsertion of nodes in the graph in reverse order, starting from x, and ending with a: When we colour a, all its 3 neighbours already have a colour and we assign it to the fourth colour.

Interestingly, under SSA, the interference graph becomes a *chordal graph*.[4] The important nice property about chordal graphs is that they can be coloured minimally in polynomial time. Even more interesting is the fact that the simplification scheme used in IRC is optimal for such graphs and will always manage to colour it with *Maxlive* colours! Thus the same colouring algorithm can be used, *without any modification*, and now becomes an exact spill test: Spilling is required if and only if the simplification scheme fails to completely simplify (hence, colour) the graph.

Tree Scan (Linear Style)

Algorithm 22.1: Tree scan

Input: T, program points in order of the dominance tree
Output: color, an assignment of live ranges to registers
1 **Function** *assign_color (p, available)*
2   **foreach** v last use at p **do**
3 | available[color(v)] ← **True**   ▷ *colors not used anymore*
4   **foreach** v defined at p **do**
5 | c ← choose_color(v, available)   ▷ *pick one of the available colors*
6 | available[c] ← **False**
7 | color(v) ← c
8   **foreach** child p' of p **do**
9 | assign_color(p', copy of available)
10 assign_color(root(T), [**True, True, …, True**])
11 **return** color

Under SSA, the live ranges are intervals that can "branch" but never "join." This allows for a simple generalization of the linear scan mentioned above that we call the *tree scan*. As we will see, the tree scan always succeeds in colouring the tree-shaped live ranges with *Maxlive* colours. This greedy assignment scans the dominance tree, colouring the variables from the root to the leaves in a top-down order. This means the variables are simply coloured in the order of appearance of their respective definition. On our example (Fig. 22.2b), tree scan would colour the variables in the following order: $p_1, a_1, b_1, x_1, y_1, x_2, y_2, x_3, y_3$. This works because branches of the tree are independent, so colouring one will not add constraints on other parts of the tree, contrary to the general non-SSA case that may expose cycles.

The pseudo-code of the tree scan is shown in Algorithm 22.1. In this pseudo-code, the program points p are processed using a depth-first search traversal of the dominance tree T. For a colour c, available[c] is a boolean that expresses if c is

---

[4] In a chordal graph, also called a triangulated graph, every cycle of length 4 or more has (at least) one chord (i.e., an edge joining two non-consecutive edges in the cycle).

available for the currently processed point. Intuitively, when the scanning arrives at the definition of a variable, the only already coloured variables are "above" it, and since there is at most $Maxlive - 1$, other variables live at this program point, and there is always a free colour. As an example, when colouring x_1, the variables live at its definition point are p_1 and b and are already coloured. So a third colour, different from the ones assigned to p_1 and b, can be given to x_1.

Conclusion

The direct consequence is that, as opposed to general form programs, and whether we consider graph-based or scan-based allocators, the only case where spilling is required is when $Maxlive > R$, i.e., when the maximum number of simultaneously live variables is greater than the number of available registers. In other words, there is no need to try to colour the interference graph to check if spilling is necessary or not: the spill test for SSA-form programs simplifies to simply computing $Maxlive$ and then comparing it to R.

This allows to design a register allocation scheme where spilling is decoupled from colouring: First, lower the register pressure to at most R everywhere in the program. Then, colour the interference graph with R colours in polynomial time.

22.2 Spilling

We have seen previously that, under SSA, it is easy to decide in polynomial time whether there is enough registers or not, simply by checking if $Maxlive \leq R$, the number of registers. The goal of this section is to present algorithms that will lower the register pressure when it is too high, i.e., when $Maxlive > R$, by *spilling* (assigning) some variables to memory.

Spilling is handled differently depending on the allocator used. For a scan-based allocator, the spilling decision happens when we are at a particular program point. Although it is actually a bit more complex, the idea when spilling a variable v is to insert a store at that point, and a load just before its next use. This process leads to spilling only a part of the live range. On the other end, a graph-based allocator has no notion of program points since the interferences have been combined in an abstract structure: the interference graph. In the graph colouring setting, spilling means removing a node of the interference graph and thus the *entire* live range of a variable. This is a called a *spill-everywhere* strategy, which implies inserting load instructions in front of every use and storing instructions after each definition of the variables. These loads and stores require temporary variables that were not present in the initial graph. These temporary variables also need to be assigned to registers. This implies that whenever the spilling/colouring is done, the interference graph has to be rebuilt and a new pass of allocation is triggered, until no variable is spilled anymore: this is where the "Iterated" comes from in the IRC name.

In this section, we will consider the two approaches: the graph-based approach with a spill-everywhere scheme, and the scan-based approach that allows partial live range spilling. In both cases, we will assume that the program was in SSA before spilling. This is important to notice that there are pros and cons of assuming so. In particular, the inability to coalesce or move the shuffle code associated to ϕ-functions can lead to spurious load and store instructions on CFG edges. Luckily, these can be handled by a post-pass of partial redundancy elimination (PRE, see Chap. 11), and we will consider here the spilling phase as a full-fledged SSA program transformation.

Suppose we have R registers, the objective is to establish $Maxlive \leq R$ (*Maxlive lowering*) by inserting loads and stores into the program. Indeed, as stated above, lowering $Maxlive$ to R ensures that a register allocation with R registers can be found in polynomial time for SSA programs. Thus, spilling should take place before registers are assigned *and* yield a program in SSA form. In such a decoupled register allocation scheme, the spilling phase is an optimization problem for which we define the following constraints and objective function:

- The *constraints* that describe the universe of possible solutions express that the resulting code should be R-colourable.
- The *objective function* expresses the fact that the (weighted) amount of inserted loads and stores should be minimized.

The constraints directly reflect the "spill test" that expresses whether more spilling is necessary or not. The objective is expressed with the profitability test: among all variables, which one is more profitable to spill? The main implication of spilling in SSA programs is that the spill test—which amounts to checking whether $Maxlive$ has been lowered to R or not—becomes precise.

The other related implication of the use of SSA form follows from this observation: consider a variable such that for any program point in its entire live range the register pressure is at most R, and then spilling this variable is useless with regard to the colourability of the code. In other words, spilling such a variable will never be profitable. We will call this yes-or-no criteria, enabled by the use of SSA form, the "usefulness test."

We will see now how to choose, among all "useful" variables (with regard to the colourability), the ones that seem most profitable. In this regard, we present in the next section how SSA allows to better account for the program structure in the spilling decision even in a graph-based allocator, thanks to the enabled capability to decouple spilling (allocation) to colouring (assignment). However, register allocation under SSA shines the most in a scan-based setting, and we present guidelines to help the spill decisions in such a scheme in Sect. 22.2.2.

22.2.1   Graph-Based Approach

In a graph-based allocator such as the IRC, a *spill-everywhere* strategy is used: a variable is either spilled completely or not at all, and loads are placed directly in front of uses and stores directly after the definition. When spilled, the live range of a variable then degenerates into small intervals: one from the definition and the store, and one from each load to its subsequent use. However, even in this simplistic setting, it is NP-complete to find the minimum number of nodes to establish $Maxlive \leq R$. In practice, heuristics such as the one in IRC spill variables (graph nodes) greedily using a weight that takes into account its node degree (the number of interfering uncoloured variables) and an estimated spill cost (estimated execution frequency of inserted loads and stores). Good candidates are high-degree nodes of low spill cost, as this means they will lessen colouring constraints on many nodes—their neighbours—while inserting few spill codes.

The node degree represents the profitability of spilling the node in terms of colourability. It is not very precise as it is only a graph property, independent of the control-flow graph. We can improve this criteria by using SSA to add a usefulness tag. We will now show how to build this criteria and how to update it.

We attach to each variable v a "useful" tag, an integer representing the number of program points that would benefit from spilling v, i.e., $v.useful$ expresses the number of program points that belong to the live range of v and for which the register pressure is strictly greater than R.

Building the "Useful" Criteria

We will now explain how the "useful" tag can be built at the same time as the interference graph. Under SSA, the interference graph can be built through a simple *bottom-up traversal* of the CFG. When encountering the last use of a variable p, p is added to the set of currently live variables (live set) and a corresponding node (that we also call p) is added in the graph. Arriving at the definition of p, it is removed from the current live set, and edges (interferences) are added to the graph: for all variables v in the live set, there is an edge $(v \rightarrow p)$. Note that, as opposed to standard interference graphs, we consider directed edges here, where the direction represents the dominance. At that point, we also associate the following fields to node p and its interferences:

- $p.pressure$ that corresponds to the number of variables live at the definition point of p, that is, $|\text{live-set}| + 1$
- $(v \rightarrow p).high$, a boolean set to true if and only if $p.pressure > R$, meaning this interference belongs to a clique of size more than R

We then create the "usefulness" field of p. If $p.pressure \leq R$, then $p.useful$ is set to 0. Otherwise, we do the following:

- $p.useful$ is set to 1.
- For all $(v \rightarrow p)$, $v.useful$ gets incremented by 1.

At the end of the build process, $v.useful$ expresses the number of program points that belong to the live range of $v.var$ and for which the register pressure is greater than R. More precisely,

$$v.useful = |\{p : (v \rightarrow p) \text{ exists } \wedge \ p.pressure > R\}| .$$

With this definition, if $v.useful = 0$, then it can be considered to be useless to spill v, as it will not help in reducing the register pressure. If not, it means that v belongs to this number of cliques of size greater than R. Since at least one of the nodes of those cliques must be spilled, spilling v is useful as it would reduce the size of each of those cliques by one. Existing colouring scheme such as the IRC can be modified to use the *useful* tag when making spill decisions, the higher the better.

Updating the "useful" Criteria
Whenever a node n is spilled (assume only useful nodes are spilled), those additional fields of the interference graph must be updated as follows:

1. If $n.pressure > R$, then for all its incoming edges $(v \rightarrow n)$, $v.useful$ is decremented by one.
2. For all its outgoing edges $(n \rightarrow p)$ such that $(n \rightarrow p).high = \text{true}$, $p.pressure$ is decremented by one; if, following this decrement, $p.pressure \leq R$; then for all the incoming edges $(v \rightarrow p)$ of p, $v.useful$ is decremented by one.

Thoughts on Graph-Based Spilling
In an existing graph-based allocation, SSA can bring information to better help the allocator in making its spill decisions. However, with the spilling and colouring fully decoupled, encoding the information using a graph does not seem as pertinent. Moreover, formulations like *spill everywhere* are often not appropriate for practical purposes, as putting the whole live range to memory is too simplistic. A variable might be spilled because at some program point the pressure is too high; however, if that same variable is later used in a loop where the register pressure is low, a spill everywhere will place a superfluous (and costly!) load in that loop. Spill-everywhere approaches try to minimize this behaviour by adding costs to variables, to make the spilling of such a variable less likely. These bring in a flow-insensitive information that a variable resides in a frequently executed area, but such approximations are often too coarse to give good performances. Hence, it is imperative to cleverly split the live range of the variable according to the *program structure* and spill only parts of it, which is why we prefer to recommend a scan-based approach of the register allocation under SSA.

22.2.2   Scan-Based Approach

In the context of a basic block, a simple algorithm that gives good results is the "furthest first" algorithm that is presented in Algorithm 22.2. The idea is to scan the block from top to bottom: whenever the register pressure is too high, we will spill the variable whose next use is the furthest away, and it is spilled only *up to this next use*. In the *evict* function of Algorithm 22.2, this corresponds to maximizing distance_to_next_use_after(p). Spilling this variable frees a register for the longest time, hence diminishing the chances to have to spill other variables later. This algorithm is not optimal because it does not take into account the fact that the first time we spill a variable is more costly than subsequent spills of the same variable (the first time, a store and a load are added, but only a load must be added afterwards). However, the general problem is NP-complete, and this heuristic, although it may produce more stores than necessary, gives good results on "straight-line codes," i.e., basic blocks.

Algorithm 22.2: Furthest first spilling algorithm for straight-line code

---

Input: B, the basic block to process
Input: B.in_regs, the set of variables in registers at entry of B
1 **foreach** program point p in B from top to bottom **do**
  ▷ *uses of v must be in a register before p*
2   protected ← ∅
3   **foreach** v used at p **do**
4   **if** $v \notin B$.in_regs **then**
5   insert a load of v from memory just before p
6   add v to B.in_regs
7   add v to protected
8   evict(p, B, protected)   ▷ *need space for vars loaded at p*
9   **foreach** v last use at p **do**
10   remove v from B.in_regs
11   protected ← ∅
12   **foreach** v defined at p **do**
13   add v to B.in_regs
14   add v to protected
15   evict(p, B, protected)   ▷ *need space for vars defined at p*
16 **Function** *evict (p, B, protected)*
  ▷ *remove variables from registers until there are enough registers*
17   **while** $|B$.in_regs$| > R$ **do**
18   let $v \in (B$.in_regs $\setminus$ protected) with maximum v.distance_to_next_use_after(p)
19   **if** $v \notin B$.in_mem **then**
20   insert a store of v to memory just before p
21   add v to B.in_mem
22   remove v from B.in_regs

We now present an algorithm that extends this "furthest first" algorithm to general control-flow graphs. The idea is to scan the CFG using a topological order and greedily evict sub-live ranges whenever the register pressure is too high. There are two main issues:

1. Generalize the priority function distance_to_next_use_after to a general CFG.
2. Find a way to initialize the "in_regs" set when starting to scan a basic block, in the situation where direct predecessor basic blocks have not been processed yet (e.g., at the entry of a loop).

Fig. 22.3 Generalization of
distance_to_next_use_after
for a CFG. Illustrating
example

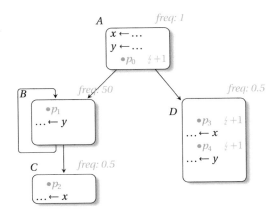

Profitability to Spill

To illustrate the generalization of the further-first priority strategy to a CFG, let us consider the example of Fig. 22.3. In this figure, the $\lightning +n$ sign denotes regions with (high) register pressure of $R + n$. At program point p_0, register pressure is too high by one variable (suppose there are other hidden variables that we do not want to spill). We have two candidates for spilling: x and y, and the classical furthest first criteria would be different depending on the chosen branch:

- If the left branch is taken, considering an execution trace $(AB^{100}C)$: In this branch, the next use of y appears in a loop, while the next use of x appears way further, after the loop has fully executed. It is more profitable to evict variable x (at distance 101).
- If the right branch is taken, the execution trace is (AD). In that case, this is variable y that has the further use (at distance 2) so we would evict variable y.

We have two opposing viewpoints, but looking at the example as a whole, we see that the left branch is not under pressure, so spilling x would only help for program point p_0, and one would need to spill another variable in block D (x is used at the beginning of D); hence, it would be preferable to evict variable y.

On the other hand, if we modify a little bit the example by assuming a high register pressure within the loop at program point p_1 (by introducing other variables), then evicting variable x would be preferred in order to avoid a load and store in a loop!

This dictates the following remarks:

1. Program points with low register pressure can be ignored.
2. Program points within loops, or more generally with higher execution frequency, should account in the computation of the "distance" more than program points with lower execution frequency.

We will then replace the notion of "distance" in the furthest first algorithm with a notion of "profitability," that is, a measure of the number of program points (weighted by frequency) that would benefit from the spilling of a variable v.

Definition 22.1 (Spill Profitability from p) Let p be a program point and v a variable live at p. Let $v.\mathrm{HP}(p)$ (high pressure) be the set of all program points q such that: 1. Register pressure at q is strictly greater than R. 2. v is live at q. 3. There exists a path from p to q that does not contain any use or definition of v. Then,

$$v.\mathrm{spill\_profitability}(p) = \sum_{q \in v.\mathrm{HP}(p)} q.\mathrm{frequency}.$$

There is an important subtlety concerning the set of points in $v.\mathrm{HP}(p)$ that is dependent on the instruction set architecture and the way spill code is inserted. Consider in our example the set $x.\mathrm{HP}(p_0)$, that is, the set of high-pressure points that would benefit from the spilling of x at point p_0. The question is the following: "does the point p_3 (just before the instruction using x) belong to this set?" If the architecture can have an operand in memory, p_3 belongs to the set. However, if the architecture requires an operand to be loaded in a register before its use, then x would be reloaded at p_3 so this point would not benefit from the spilling of x.

Let us see how the profitability applies to our running example with two scenarios: low pressure and high pressure in the B loop at program point p_1. We also assume that an operand should be loaded in a register before its use. In particular, p_3 must be excluded from $x.\mathrm{HP}(p_0)$.

	Loop B	Low pressure at p_1 scenario	High pressure at p_1 scenario
	$x.\mathrm{HP}(p_0)$	$\{p_0\}$	$\{p_0, p_1\}$
$x.\mathrm{spill\_profitability}(p_0)$		1	**51**
	$y.\mathrm{HP}(p_0)$	$\{p_0, p_3\}$	$\{p_0, p_3\}$
$y.\mathrm{spill\_profitability}(p_0)$		**1.5**	1.5

In the first scenario, we would evict y with a profitability of 1.5. In the second, we would evict x with a profitability of 51 (plus another variable later, when arriving at p_3), which is the behaviour we wanted in the first place.

Initial Register Filling at the Beginning of a Basic Block

When visiting basic block B, the set of variables that must reside in a register is stored in $B.\mathrm{in\_regs}$. For each basic block, the initial value of this set has to be computed before we start processing it. The heuristic for computing this set is different for a "regular" basic block and for a loop entry. For a regular basic block, as we assume a topological order traversal of the CFG, all its direct predecessors will have been processed. Live-in variables fall into three sets:

1. The ones that are available at the exit of all direct predecessor basic blocks:

$$\text{allpreds\_in\_regs} = \bigcap_{P \in \text{directpred}(B)} P.\text{in\_regs}$$

2. The ones that are available in some of the direct predecessor basic blocks:

$$\text{somepreds\_in\_regs} = \bigcup_{P \in \text{directpred}(B)} P.\text{in\_regs}$$

3. The ones that are available in none of them.

As detailed in Algorithm 22.3, $B.\text{in\_regs}$ is initialized with `allpreds_in_regs` plus, as space allows, elements of `somepreds_in_regs` sorted in decreasing order of their `spill_profitability` metric. Note that this will add shuffle code (here, loads) on the previous basic blocks or on the corresponding incoming edges that are critical. In practice, this can be handled in a post-pass, in a similar fashion as during the SSA destruction presented in Chap. 21: indeed, lowering the ϕ-functions will create shuffle code that must be executed prior to entering the basic block.

Algorithm 22.3: Initial value of $B.\text{in\_regs}$ for a regular basic block

---

Input: B, the basic block to process
1 allpreds_in_regs $\leftarrow \bigcap_{P \in \text{directpred}(B)} P.\text{in\_regs}$
2 somepreds_in_regs $\leftarrow \bigcup_{P \in \text{directpred}(B)} P.\text{in\_regs}$
3 $B.\text{in\_regs} \leftarrow$ allpreds_in_regs
4 **while** $|B.\text{in\_regs}| < R$ and $|\text{somepreds\_in\_regs}| > |B.\text{in\_regs}|$ **do**
5 $\quad$ let $v \in (\text{somepreds\_in\_regs} \setminus B.\text{in\_regs})$ with mininum $v.\text{spill\_profitability}(B.\text{entry})$
6 $\quad$ add v to $B.\text{in\_regs}$

For a basic block at the entry of a loop, as illustrated by the example of Fig. 22.4, one does not want to account for allocation on the direct predecessor basic block but starts from scratch instead. Here, we assume the first basic block has already been processed, and one wants to compute $B.\text{in\_regs}$:

1. Example (a): Even if at the end of the direct predecessor basic block, at p_0, x is not available in a register, one wants to insert a reload of x at p_1, that is, include x in $B.\text{in\_regs}$. Not doing so would involve a reload at every iteration of the loop at p_2.
2. Example (b): Even if at the entry of the loop, x is available in a register, one wants to spill it at p_1 and restore at p_4 so as to lower the register pressure that

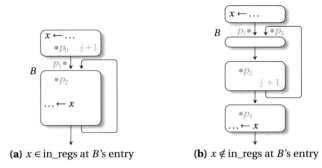

(a) $x \in$ in_regs at B's entry   **(b)** $x \notin$ in_regs at B's entry

Fig. 22.4 Initial values of B.in_regs at loop entry. Illustrating examples

is too high within the loop. Not doing so would involve a store in the loop at p_2 and a reload on the back edge of the loop at p_3. This means excluding x from B.in_regs.

This leads to Algorithm 22.4 where B.livein represents the set of live-in variables of B and L.*Maxlive* is the maximal register pressure in the whole loop L. Init_inregs first fills B.in_regs with live-in variables that are used within the loop L. Then, we add live-through variables to B.in_regs, but only those that can survive the loop: If L.*Maxlive* $>$ R, then L.*Maxlive* $- R$ variables will have to be spilled (hopefully some live-through variables), so no more than $|B.\texttt{livein}| - (L.Maxlive - R)$ are allocated to a register at the entry of B.

Algorithm 22.4: Initial value of B.in_regs at the entry of a loop L

---

1 B.in_regs $\leftarrow \emptyset$
2 **while**
  $|B.\text{in\_regs}| < R$ and
3    $|B.\text{livein}| > |B.\text{in\_regs}|$ and
  $|B.\text{in\_regs}| < R + |B.\text{livein}| - L.Maxlive$
4 **do**
5   let $v \in (B.\text{livein} \setminus B.\text{in\_regs})$ with minimum v.spill_profitability(B.entry)
6   add v to B.in_regs

Putting All Together

The overall algorithm for spilling comprises several phases: First, we pre-compute both liveness and profitability metrics. Then we traverse the CFG in topological order, and each basic block is scanned using the initial value of B.in_regs as explained above. During this phase, we maintain the sets of live variables available in register and in memory at basic block boundaries. The last phase of the algorithms handles the insertion of shuffle code (loads and stores) where needed.

22.3   Colouring and Coalescing

We advocate here a decoupled register allocation: First, lower the register pressure so that *Maxlive* $\leq$ R; second, assign variable to registers. Live range splitting ensures that, after the first phase is done, no more spilling will be required as R will be sufficient, possibly at the cost of inserting register-to-register copies.

We already mentioned in Sect. 22.1.3 that the well-known "Iterated Register Coalescing" (IRC) allocation scheme, which uses a simplification scheme, can take advantage of the SSA-form property. We will show here that, indeed, the underlying structural property makes a graph colouring simplification scheme (recalled below) an "optimal" scheme. This is especially important because, besides minimizing the amount of spill code, the second objective of register allocation is to perform a "good coalescing," that is, try to minimize the amount of register-to-register copies: a decoupled approach is practically viable if the coalescing phase is effective in merging most of the live ranges, introduced by the splitting from SSA, by assigning live ranges linked by copies to the same register.

In this section, we will first present the traditional graph colouring heuristic, based on a simplification scheme, and show how it successfully colours programs under SSA form. We will then explain in greater detail the purpose of coalescing and how it translates when performed on SSA-form program. Finally, we will show how to extend the graph-based (from IRC) and the scan-based (of Algorithm 22.1) greedy colouring schemes to perform efficient coalescing.

22.3.1   Greedy Colouring Scheme

In traditional graph colouring register allocation algorithms, the assignment of registers to variables is done by colouring the interference graph using a simplification scheme. This result in what is called a *greedy colouring scheme*: this scheme is based on the observation that, given R colours—representing the registers, if a node in the graph has at most $R - 1$ neighbours, there will always be one colour available for this node whatever colours the remaining nodes have. Such a node can be *simplified*, that is, removed from the graph and placed on a stack. This process can be iterated with the remaining nodes, whose degree may have decreased. If the graph becomes empty, we know it is possible to colour the graph with R colours, by assigning colours to nodes in the reverse order of their simplification, that is, popping nodes from the stack and assigning them one available colour. This is always possible since they have at most $R - 1$ coloured neighbours. We call the whole process the *greedy colouring scheme*: Its first phase, whose goal is to eliminate nodes from the graph to create the stack, is presented in Algorithm 22.5, while the second phase, which assigns variables to registers, is presented in Algorithm 22.6.

The greedy colouring scheme is a colouring heuristic for general graphs, and as such, it can fail. The simplification can get stuck whenever all remaining nodes have

---

Algorithm 22.5: Greedy simplification scheme used by the greedy colouring scheme to create the stack. If this stage succeeds, the graph is said to be greedy-R-colourable

---

Input: $G = (V, E)$ an undirected graph
Input: R: number of colors (registers)
Input: For each vertex v, degree[v], the number of neighbours in G

1 **Function** *Simplify(G)*
2   stack $\leftarrow \emptyset$
3   worklist $\leftarrow \{v \in V \mid$ degree[v] $< R\}$
4   **while** worklist $\neq \emptyset$ **do**
5   let $v \in$ worklist
6   **foreach** w neighbour of v **do**
7   degree[w] $\leftarrow$ degree[w]-1
8   **if** degree[w] $= R - 1$ **then**
9   add w to worklist
10   push v on stack
11   remove v from worklist
12   remove v from V ▷ *Remove v from the graph*
13   **if** $V \neq \emptyset$ **then**
14   Failure "The graph is not simplifiable"
15   **return** stack

Algorithm 22.6: Greedy colouring assignment used in the greedy colouring scheme. The stack is the output of the Simplify function described in Algorithm 22.5

---

1 **Function** *Assign_colors(G)*
2   available $\leftarrow$ new array of size R, with initialized values to **True**
3   **while** stack $\neq \emptyset$ **do**
4         $v \leftarrow$ pop stack
5   **foreach** neighbour w in G **do**
6   available[color(w)] $\leftarrow$ **False**   ▷ *color already used by neighbour*
7   **foreach** color c **do**
8   **if** available[c] **then**
9   col $\leftarrow c$
10   available[c] $=$ **True**   ▷ *prepare for next round*
11   color(v) $\leftarrow$ col
12   add v to G

---

degree at least R. In that case, we do not know whether the graph is R-colourable or not. In traditional register allocation, this is the trigger for spilling some variables so as to unstuck the simplification process. However, under the SSA form, if spilling has already been done so that the maximum register pressure is at most R, the greedy colouring scheme can never get stuck! We will not formally prove this fact here but will nevertheless try to give insight as to why this is true.

The key to understanding that property is to picture the dominance tree, with live ranges are sub-trees of this tree, such as the one in Fig. 22.2b. At the end of each dangling branch, there is a "leaf" variable: the one that is defined last in this branch. These are the variables y_1, y_2, y_3, and x_3 in Fig. 22.2b. We can visually see that this variable will not have many intersecting variables: those are the variables alive at its definition point, i.e., no more than $Maxlive - 1$, hence less than $R - 1$. In Fig. 22.2b, with $Maxlive = 3$, we see that each of them has no more than two neighbours.

Considering again the greedy scheme, this means each of them is a candidate for simplification. Once removed, another variable will become the new leaf of that particular branch (e.g., x_1 if y_1 is simplified). This means simplification can always happens at the end of the branches of the dominance tree, and the simplification process can progress upward until the whole tree is simplified.

In terms of graph theory, the general problem is knowing whether a graph is k-colourable or not. Here, we can define a new class of graphs that contains the graphs that can be coloured with this simplification scheme.

Definition 22.2 A graph is greedy-k-colourable if it can be simplified using the Simplify function of Algorithm 22.5.

We could prove the following theorem:

Theorem 1 *Setting $k = Maxlive$, the interference graph of a code under SSA form is always greedy-k-colourable.*

This tells us that, if we are under SSA form and the spilling has already been done so that $Maxlive \leq R$, the classical greedy colouring scheme is *guaranteed* to perform register allocation with R colours without any additional spilling, as the graph is greedy-R-colourable.[5]

22.3.2 Coalescing Under SSA Form

The goal of *coalescing* is to minimize the number of register-to-register *copies* instructions in the final code. In a graph-based allocation approach where the colour-everywhere strategy is applied, each node of the interference graph corresponds to a unique variable that will be allocated to a register along its entire live range. Coalescing two nodes, or by extension coalescing the two corresponding variables, means merging the nodes in the graph, thus imposing allocating the same register to those variables. Any copy instruction between those two variables becomes useless and can be safely removed.

---

[5] Observe that, with a program originally under SSA form, practical implementation may still choose to interleave the process of spilling and coalescing/colouring. Result will be unchanged, but speed might be impacted.

This way, coalescing, when done during the register allocation phase, is used to minimize the amount of register-to-register copies in the final code. While there may not be so many such copies at the high level (e.g., instructions "$a \leftarrow b$")— especially after a phase of copy propagation under SSA (see Chap. 8), many such instructions are added in different compiler phases by the time compilation reaches the register allocation phase. For instance, adding copies is a common way to deal with register constraints (see the practical discussion in Sect. 22.4).

An even more obvious and unavoidable reason in our case is the presence of ϕ-functions due to the SSA form: The semantic of a ϕ-function corresponds to parallel copies on incoming edges of basic blocks, and destructing SSA, that is, getting rid of ϕ-functions that are not machine instructions, is done through the insertion of copy instructions. It is thus better to assign variables linked by a ϕ-function to the same register, so as to "remove" the associated copies between subscripts of the same variable. As already formalized (see Sect. 21.2 of Chap. 21) for the aggressive coalescing scheme, we define a notion of *affinity*, acting as the converse of the relation of interference and expressing how much two variables "want" to share the same register. By adding a metric to this notion, it measures the benefit one could get if the two variables were assigned to the same register: the weight represents how many instructions coalescing would save at execution.

22.3.3   Graph-Based Approach

Coalescing comes with several flavours, which can be either aggressive or conservative. Aggressively coalescing an interference graph means coalescing non-interfering nodes (that is, constraining the colouring) regardless of the chromatic number of the resulting graph. An aggressive coalescing scheme is presented in Chap. 21. Conservatively coalescing an interference graph means coalescing non-interfering nodes without increasing the chromatic number of the graph. In both cases, the objective function is the maximization of satisfied affinities, that is, the maximization of the number of (weighted) affinities between nodes that have been coalesced together. In the current context, we will focus on the conservative scheme, as we do not want more spilling.

Obviously, because of the reducibility to graph-k-colouring, both coalescing problems are NP-complete. However, graph colouring heuristics such as the Iterated Register Coalescing use incremental coalescing schemes where affinities are considered one after another. Incrementally, for two nodes linked by an affinity, the heuristic will try to determine whether coalescing those two nodes will, with regard to the colouring heuristic, increase the chromatic number of the graph or not. If not, then the two corresponding nodes are (conservatively) coalesced. The IRC considers two conservative coalescing rules that we recall here. Nodes with degree strictly less than R are called *low-degree* nodes (those are simplifiable), while others are called *high-degree* nodes.

Briggs rule   merges u and v if the resulting node has less than R neighbours of high degree. This node can always be simplified after its neighbours, low-degree neighbours, are simplified; thus the graph remains greedy-R-colourable.

George rule   merges u and v if all neighbours of u with high degree are also neighbours of v. After coalescing and once all low-degree neighbours are simplified, one gets a sub-graph of the original graph, thus greedy-R-colourable too.

The Iterated Register Coalescing algorithm normally also performs spilling and includes many phases that are interleaved with colouring and coalescing, called in the literature "freezing," "potential spills," "select," and "actual spill" phases.

A pruned version of the coalescing algorithm used in the IRC can be obtained by removing the freezing mechanism (explained below) and the spilling part. It is presented in Algorithm 22.7. In this code, both the processes of coalescing and simplification are combined. It works as follows:

1. Low-degree nodes that are not copy-related (no affinities) are simplified as much as possible.
2. When no more nodes can be simplified this way, an affinity is chosen. If one of the two rules (Briggs or George) succeeds, the corresponding nodes are merged. If not, the affinity is erased.
3. The process iterates (from stage 1) until the graph is empty.

Originally, those rules were used for any graph, not necessarily greedy-R-colourable, and with an additional clique of pre-coloured nodes—the physical machine registers. With such general graphs, some restrictions on the applicability of those two rules had to be applied when one of the two nodes was a pre-coloured one. But in the context of greedy-R-colourable graphs, we do not need such restrictions.

However, in practice, those two rules give insufficient results to coalesce the many copies introduced, for example, by a basic SSA destruction conversion. The main reason is because the decision is too local: it depends on the degree of neighbours only. But these neighbours may have a high degree just because their neighbours are not simplified yet, that is, the coalescing test may be applied too early in the simplify phase.

This is the reason why the IRC actually iterates: instead of giving up coalescing when the test fails, the affinity is "frozen," that is, placed in a sleeping list and "awakened" when the degree of one of the nodes implied in the rule changes. Thus, affinities are in general tested several times, and copy-related nodes—nodes linked by affinities with other nodes—should not be simplified too early to ensure the affinities get tested.

The advocated scheme corresponds to the pseudo-code of Algorithm 22.7 and is depicted in Fig. 22.5. It tries the coalescing of a given affinity only once and thus does not require any complex freezing mechanism as done in the original IRC. This is made possible thanks to the following enhancement of the conservative coalescing rule: Recall that the objective of Briggs and George rules is to test

---

Algorithm 22.7: Pruned version of the Iterated Register Coalescing.
Does not handle spilling but only colouring plus coalescing. It can be used
instead of the Simplify function described in Algorithm 22.5, as it also produces
a stack useable by the Assign_colour function of Algorithm 22.6

Input: Undirected greedy-R-colorable graph $G = (V, I)$
Data: R: number of colors
Data: A: the affinity edges of G
Data: For all v, degree[v] denotes the number of interfering neighbours of v in G
Data: For all v, affinities[v] denotes $\{(v, w) \in A\}$

```
 1  Function Simplify_and_Coalesce(G)
 2  |    stack ← ∅
 3  |    simplifiable ← {v ∈ V | degree[v] < R and affinities[v] = ∅}
 4  |    while V ≠ ∅ do
 5  |    |    while simplifiable ≠ ∅ do
 6  |    |    |    let v ∈ simplifiable
 7  |    |    |    push v on stack
 8  |    |    |    remove v from G: update V, I, and simplifiable sets accordingly
 9  |    |    let a = (v, u) ∈ A of highest weight
    |    |    ▷ verify there is no interference (can happen after other merges)
10  |    |    if a exists and a ∉ I and can_be_coalesced(a) then
11  |    |    |    merge u and v into uv
12  |    |    |    update V, I, and simplifiable sets accordingly
13  |    |    remove a from A
14  |    return stack
```

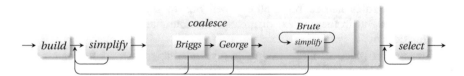

Fig. 22.5 Combined coalescing and colouring simplification scheme including Brute rule

whether the coalescing v and u breaks the greedy-R-colourable property of G
or not. If the Briggs and George tests fail, testing this property can be done
by running function *Simplify(G)* itself! Theoretically, this greatly increases the
complexity, as for each affinity, a full simplification process could be potentially
performed. However, experience shows that the overhead is somewhat balanced
by the magnitude lowering of calls to "can_be_coalesced." This approach still
looks quite rough; hence, we named it the *Brute* coalescing heuristic. While this
is more costly than only using Briggs and George rules, adding this "brute"
rule improves substantially the quality of the result, in terms of suppressed copy
instructions.

22.3.4 Scan-Based Approach

The most natural way to perform coalescing using a scan-based (linear scan or tree scan) approach would be to simply do biased colouring. Consider again the "tree scan" depicted in Algorithm 22.1. Let us suppose the current program point p is a copy instruction from variable v to v'. The colour of v that is freed at line 3 can be reused for v' at line 5. In practice, this extremely local strategy does not work that well for ϕ-functions. Consider as an example variables x_1, x_2, and x_3 in the program Fig. 22.2b. As there is no specific reason for the greedy allocation to assign the same register to both x_1 and x_2, when it comes to assigning one to x_3, the allocator will often be able to only satisfy one of its affinities.

To overcome this limitation, the idea is to use an aggressive pre-coalescing as a pre-pass of our colouring phase. We can use one of the algorithms presented in Chap. 21, but the results of aggressive coalescing should only be kept in a separate structure and *not applied* to the program. The goal of this pre-pass is to put copy-related variables into *equivalence classes*. In a classical graph colouring allocator, the live ranges of the variables in a class are fused. We do not do so but use the classes to bias the colouring. Each equivalence class has a colour, which is initially unset, and is set as soon as one variable of the class is assigned to register. When assigning a colour to a variable, tree scan checks if the colour of the class is available, and picks it if it is. If not, it chooses a different colour (based on the other heuristics presented here) and updates the colour of the class.

22.4 Practical Discussions and Further Reading

The amount of papers on register allocation is humongous. The most popular approach is based on graph colouring, including many extensions [71, 190, 219] of the seminal paper of Chaitin et al. [61]. The Iterated Register Coalescing, mentioned several times in this chapter, is from George et al. [124]. The linear scan approach is also extremely popular in particular in the context of just-in-time compilation. This elegant idea goes back to Traub et al. [287] and Poletto and Sarkar [230]. If this original version of linear scan contains a lot of hidden subtleties (such as the computation of live ranges, handling of shuffle code in the presence of critical edges, etc.), its memory footprint is smaller than a graph colouring approach, and its practical complexity is smaller than the IRC. However, mostly due to the highly over-approximated live ranges, its apparent simplicity comes with a poor-quality resulting code. This leads to the development of interesting but more complex extensions such as the works of Wimmer and Sarkar [252, 305]. All those approaches are clearly subsumed by the tree scan approach [80], both in terms of simplicity, complexity, and quality of result. On the other hand, the linear scan extension proposed by Barik in his thesis [17] is an interesting competitor, its footprint being, possibly, more compact than the one used by a tree scan.

There exists an elegant relationship between tree scan colouring and graph colouring: Back in 1974, Gavril [122] showed that the intersection graphs of subtrees are the *chordal graphs*. By providing an elimination scheme that exposes the underlying tree structure, this relationship allowed to prove that chordal graphs can be optimally coloured in linear time with respect to the number of edges in the graph. This is the rediscovering of this relationship in the context of live ranges of variables for SSA-form programs that motivated different research groups [39, 51, 136, 221] to revisit register allocation in the light of this interesting property. Indeed, at that time, most register allocation schemes were incomplete assuming that the assignment part was hard by referring to the NP-completeness reduction of Chaitin et al. to graph colouring. However, the observation that SSA-based live range splitting allows to decouple the allocation and assignment phases was not new [87, 111]. Back in the nineties, the LaTTe [314] just-in-time compiler already implemented the ancestor of our tree scan allocator. The most aggressive live range splitting that was proposed by Appel and George [10] allowed to stress the actual challenge that past approaches were facing when splitting live range to help colouring, which is coalescing [40]. The PhD theses of Hack [135], Bouchez [38], and Colombet [79] address the difficult challenge of making a neat idea applicable to real life but without trading the elegant simplicity of the original approach. For some more exhaustive related work references, we refer to the bibliography of those documents.

Looking Under the Carpet

As done in (too) many register allocation papers, the heuristics described in this chapter assume a simple non-realistic architecture where all variables or registers are equivalent and where the instruction set architecture does not impose any specific constraint on register usage. Reality is different, including: 1. register constraints such as 2-address mode instructions that impose two of the three operands to use the same register, or instructions that impose the use of specific registers; 2. registers of various sizes (vector registers usually leading to register aliasing), historically known as register pairing problem; 3. instruction operands that cannot reside in memory. Finally, SSA-based—but also any scheme that relies on live range splitting—must deal with critical edges possibly considered abnormal (i.e., that cannot be split) by the compiler.

In the context of graph colouring, *register constraints* are usually handled by adding an artificial clique of pre-coloured node in the graph and splitting live ranges around instructions with pre-coloured operands. This approach has several disadvantages. First, it substantially increases the number of variables. Second, it makes coalescing much harder. This motivated Colombet et al. [80] to introduce the notion of antipathies (affinities with negative weight) and extend the coalescing rules accordingly. The general idea is, instead of enforcing architectural constraints, to simply express the cost (through affinities and antipathies) of shuffle code inserted by a post-pass repairing. In a scan-based context, handling of register constraints is usually done locally [203, 252]. The biased colouring strategy used in this chapter is proposed by Braun et al. and Colombet et al. [46, 80] and allows to reduce need for shuffle code.

In the context of graph-based heuristics, *vector registers* are usually handled through a generalized graph colouring approach [262, 281]. In the context of scan-based heuristics, the puzzle solver [222] is an elegant formulation that allows to express the local constraints as a puzzle.

One of the main problems of the graph-based approach is its underlying assumption that the live range of a variable is atomic. But when spilled, not all instructions can access it through a *memory operand*. In other words, spilling has the effect of removing the majority of the live ranges, but leaving "chads" that correspond to shuffle code around instructions that use the spilled variable. These replace the removed node in the interference graph, which is usually re-built at some point. Those subtleties and associated complexity issues are exhaustively studied by Bouchez et al. [41].

As already mentioned, the semantics of ϕ-functions correspond to parallel copies on the incoming edges of the basic block where they textually appear. When lowering ϕ-functions, the register for the def operand may not match the register of the use operand. If done naively, this imposes the following:

1. *Splitting* of the corresponding control-flow edge
2. Inserting of copies on this freshly created basic block
3. Using of spill code in case parallel copy requires a temporary register but none is available

This issue shares similarities with the SSA destruction that was described in Chap. 21. The advocated approach is, just as for the register constraints, to express the cost of an afterward repairing in the objective function of the register allocation scheme. Then, locality, the repairing can be expressed as a standard graph colouring problem on a very small graph containing the variables created by ϕ-nodes isolation (see Fig. 21.1). However, it turns out that most of the associated shuffle codes can usually be moved (and even annihilated) to and within the surrounding basic blocks. Such post-pass optimizations correspond to the optimistic copy insertion of Braun et al. [46] or the parallel copy motion of Bouchez et al. [43].

Decoupled Load/Store and Coalescing Optimization Problems

The scan-based spilling heuristic described in this chapter is inspired from the heuristic developed by Braun and Hack [45]. It is an extension of Belady's algorithm [22], which was originally designed for page eviction, but can easily be shown to be optimal for interval graphs (straight-line code and single use variables). For straight-line code but multiple uses per variable, Farrach and Liberatore [114] showed that if this further-first strategy works reasonably well, it can also be formalized as a flow problem. For a deeper discussion about the complexity of the spilling problem under different configurations, we refer to the work of Bouchez et al. [41]. To conclude on the spilling part of the register allocation problem, we need to mention the important problem of load/store placement. As implicitly done by our heuristic given in Algorithm 22.4, one should, whenever possible, hoist shuffle code outside of loops. This problem corresponds to the global code motion addressed in Chap. 11 that should idealistically be coupled to the register allocation problem. Another related problem, not mentioned in this chapter, is rematerialization [48]. Experience shows that: 1. Rematerialization is one of the main sources of performance improvement for register allocation. 2. In the vast majority of cases, rematerialization simply amounts to rescheduling some of the instructions. This remark allows to highlight one of the major limitations of the presented approach common to almost all papers in the area of register allocation: while scheduling and register allocation are clearly highly coupled problems, all those approaches only consider a fixed schedule and only few papers try to address the coupled problem [27, 78, 204, 209, 226, 242, 286, 301].

Static Single Assignment has not been adopted by compiler designers for a long time. One of the reasons is related to the numerous copy instructions inserted by SSA destruction that the compiler could not get rid of afterwards. The coalescing heuristics were not effective enough in removing all such copies, so even if it was clear that live range splitting was useful for improving colourability, that is, avoiding spill code, people were reluctant to split live ranges, preferring to do it on demand [82]. The notion of value-based interference described in Paragraph 21.2 of Chap. 21 is an important step for improving the quality of existing schemes. Details on the brute-force approach that extends the conservative coalescing scheme depicted in this chapter can be found in Bouchez et al. [42]. However, there exist several other schemes whose complexities are studied by Bouchez et al. [40]. Among others, the optimistic approach of Park and Moon [219] has been shown to be extremely competitive [42, 132] in the context of the, somehow restricted, "Optimal Coalescing Challenge" initiated by Appel and George, designers of the Iterated Register Coalescing [124].

Chapter 23
Hardware Compilation Using SSA

Pedro C. Diniz and Philip Brisk

This chapter describes the use of SSA-based high-level program representations for the realization of the corresponding computations using hardware digital circuits. We begin by highlighting the benefits of using a compiler SSA-based intermediate representation in this hardware mapping process, using an illustrative example. The subsequent sections describe hardware translation schemes for discrete hardware logic structures or datapaths of hardware circuits and outline several compiler transformations that benefit from SSA. We conclude with a brief survey of various hardware compilation efforts from both academia and industry that have adopted SSA-based internal representations.

23.1 Brief History and Overview

Hardware compilation is the process by which a high-level behavioural description of a computation is translated into a hardware-based implementation, that is, a circuit expressed in a hardware design language such as VHDL, or Verilog which can be directly realized as an electrical (often digital) circuit.

Hardware-oriented languages such as VHDL or Verilog allow programmers to develop such digital circuits either by structural composition of blocks using abstractions such as wires and ports or behaviourally by definition of the input-output relations of signals in these blocks. A mix of both design approaches is often found in medium to large designs. Using a structural approach, a circuit

P. C. Diniz (✉)
University of Porto, Porto, Portugal
e-mail: peddiniz@gmail.com

P. Brisk
UC Riverside, Riverside, CA, USA
e-mail: philip@cs.ucr.edu

© The Author(s), under exclusive license to Springer Nature Switzerland AG 2022
F. Rastello, F. Bouchez Tichadou (eds.), *SSA-based Compiler Design*,
https://doi.org/10.1007/978-3-030-80515-9_23

description will typically include discrete elements such as registers (flip-flops) that capture the state of the computation at specific events, such as clock edges, and combinatorial elements that transform the values carried by wires. The composition of these elements allows programmers to build finite-state machines (FSMs) that orchestrate the flow and the processing of data stored in internal registers or RAM structures. These architectures support an execution model with operations akin to assembly instructions found in common processors that support the execution of high-level programs.

A common vehicle for the realization of hardware designs is an FPGA or field-programmable gate array. These devices include a large number of configurable logic blocks (or CLBs), each of which can be individually programmed to realize an arbitrary combinatorial function of k inputs whose outputs can be latched in flip-flops and connected via an internal interconnection network to any subset of the CLBs in the device. Given their design regularity and simple structure, these devices, popularized in the 1980s as fast hardware prototyping vehicles, have taken advantage of Moore's law to grow to large sizes with which programmers can define custom architectures capable of TFlops/Watt performance, thus making them the vehicle of choice for very power-efficient custom computing machines.

While developers were initially forced to design hardware circuits exclusively using schematic capture tools, over the years, high-level behavioural synthesis allowed them to leverage a wealth of hardware mapping and design exploration techniques to realize substantial productivity gains.

As an example, Fig. 23.1 illustrates these concepts of hardware mapping for the computation expressed as $x \leftarrow (a \times b) - (c \times d) + f$. Figure 23.1b depicts a graphical representation of a circuit that directly implements this computation. Here, there is a direct mapping between hardware operators such as adders and multipliers and the operations in the computation. Input values are stored in the registers at the top of the diagram, and the entire computation is carried out during a single (albeit long) clock cycle, at the end of which the results propagated through the various hardware operators are captured (or latched) in the registers at the bottom of the diagram. In all, this direct implementation uses two multipliers, two adders/subtractors, and six registers, five registers to hold the computation's input values and one to capture the computation's output result. This hardware implementation requires a simple control scheme, as it just needs to record the input values, and wait for a single clock cycle at the end of which it stores the outputs of the operations in the output register. Figure 23.1c depicts a different implementation variant of the same computation, this time using nine registers and the same number of adders and subtractors.[1] The increased number of registers allows the circuit to be clocked at a higher frequency as well as to be executed in a pipelined fashion. Lastly, Fig. 23.1d depicts yet another possible implementation of the same computation,

---

[1] It may be apparent that the original computation lacks any temporal specification in terms of the relative order in which data-independent operations can be carried out. Implementation variants exploit this property.

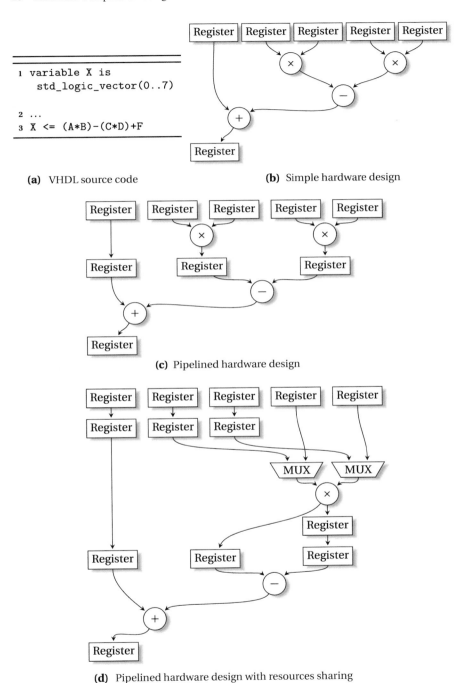

```
1 variable X is
   std_logic_vector(0..7)

2 ...
3 X <= (A*B)-(C*D)+F
```

(a) VHDL source code

(b) Simple hardware design

(c) Pipelined hardware design

(d) Pipelined hardware design with resources sharing

Fig. 23.1 Different variants of mapping a computation to hardware

but using a single multiplier operator. This last version allows for the reuse in time of the multiplier operator and the required thirteen registers as well as multiplexers to route the inputs to the multiplier in two distinct control steps. As is apparent, the reduction in the number of operators, in this particular case the multipliers, carries a penalty in the form of an increased number of registers and multiplexers.[2] It can be viewed as a hardware implementation of the C programming language selection operator: $out = q \ ? \ in_1 \ : \ in_2$ and increased complexity of the control scheme.

This example illustrates the many degrees of freedom in high-level behavioural hardware synthesis. Synthesis techniques perform the classical tasks of allocation, binding, and scheduling of the various operations in a computation given specific target hardware resources. For instance, a designer can use a behavioural synthesis tool (e.g., Xilinx's Vivado) to automatically derive a hardware implementation for a computation, as expressed in the example in Fig. 23.1a. This high-level description suggests a simple, direct hardware implementation using a single adder and two multipliers, as depicted in Fig. 23.1b. Alternative hardware implementations taking advantage of pipelined execution and the sharing of a multiplier to reduce hardware resources are respectively illustrated in Fig. 23.1c and Fig. 23.1d. The tool then derives the control scheme required to route the data from registers to the selected units so as to meet the designers' goals.

Despite the introduction of high-level behavioural synthesis techniques into commercially available tools, hardware synthesis, and thus hardware compilation, has never enjoyed the same level of success as traditional software compilation. Sequential programming paradigms popularized by programming languages such as C/C++ and, more recently, by Java, allow programmers to easily reason about program behaviour as a sequence of program memory state transitions. The underlying processors and the corresponding system-level implementations present a number of simple unified abstractions—such as a unified memory model, a stack, and a heap that do not exist (and often do not make sense) in customized hardware designs.

Hardware compilation, in contrast, has faced numerous obstacles that have hampered its wide adoption. When developing hardware solutions, designers must understand the concept of spatial concurrency that hardware circuits offer. Precise timing and synchronization between distinct hardware components are key abstractions in hardware. Solid and robust hardware design implies a detailed understanding of the precise timing of specific operations, including I/O, that simply cannot be expressed in languages such as C, C++, or Java. Alternatives such as SystemC have emerged in recent years, giving the programmer considerably more control over these issues. The inherent complexity of hardware designs has hampered the development of robust synthesis tools that can offer high-level programming abstractions enjoyed by tools that target traditional architecture and software systems, thus substantially raising the barrier of entry for hardware design-

---

[2] A 2×1 multiplexer is a combinatorial circuit with two data inputs, a single output and a control input, where the control input selects which of the two data inputs is transmitted to the output.

ers in terms of productivity and robustness of the generated hardware solutions. At best, hardware compilers today can only handle certain subsets of mainstream high-level languages and at worst are limited to purely arithmetic sequences of operations with strong restrictions on control flow.

Nevertheless, the emergence of multi-core processing has led to the introduction of new parallel programming languages and parallel programming constructs that are more amenable to hardware compilation than traditional languages. For example, MapReduce, originally introduced by Google to spread parallel jobs across clusters of servers, has been an effective programming model for FPGAs as it naturally exposes task-level concurrency with data independence. Similarly, high-level languages based on parallel models of computation such as synchronous data flow, or functional single-assignment languages, have also been shown to be good choices for hardware compilation, as not only do they make data independence obvious, but in many cases, the natural data partitioning they expose is a natural match for the spatial concurrency of FPGAs.

Although in the remainder of this chapter we will focus primarily on the use of SSA representations for hardware compilation of imperative high-level programming languages, many of the emerging parallel languages, while including sequential constructs (such as control-flow graphs), also support true concurrent constructs. These languages can be a natural fit to exploit the spatial and customization opportunities of FPGA-based computing architectures. While the extension of SSA form to these emerging languages is an open area of research, the fundamental uses of SSA for hardware compilation, as discussed in this chapter, are likely to remain a solid foundation for the mapping of these parallel constructs to hardware.

23.2   Why Use SSA for Hardware Compilation?

Hardware compilation, unlike its software counterpart, offers a spatially oriented computational infrastructure that presents opportunities to leverage information exposed by the SSA representation. We illustrate the direct connection between SSA representation form and hardware compilation using the example of computation mapping in Fig. 23.2a. Here the value of a variable v depends on the control flow of the computation, as the temporary variable t can be assigned different values depending on the value of the p predicate. The representation of this computation is depicted in Fig. 23.2b, where a ϕ-function is introduced to capture the two possible assignments to the temporary variable t in both control branches of the if-then-else construct. Lastly, in Fig. 23.2c, we illustrate the corresponding mapping to hardware.

The basic observation is that the confluence of values for a given program variable leads to the use of a ϕ-function. This ϕ-function abstraction thus corresponds in terms of hardware implementation to the insertion of a multiplexer logic circuit. This logic circuit uses the boolean value of a control input to select which of its input's values is to be propagated to its output. The selection or control input of

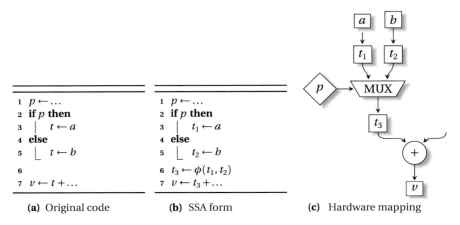

| (a) Original code | (b) SSA form | (c) Hardware mapping |

(a) Original code

```
1  p ← …
2  if p then
3  |   t ← a
4  else
5  |_  t ← b
6
7  v ← t + …
```

(b) SSA form

```
1  p ← …
2  if p then
3  |   t₁ ← a
4  else
5  |_  t₂ ← b
6  t₃ ← φ(t₁, t₂)
7  v ← t₃ + …
```

Fig. 23.2 Basic hardware mapping using SSA representation

a multiplexer thus acts as a gated transfer of value that parallels the actions of an `if-then-else` construct in software. Note also that in the case of a backward control flow (e.g., associated with a back edge of a loop), the possible indefinition of one of the ϕ-function's inputs is transparently ignored by the fact that in a correct execution the predicate associated with the true control-flow path will yield the value associated with a defined input of the SSA representation.

Equally important in this mapping is the notion that the computation in hardware can now take a spatial dimension. In the hardware circuit in Fig. 23.2c, the computation derived from the statement in both branches of the `if-then-else` construct can be evaluated concurrently by distinct logic circuits. After the evaluation of both circuits, the multiplexer will define which set of values is used based on the value of its control input, in this case of the value of the computation associated with p.

In a sequential software execution environment, the predicate p would be evaluated first, and then either branches of the `if-then-else` construct would be evaluated, based on the value of p; as long as the register allocator is able to assign t_1, t_2, and t_3 to the same register, then the ϕ-function is executed implicitly; if not, it is executed as a register-to-register copy.

There have been some efforts to automatically convert the sequential program above into a semi-spatial representation that could obtain some speedup if executed on a VLIW (very long instruction word) type of processor. For example, if-conversion (see Chap. 20) would convert the control dependency into a data dependency: statements from the `if` and `else` blocks could be interleaved, as long as they do not overwrite one another's values, and the proper result (the ϕ-function) could be selected using a conditional move instruction. In the worst case, however, this approach would effectively require the computation of both branch sides, rather than one, so it could increase the computation time. In contrast, in a spatial representation, the correct result can be output as soon as two of the three inputs to the multiplexer are known (p, and one of t_1 or t_2, depending on the value of p).

While the Electronic Design Automation (EDA), community has for decades now exploited similar information regarding data and control dependencies for the generation of hardware circuits from increasingly higher-level representations (e.g., Behavioural HDL), SSA-based representations make these dependencies explicit in the intermediate representation itself. Similarly, the more classical compiler representation, using three-address instructions augmented with the def-use chains, already exposes the data-flow information as for the SSA-based representation. However, as we will explore in the next section, the latter facilitates the mapping and selection of hardware resources.

23.3   Mapping a Control-Flow Graph to Hardware

In this section, we focus on hardware implementations of circuits that are spatial in nature. We therefore do not address the mapping to architectures such as VLIW or Systolic Arrays. While these architectures raise interesting and challenging issues, namely scheduling and resource usage, we are more interested in exploring and highlighting the benefits of SSA representation which, we believe, are more naturally (although not exclusively) exposed in the context of spatial hardware computations.

23.3.1   Basic Block Mapping

As a basic block is a straight-line sequence of three-address instructions, a simple hardware mapping approach consists in composing or evaluating the operations in each instruction as a data-flow graph. The inputs and outputs of the instructions are transformed into registers[3] connected by nodes in the graph that represent the operators.

As a result of the "evaluation" of the instructions in the basic block, this algorithm constructs a hardware circuit that has as input registers that will hold the values of the input variables to the various instructions and will have as outputs registers that hold only variables that are live outside the basic block.

---

[3] As a first approach, these registers are virtual, and then, after synthesis, some of them are materialized to physical registers in a process similar to register allocation in software-oriented compilation.

23.3.2   Basic Control-Flow Graph Mapping

One can combine the various hardware circuits corresponding to a control-flow graph in two basic approaches, respectively, *spatial* and *temporal*. The spatial form of combining hardware circuits consists in laying out the various circuits spatially by connecting variables that are live at the output of a basic block, and therefore the output registers of the corresponding hardware circuit, to the registers that will hold the values of those same variables in subsequent hardware circuits of the basic blocks that execute in sequence.

In the temporal approach, the hardware circuits corresponding to the various CFG basic blocks are not directly interconnected. Instead, their input and output registers are connected via dedicated buses to a local storage module. An execution controller "activates" a basic block or a set of basic blocks by transferring data between the storage and the input registers of the hardware circuits to be activated. Upon execution completion, the controller transfers the data from the output registers of each hardware circuit to the storage module. These data transfers do not necessarily need to be carried out sequentially, but instead can leverage the aggregation of the outputs of each hardware circuit to reduce transfer time to and from the storage module via dedicated wide buses.

The temporal approach described above is well suited to the scenario where the target hardware architecture does not have sufficient resources to simultaneously implement the hardware circuits corresponding to all basic blocks of interest, as it trades off execution time for hardware resources.

In such a scenario, where hardware resources are very limited or the hardware circuit corresponding to a set of basic blocks is exceedingly large, one could opt for partitioning a basic block or a set of basic blocks into smaller blocks until the space constraints for the realization of each hardware circuit are met. In reality, this is the common approach in every processor today. It limits the hardware resources to those required for each of the ISA instructions and schedules them in time at each step, saving the states (registers) that were the output of the previous instruction. The computation thus proceeds as described above by saving the values of the output registers of the hardware circuit corresponding to each smaller block.

These two approaches, illustrated in Fig. 23.3, can obviously be merged in a hybrid implementation. As they lead to distinct control schemes for the orchestration of the execution of computation in hardware, their choice depends heavily on the nature and granularity of the target hardware architecture. For fine-grain hardware architectures such as FPGAs, a spatial mapping can be favoured, while for coarse-grain architectures a temporal mapping is common.

While the overall execution control for the temporal mapping approach is simpler, as the transfers to and from the storage module are done upon the transfer of control between hardware circuits, a spatial mapping approach makes it easier to

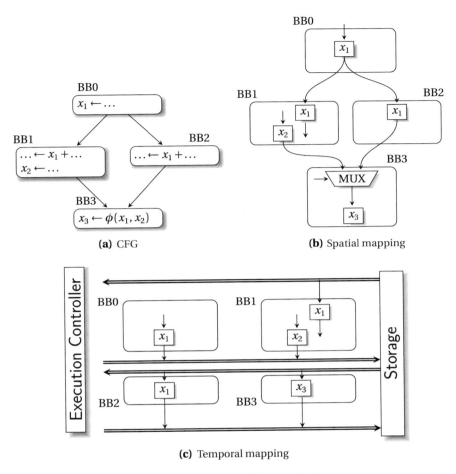

(a) CFG

(b) Spatial mapping

(c) Temporal mapping

Fig. 23.3 Combination of hardware circuits for multiple basic blocks

take advantage of pipelining execution techniques and speculation.[4] The temporal mapping approach can, however, be area-inefficient, as often only one basic block will execute at any point in time. This issue can nevertheless be mitigated by exposing additional amounts of instruction-level parallelism by merging multiple basic blocks into a single *hyper-block* and combining this aggregation with loop unrolling. Still, as these transformations and their combination can lead to a substantial increase in the required hardware resources, a compiler can exploit resource sharing between the hardware units corresponding to distinct basic blocks

---

[4] Speculation is also possible in the temporal mode by activating the inputs and execution of multiple hardware blocks and is only limited by the available storage bandwidth to restore the input context in each block, which, in the spatial approach, is trivial.

to reduce the pressure on resource requirements and thus lead to feasible hardware implementation designs. As these optimizations are not specific to the SSA representation, we will not discuss them further here.

23.3.3 Control-Flow Graph Mapping Using SSA

In the spatial mapping approach, the SSA form plays an important role in the minimization of multiplexers and thus in the simplification of the corresponding data-path logic and execution control.

Consider the illustrative example in Fig. 23.4a. Here, basic block BB0 defines a value for the variables x and y. One of the two subsequent basic blocks BB1 redefines the value of x, whereas the other basic block BB2 only reads x.

A naive implementation based exclusively on liveness analysis (see Chap. 9) would use multiplexers for both variables x and y to merge their values as inputs to the hardware circuit implementing basic block BB3, as depicted in Fig. 23.4b. As can be observed, however, the SSA form representation captures the fact that such a multiplexer is only required for variable x. The value for variable y can be propagated either from the output value in the hardware circuit for basic block BB0 (as shown in Fig. 23.4c) or from any other register that has a valid copy of the y variable. The direct flow of the single definition point to all its uses, across the hardware circuits corresponding to the various basic blocks in the SSA form, thus allows a compiler to use the minimal number of multiplexers strictly required.[5]

An important aspect of the implementation of a multiplexer associated with a ϕ-function is the definition and evaluation of the predicate associated with each multiplexer's control (or selection) input signal. In the basic SSA representation, the selection predicates are not explicitly defined, as the execution of each ϕ-function is implicit when the control flow reaches it. When mapping a computation to hardware, however, a ϕ-function clearly elicits the need to define a predicate to be included as part of the hardware logic circuit that defines the value of the multiplexer circuit's selection input signal. To this effect, hardware mapping must rely on a variant of SSA, named gated SSA (see Chap. 14), which explicitly captures the symbolic predicate information in the representation.[6] The generation of the hardware circuit simply uses the register that holds the corresponding variable's version value of the predicate. Figure 23.5 illustrates an example of a mapping using the information provided by the gated SSA form.

When combining multiple predicates in the gated SSA form, it is often desirable to leverage the control-flow representation in the form of the program dependence graph (PDG) described in Chap. 14. In the PDG representation, basic blocks that

---

[5] Under the scenarios of a spatial mapping and with the common disclaimers about static control-flow analysis.

[6] As with any SSA representation, variable names fulfil the referential transparency.

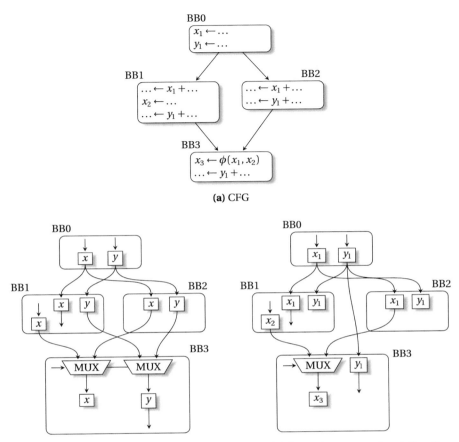

(a) CFG

(b) Naive multiplexer placement using liveness   **(c)** Multiplexer placement using ϕ-function

Fig. 23.4 Mapping of variable values across hardware circuit using spatial mapping

share common execution predicates (i.e., both execute under the same predicate conditions) are linked to the same *region* nodes. Nested execution conditions are easily recognized as the corresponding nodes are hierarchically organized in the PDG representation. As such, when generating code for a given basic block, an algorithm will examine the various region nodes associated with a given basic block and compose (using AND operators) the outputs of the logic circuits that implement the predicates associated with these nodes. If a hardware circuit already exists that evaluates a given predicate that corresponds to a given region, the implementation can simply reuse its output signal. This lazy code generation and predicate composition achieves the goal of hardware circuit sharing, as illustrated by the example in Fig. 23.6 where some of the details were omitted for simplicity. When using the PDG representation, however, care must be taken regarding the

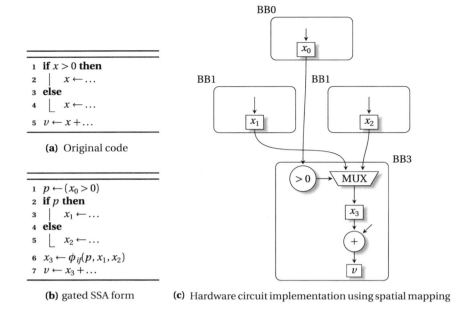

(a) Original code

(b) gated SSA form   **(c)** Hardware circuit implementation using spatial mapping

Fig. 23.5 Hardware generation example using gated SSA form

potential lack of referential transparency. To this effect, it is often desirable to combine the SSA information with the PDG's regions to ensure correct reuse of the hardware that evaluates the predicates associated with each control-dependence region.

23.3.4 φ-Function and Multiplexer Optimizations

We now describe a set of hardware-oriented transformations that can be applied to potentially reduce the amount of hardware resources devoted to multiplexer implementation or to use multiplexers to change the temporal features of the execution and thus enable other aspects of hardware execution to be more effective (e.g., scheduling). Although these transformations are not specific to the mapping of computations to hardware, the explicit representation of the selection constructs in SSA makes it very natural to map and therefore manipulate/transform the resulting hardware circuit using multiplexers. Other operations in the intermediate representation (e.g., predicated instructions) can also yield multiplexers in hardware without the explicit use of SSA form.

A first transformation is motivated by a well-known result in computer arithmetic: integer addition scales with the number of operands. Building a large, unified

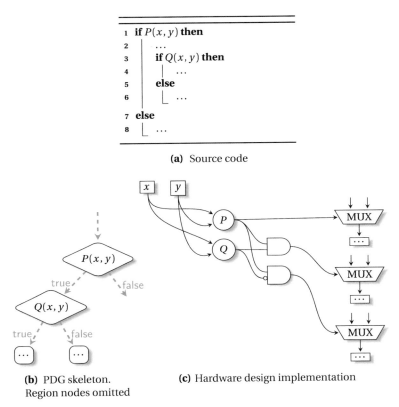

(a) Source code

(b) PDG skeleton.
Region nodes omitted

(c) Hardware design implementation

Fig. 23.6 Use of the predicates in region nodes of the PDG for mapping into the multiplexers associated with each ϕ-function

k-input integer addition circuit is more efficient than adding k integers two at a time. Moreover, hardware multipliers naturally contain multi-operand adders as building blocks: a partial product generator (a layer of AND gates) is followed by a multi-operand adder called a *partial product reduction tree*. For these reasons, there have been several efforts in recent years to apply high-level algebraic transformations to source code with the goal of merging multiple addition operations with partial product reduction trees of multiplication operations. The basic flavour of these transformations is to push the addition operators towards the outputs of a data-flow graph, so that they can be merged at the bottom. Examples of these transformations that use multiplexers are depicted in Fig. 23.7a and b. In the case of Fig. 23.7a, the transformation leads to the fact that an addition is always executed, unlike in the original hardware design. This can lead to more predictable timing or more uniform power draw signatures.[7] Figure 23.7c depicts a similar transformation that

---

[7] An important issue in security-related aspects of the execution.

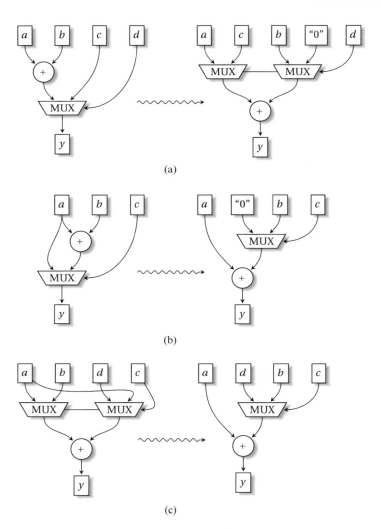

Fig. 23.7 Multiplexer-operator transformations: juxtaposition of a multiplexer and an adder (**a** and **b**); reducing the number of multiplexers placed on the input of an adder (**c**)

merges two multiplexers sharing a common input, while exploiting the commutative property of the addition operator. The SSA-based representation facilitates these transformations as it explicitly indicates which values (by tracing backward in the representation) are involved in the computation of the corresponding values. For the example in Fig. 23.7b, a compiler could quickly detect the variable a to be common to the two expressions associated with the ϕ-function.

A second transformation that can be applied to multiplexers is specific to FPGA, whose basic building block consists of a k-input lookup table (LUT) logic element which can be programmed to implement any k-input logic function.[8] For example, a 3-input LUT (3-LUT) can be programmed to implement a multiplexer with two data inputs and one selection bit. Similarly, a 6-LUT can be programmed to implement a multiplexer with four data inputs and two selection bits, thus enabling the implementation of a tree of 2-input multiplexers.

23.3.5 Implications of Using SSA Form in Floor Planning

For spatial oriented hardware circuits, moving a ϕ-function from one basic block to another can alter the length of the wires that are required to transmit data from the hardware circuits corresponding to the various basic blocks. As the boundaries of basic blocks are natural synchronization points, where values are captured in hardware registers, the length of wires dictates the maximum allowed hardware clock rate for synchronous designs. We illustrate this effect via an example as depicted in Fig. 23.8. In this figure, each basic block is mapped to a distinct hardware unit, whose spatial implementation is approximated by a rectangle. A floor planning algorithm must place each of the units in a two-dimensional plane while ensuring that no two units overlap. As can be seen in Fig. 23.8a, placing block 4 on the right-hand side of the plane will result in several mid-range and one long-range wire connections. However, placing block 4 at the centre of the design would virtually eliminate all mid-range connections as all connections corresponding to the transmission of the values for variable x are now next-neighbouring connections.

The same result could be obtained by moving the ϕ-function to block 5, which is illustrated in Fig. 23.8b. But moving a multiplexer from one hardware unit to another can significantly change the dimensions of the resulting unit, which is not under the control of the compiler. Changing the dimensions of the hardware units fundamentally changes the placement of modules, so it is very difficult to predict whether moving a ϕ-function will actually be beneficial. For this reason, compiler optimizations that attempt to improve the physical layout must be performed using a feedback loop so that the results of the lower-level CAD tools that produce the layout can be reported back to the compiler.

---

[8] A typical k-input LUT will include an arbitrary combinatorial functional block of those k inputs followed by an optional register element (e.g., Flip-Flop).

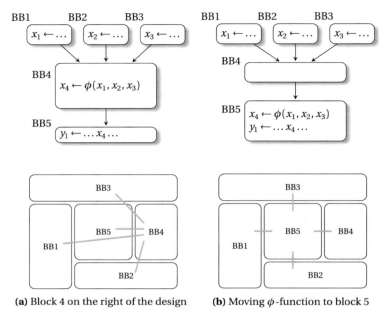

(a) Block 4 on the right of the design   **(b)** Moving ϕ-function to block 5

Fig. 23.8 Example of the impact of ϕ-function movement in reducing hardware wire length

23.4   Existing SSA-Based Hardware Compilation Efforts

Several research projects have relied on SSA-based intermediate representations that leverage control- and data-flow information to exploit fine-grain parallelism. Often, but not always, these efforts have been geared towards mapping computations to fine-grain hardware structure, such as the ones offered by FPGAs.

The standard approach followed in these compilers has been to translate a high-level programming language such as Java, in the case of the Sea Cucumber [288] compiler, or C, in the case of the DEFACTO [105] and ROCCC [133] compilers that translate C code to sequences of intermediate instructions. These sequences are then organized in basic blocks that compose the control-flow graph (CFG). For each basic block a data-flow graph (DFG) is typically extracted followed by a conversion into SSA representation, possibly using predicates associated with the control flow in the CFG, thus explicitly using Predicated SSA representation (see the gated SSA representation [290] and the Predicate SSA [58, 104, 275]).

As an approach to increase the potential amount of exploitable ILP at the instruction level, many of these efforts (as well as others such as the earlier Garp compiler [55]) restructure the CFG into hyper-blocks [191]. A hyper-block consists in single-entry multi-exit regions derived from the aggregation of multiple basic blocks, thus serializing longer sequences of instruction. As not all instructions are executed in a hyper-block (due to early exit of a block), hardware circuit implementation must rely on predication to exploit the potential for additional ILP.

The CASH compiler [54] uses an augmented predicated SSA representation with tokens to explicitly express synchronization and handle *may-dependencies*, thus supporting speculative execution. This fine-grain synchronization mechanism is also used to serialize the execution of consecutive hyper-blocks, thus greatly simplifying the code generation. Other efforts also exploit instruction-level optimizations or algebraic properties of the operators for minimization of expensive hardware resources such as multipliers, adders, and in some cases even multiplexers [198, 298]. For a comprehensive description of a wide variety of hardware-oriented high-level program transformations, the reader is referred to [57].

Further Reading

Despite their promise in terms of high performance and high computational efficiency, hardware devices such as FPGAs have long been beyond the reach of the "average" software programmer. To effectively program them using hardware-oriented programming languages such as VHDL [12], Verilog [283], OpenCL [173], or SystemC [218], the developers must assume the role of both software and hardware designs.

To address the semantic gap between a hardware-oriented programming model and high-level software programming models, various research projects, first in academia and later in industry, developed prototype tools that could bridge this gap and make the promising technology of configurable logic approachable among a wider audience of programmers. In these efforts, loosely labeled C-to-Gates, compilers performed the traditional phases of program data- and control-dependence analysis to uncover opportunities for concurrent and/or pipelined execution and directly translated the underlying data flow to Verilog/VHDL description alongside the corresponding control logic. In practice, these compilers, of which Vivado HLS [310] and LegUp [56] are notable examples, focus on loop constructs with significant execution time weights (the so-called hot-spots) for which they automatically derive hardware pipelined implementations (often guided by user-provided compilation directives) that efficiently executed them. When the target architecture is a "raw" FPGA (rather than an overlay architecture), this approach invariably leads to long compilation and synthesis times.

The inherent difficulties and limitations in extracting enough instruction-level Parallelism in these approaches, coupled with the increase in the devices' capacities (e.g. Intel's Arria [147] and Xilinx's Virtex UltraScale+ [311]), have prompted a search for programming models with a more natural concurrency that would facilitate the mapping of high-level computation to hardware. One such example is MapReduce[95], originally introduced by Google to naturally distribute concurrent jobs across clusters of servers [95], which has been an effective programming model for FPGAs [315]. Similarly, high-level languages based on parallel models of computation such as synchronous data flow [179] or functional single-assignment languages have also been shown to be good choices for hardware compilation [33, 137, 145].

Chapter 24
Building SSA in a Compiler for PHP

Paul Biggar and David Gregg

Dynamic scripting languages such as PHP, Python, and JavaScript are among the most widely used programming languages.

Dynamic scripting languages provide flexible high-level features, a fast modify-compile-test environment for rapid prototyping, strong integration with popular strongly typed programming languages, and an extensive standard library. Dynamic scripting languages are widely used in the implementation of many of the best-known web applications of the last decade such as Facebook, Wikipedia, and Gmail. Most web browsers support scripting languages such as JavaScript for client-side applications that run within the browser. Languages such as PHP and Ruby are popular for server-side web pages that generate content dynamically and provide close integration with back-end databases.

One of the most widely used dynamic scripting languages is PHP, a general-purpose language that was originally designed for creating dynamic web pages. PHP has many of the features that are typical of dynamic scripting languages. These include a simple syntax, dynamic typing and the ability to dynamically generate new source code during runtime, and execute that code. These simple flexible features facilitate rapid prototyping, exploratory programming, and in the case of many non-professional websites, a copy-paste-and-modify approach to scripting by non-expert programmers.

Constructing SSA form for languages such as C/C++ and Java is a well-studied problem. Techniques exist to handle the most common features of static languages, and these solutions have been tried and tested in production-level compilers over many years. In these static languages, it is not difficult to identify a set of scalar

P. Biggar (✉)
Darklang, New York, NY, USA
e-mail: paul.biggar@gmail.com

D. Gregg
Trinity College, Dublin, Ireland
e-mail: David.Gregg@tcd.ie

variables that can be safely renamed. Better analysis may lead to more such variables being identified, but significant numbers of such variables can be found with very simple analysis.

In our study of optimizing dynamic scripting languages, specifically PHP, we find this is not the case. The information required to build SSA form—that is, a conservatively complete set of unaliased scalars, and the locations of their uses and definitions—is not available directly from the program source and cannot be derived from a simple analysis. Instead, we find a litany of features whose presence must be ruled out, or heavily analysed, in order to obtain a non-pessimistic estimate.

Dynamic scripting languages commonly feature runtime generation of source code, which is then executed, built-in functions, and variable-variables, all of which may alter arbitrary unnamed variables. Less common—but still possible—features include the existence of object handlers which have the ability to alter any part of the program state, most dangerously a function's local symbol table. The original implementations of dynamic scripting languages were all interpreted, and in many cases, this led to their creators including very dynamic language features that are easy to implement in an interpreter but make program analysis and compilation very difficult.

Ruling out the presence of these features requires precise, inter-procedural, whole-program analysis. We discuss the futility of the pessimistic solution, the analyses required to provide a precise SSA form, and how the presence of variables of unknown types affects the precision of SSA.

Note that dynamic scripting languages share similar features of other kinds of dynamic programming languages, such as Lisp and Smalltalk. The main distinguishing feature of dynamic languages is that they resolve at runtime behaviours that many other languages perform at compile time. Thus, languages such as Lisp support dynamic typing and runtime generation of source code which can then be executed. The main difference between scripting languages and other dynamic languages is their intended use, rather than the features of the language. Scripting languages were originally designed for writing short scripts, often to invoke other programs or services. It was only later that developers started to write large complex programs using scripting languages.

The rest of this chapter describes our experiences with building SSA in PHC, an open source compiler for PHP. We identify the features of PHP that make building SSA difficult, outline the solutions we found to these some of these challenges, and draw some lessons about the use of SSA in analysis frameworks for PHP.

24.1 SSA Form in Statically Typed Languages

In statically typed, non-scripting languages, such as C, C++, and Java, it is straightforward to identify which variables may be converted into SSA form. In Java, all scalars may be renamed. In C and C++, any scalar variables that do not have their address taken may be renamed. Figure 24.1 shows a simple C function,

Fig. 24.1 Simple C code

```
1: int factorial(int num) {
2:    int result = 1, i;
3:
4:    while ( num > 1 ) {
5:       result = result * num;
6:       num --;
7:    }
8:    return result;
9: }
```

Fig. 24.2 C code with
pointers

```
1:   void func() {
2:      int x=5, *y;
3:      y = &x;
4:      *y = 7;
5:      return x;
6:   }
```

whose local variables may all be converted into SSA form. In Fig. 24.1, it is trivial
to convert each variable into SSA form. For each statement, the list of variables that
are used and defined is immediately obvious from a simple syntactic check.

By contrast, Fig. 24.2 contains variables that cannot be trivially converted into
SSA form. On Line 3, the variable x has its address taken. As a result, to convert x
into SSA form, we must know that Line 4 modifies x and introduce a new version
of x accordingly. Chapter 16 describes HSSA form, a powerful way to represent
indirect modifications to scalars in SSA form.

To discover the modification of x on Line 4 requires an *alias analysis*. Alias
analyses detect when multiple program names (*aliases*) represent the same memory
location. In Fig. 24.2, x and $*y$ alias each other.

There are many variants of alias analysis, of varying complexity. The most
complex ones analyse the entire program text, taking into account the control flow,
the function calls, and multiple levels of pointers. However, it is not difficult to
perform a very simple alias analysis. *Address-taken alias analysis* identifies all
variables whose addresses are taken by a referencing assignment. All variables
whose address has been taken anywhere in the program are considered to alias each
other (that is, all address-taken variables are in the same *alias set*).

When building SSA form with address-taken alias analysis, variables in the alias
set are not renamed into SSA form. All other variables are converted. Variables not
in SSA form do not possess any SSA properties, and pessimistic assumptions must
be made. As a result of address-taken alias analysis, it is straightforward to convert
any C program into SSA form, without sophisticated analysis. In fact, this allows
more complex alias analyses to be performed on the SSA form.

24.2 PHP and Aliasing

In PHP, it is not possible to perform an address-taken alias analysis. The syntactic
clues that C provides—notably as a result of static typing—are not available in PHP.
Indeed, statements that may appear innocuous may perform complex operations
behind the scenes, affecting any variable in the function. Figure 24.3 shows a small
piece of PHP code. Note that in PHP, variable names start with the $ sign.

The most common appearance of aliasing in PHP is due to variables that store
references. Creating references in PHP does not look so very different from C. In
Fig. 24.3, the variable y becomes a reference to the variable x. Once y has become
an alias for x, the assignment to y (in Line 3) also changes the value of x.

On first glance, the PHP code in Fig. 24.3 is not very different from similar C
code in Fig. 24.2. From the syntax, it is easy to see that a reference to the variable x
is taken. Thus, it is clear that x cannot be easily renamed. However, the problem is
actually with the variable y which contains the reference to the variable x.

There is no type declaration to say that y in Fig. 24.3 is a reference. In fact, due
to dynamic typing, PHP variables may be references at one point in the program
and stop being references a moment later. Or at a given point in a program, a given
variable may be a reference variable or a non-reference variable depending upon
the control flow that preceded that point. PHP's dynamic typing makes it difficult to
simply identify when this occurs, and a sophisticated analysis over a larger region
of code is essential to building a more precise conservative SSA form.

The size of this larger region of code is heavily influenced by the semantics of
function parameters in PHP. Consider the PHP function in Fig. 24.4a. Superficially,
it resembles the C function in Fig. 24.4b. In the C version, we know that x and y are
simple integer variables and that no pointer aliasing relationship between them is
possible. They are separate variables that can be safely renamed. In fact, a relatively
simple analysis can show that the assignment to x in Line 2 of the C code can be
optimized away because it is an assignment to a dead variable.

In the PHP version in Fig. 24.4a, x may alias y upon function entry. This can
happen if x is a reference to y, or vice versa, or if both x and y are references to a

```
1:    $x = 5;
2:    $y =& $x;
3:    $y = 7;
```

Fig. 24.3 Creating and using a reference in PHP

```
1:    function foo($x, $y) {        1:    int foo(int x, int y) {
2:        $x = $x + $y;             2:        x = x + y;
3:        return $y;                3:        return y;
4:    }                            4:    }
              (a)                             (b)
```

Fig. 24.4 Similar (**a**) PHP and (**b**) C functions with parameters

third variable. It is important to note, however, that the possibility of such aliasing is not apparent from the function prototype or any type declarations. Instead, whether the formal parameter x and/or y are references depends on the types of actual parameters that are passed when the function is invoked. If a reference is passed as a parameter to a function in PHP, the corresponding formal parameter in the function also becomes a reference.

The addition operation in Line 2 of Fig. 24.4a may therefore change the value of y, if x is a reference to y or vice versa. In addition, recall that dynamic typing in PHP means that whether or not a variable contains a reference can depend on control flow leading to different assignments. Therefore, on some executions of a function, the passed parameters may be references, whereas on other executions they may not.

In the PHP version, there are no syntactic clues that variables may alias. Furthermore, as we show in Sect. 24.4, there are additional features of PHP that can cause the values of variables to be changed without simple syntactic clues. In order to be sure that no such features can affect a given variable, an analysis is needed to detect such features. As a result, a simple conservative aliasing estimate that does not take account of PHP's references and other difficult features—similar to C's address-taken alias analysis—would need to place all variables in the alias set. This would leave no variables available for conversion to SSA form. Instead, an inter-procedural analysis is needed to track references between functions.

24.3   Our Whole-Program Analysis

PHP's dynamic typing means that program analysis cannot be performed a function at a time. As function signatures do not indicate whether parameters are references, this information must be determined by inter-procedural analysis. Furthermore, each function must be analysed with full knowledge of its calling context. This requires a whole-program analysis. We present an overview of the analysis below. A full description is beyond the scope of this chapter.

24.3.1   The Analysis Algorithm

The analysis is structured as a symbolic execution. This means the program is analysed by processing each statement in turn and modelling the effect of the statement on the program state. This is similar in principle to the sparse conditional constant propagation (SSA-CCP) algorithm described in Sect. 8.3.[1]

The SCCP algorithm models a function at a time. Instead, our algorithm models the entire program. The execution state of the program begins empty, and the

---

[1] The analysis is actually based on a variation, conditional constant propagation.

analysis begins at the first statement in the program, which is placed in a worklist. The worklist is then processed a statement at a time.

For each analysed statement, the results of the analysis are stored. If the analysis results change, the statement's direct successors (in the control-flow graph) are added to the worklist. This is similar to to the treatment of CFG edges (through *CFGWorkList*) in the SCCP algorithm. There is no parallel to the treatment of SSA edges (*SSAWorkList*), since the analysis is not performed on the SSA form. Instead, loops must be fully analysed if their headers change.

This analysis is therefore less efficient than the SCCP algorithm, in terms of time. It is also less efficient in terms of space. SSA form allows results to be compactly stored in a single array, using the SSA index as an array index. This is very space efficient. In our analysis, we must instead store a table of variable results at all points in the program.

Upon reaching a function or method call, the analysis begins analysing the callee function, pausing the caller's analysis. A new worklist is created and initialized with the first statement in the callee function. The worklist is then run until it is exhausted. If another function call is analysed, the process recurses.

Upon reaching a callee function, the analysis results are copied into the scope of the callee. Once a worklist has ended, the analysis results for the exit node of the function are copied back to the calling statement's results.

24.3.2 Analysis Results

Our analysis computes and stores three different kinds of results. Each kind of result is stored at each point in the program.

The first models the alias relationships in the program in a *points-to graph*. The graph contains variable names as nodes, and the edges between them indicate aliasing relationships. An aliasing relationship indicates that two variables either *must*-alias or *may*-alias. Two unconnected nodes cannot alias. A points-to graph is stored for each point in the program. Graphs are merged at CFG join nodes.

Secondly, our analysis also computes a conservative estimate of the types of variables in the program. Since PHP is an object-oriented language, polymorphic method calls are possible, and they must be analysed. As such, the set of possible types of each variable is stored at each point in the program. This portion of the analysis closely resembles using SCCP for type propagation.

Finally, like the SCCP algorithm, constants are identified and propagated through the analysis of the program. Where possible, the algorithm resolves branches statically using propagated constant values. This is particularly valuable because our PHC ahead-of-time compiler for PHP creates many branches in the intermediate representation during early stages of compilation. Resolving these branches statically eliminates unreachable paths, leading to significantly more precise results from the analysis algorithm.

24.3.3 Building Def-Use Sets

To build SSA form, we need to be able to identify the set of points in a program where a given variable is defined or used. Since we cannot easily identify these sets due to potential aliasing, we build them as part of our program analysis. Using our alias analysis, any variables that may be written to or read from during a statement's execution are added to a set of defs and uses for that statement. These are then used during construction of the SSA form.

For an assignment by copy, $x = $y:

1. $x's value is defined.
2. $x's reference is used (by the assignment to $x).
3. For each alias $x_alias of $x, $x_alias's value is defined. If the alias is possible, $x_alias's value is may-defined instead of defined. In addition, $x_alias's reference is used.
4. $y's value is used.
5. $y's reference is used.

For an assignment by reference, $x =& $y:

1. $x's value is defined.
2. $x's reference is defined (it is not used—$x does not maintain its previous reference relationships).
3. $y's value is used.
4. $y's reference is used.

24.3.4 HSSA

Once the set of locations where each variable is defined or used has been identified, we have the information needed to construct SSA. However, it is important to note that due to potential aliasing and potential side effects of some difficult-to-analyse PHP features (see Sect. 24.4), many of the definitions we compute are may-definitions. Whereas a normal definition of a variable means that the variable will definitely be defined, a may-definition means that the variable may be defined at that point.[2]

In order to accommodate these may-definitions in our SSA form for PHP, we adapt a number of features from Hashed SSA form.

---

[2] Or more precisely, a may-definition means that there exists at least one possible execution of the program where the variable is defined at that point. Our algorithm computes a conservative approximation of may-definition information. Therefore, our algorithm reports a may-definition in any case where the algorithm cannot prove that no such definition can exist on any possible execution.

After performing our alias analysis, we convert our intermediate representation into SSA form. As in HSSA form, we give names to all memory locations in our program, including locations inside arrays and hash tables, if they are directly referred to in source being analysed.

We augment our SSA form with μ and χ functions, representing may-uses and may-definitions, respectively.

Principally, the difference between our SSA form and traditional SSA is that all names in the program are included in the SSA representation. In addition, for each statement in the program, we maintain a set of names which must be defined, which might be defined, and which are used (either possibly or definitely). Names that may be defined are the same concept as χ variables in HSSA, but there may also be more than one name that *must* be defined.

We also add ϕ-functions for names that are only used in χ operations or which must be defined but are not syntactically present in the statement. As these ϕ-functions are not for real variables, we cannot add copies to represent them when destructing SSA form and must instead drop all SSA indices. While this is a valid means of destructing SSA form, it slightly limits the optimizations that can be performed. For example, copy-propagation cannot move copies past the boundary of a variable's live range.

We do not use virtual variables, zero versioning, or global value numbering (the *hash* in *Hashed SSA*). However, in order to scale our SSA form in the future, it would be necessary to incorporate these features.

The final distinction from traditional SSA form is that we do not only model scalar variables. All names in the program, such as fields of objects or the contents of arrays, can be represented in SSA form.

24.4 Other Challenging PHP Features

For simplicity, the description so far of our algorithm has only considered the problems arising from aliasing due to PHP reference variables. However, in the process of constructing SSA in our PHP compiler several other PHP language features make it difficult to identify all the points in a program where a variable may be defined. In this section, we briefly describe these language features and how they may be dealt with in order to conservatively identify all may-definitions of all variables in the program.

24.4.1 Runtime Symbol Tables

Figure 24.5 shows a program which accesses a variable indirectly. On line 2, a string value is read from the user and stored in the variable *var_name*. On line 3, some variable—whose name is the value in *var_name*—is set to 5. That is, any variable

Fig. 24.5 PHP code using a
variable-variable

```
1:  $x = 5;
2:  $var_name = readline( );
3:  $$var_name = 6;
4:  print $x;
```

can be updated, and the updated variable is chosen by the user at runtime. It is not
possible to know whether the user has provided the value "*x*" and so know whether
x has the value 5 or 6.

This feature is known as *variable-variables*. They are possible because a
function's symbol table in PHP is available at runtime. Variables in PHP are not
the same as variables in C. A C local variable is a name which represents a memory
location on the stack. A PHP local variable is the domain of a map from strings
to values. The same runtime value may be the domain of multiple string keys
(references, discussed above). Similarly, variable-variables allow the symbol table
to be accessed dynamically at runtime, allowing arbitrary values to be read and
written.

Upon seeing a variable-variable, all variables may be written to. In HSSA form,
this creates a χ-function for every variable in the program. In order to reduce the set
of variables that might be updated by assigning to a variable-variable, the contents of
the string stored in *var_name* may be modelled using a string analysis by Wasserman
and Su.

String analysis is a static program analysis that models the structure of strings.
For example, string analysis may be able to tell us that the name stored in the
variable-variable (that is the name of the variable that will be written to) begins with
the letter "x." In this case, all variables that do not begin with "x" do not require a
χ-function, leading to a more precise SSA form.

24.4.2   Execution of Arbitrary Code

PHP provides a feature that allows an arbitrary string to be executed as a code. This
eval statement simply executes the contents of any string as if the contents of the
string appeared inline in the location of the *eval* statement. The resulting code can
modify local or global variables in arbitrary ways. The string may be computed by
the program or read in from the user or from a web form.

The PHP language also allows the contents of a file to be imported into the
program text using the *include* statement. The name of the file is an arbitrary
string which may be computed or read in at runtime. The result is that any file
can be included in the text of the program, even one that has just been created by
the program. Thus, the *include* statement is potentially just as flexible as the *eval*
statement for arbitrarily modifying program state.

Both of these may be modelled using the same string analysis techniques as
discussed in Sect. 24.4.1. The *eval* statement may also be handled using profiling,

which restricts the set of possible *eval* statements to those which actually are used in practice.

24.4.3   Object Handlers

PHP's reference implementation allows classes to be partially written in C. Objects that are instantiations of these classes can have special behaviour in certain cases. For example, the objects may be dereferenced using array syntax. Or a special handler function may be called when the variable holding the object is read or written to.

The handler functions for these special cases are generally unrestricted, meaning that they can contain arbitrary C code that does anything the programmer chooses. Being written in C, they are given access to the entire program state, including all local and global variables.

These two characteristics of handlers make them very powerful but break any attempts at function-at-a-time analysis. If one of these objects is passed into a function and is read, it may overwrite all variables in the local symbol table. The overwritten variables might then have the same handlers. These can then be returned from the function or passed to any other called functions (indeed, it can also call any function). This means that a single unconstrained variable in a function can propagate to any other point in the program, and we need to treat this case conservatively.

24.5   Concluding Remarks and Further Reading

In this chapter, we have described our experience of building SSA in PHC, an ahead-of-time compiler for PHP. PHP is quite different to static languages such as C/C++ and Java. In addition to dynamic typing, it has other dynamic features such as very flexible and powerful reference types, variable-variables, and features that allow almost arbitrary code to be executed.

The main result of our experience is to show that SSA cannot be used as an end-to-end intermediate representation (IR) for a PHP compiler. The main reason is that in order to build SSA, significant analysis of the PHP program is needed to deal with aliasing and to rule out potential arbitrary updates of variables. We have found that in real PHP programs these features are seldom used in ways that make analysis really difficult [28]. But analysis is nonetheless necessary to show the absence of the bad use of these features.

In principle, our analysis could perform only the alias analysis prior to building SSA and perform type analysis and constant propagation in SSA. But our experience is that combining all three analyses greatly improves the precision of alias analysis.

In particular, type analysis significantly reduces the number of possible method callees in object-oriented PHP programs.

Once built, the SSA representation forms a platform for further analysis and optimization. For example, we have used it to implement an aggressive dead-code elimination pass which can eliminate both regular assignments and reference assignments.

In recent years, there has been an increasing interest in the static analysis of scripting languages, with a particular focus on detecting potential security weaknesses in web-based PHP programs. For example, Jovanovic et al. [156] describe an analysis of PHP to detect vulnerabilities that allow the injection of executable code into a PHP program by a user or web form. Jensen et al. describe an alias analysis algorithm for JavaScript, which works quite differently to ours [28] but works well for JavaScript [150].

To our knowledge, we are the first to investigate building SSA in a PHP compiler. Our initial results show that building a reasonably precise SSA for PHP requires a great deal of analysis. Nonetheless, as languages such as PHP are used to build a growing number of web applications, we expect that there will be an increasing interest in tools for the static analysis of scripting languages.

Further Reading

The HSSA algorithm is described further in Chap. 16. Array SSA form—as described in Chap. 17—was also considered for our algorithm. While we deemed HSSA a better match for the problem space, we believe we could also have adapted Array SSA for this purpose.

Our propagation algorithm is based on the Conditional Constant Propagation algorithm by Pioli and Hind [227]. It was also based on the SSA-based cousin of CCP, *Sparse* Conditional Constant Propagation (SSA-CCP). SSA-CCP is described earlier, in Chaps. 8 and 17.

Our type-inference algorithm is based on the work by Lenart [181], which itself is based on SSA-CCP.

The variant of CCP on which we based our analysis considers alias analysis [110] alongside constant propagation. Our alias analysis was based on "points-to" analysis by Emami et al. [110]. A more detailed analysis of our alias analysis and how it interacts with constant propagation is available [28]. Previous work also looked at alias analysis in PHP, taking a simpler approach to analyse a subset of PHP [156].

Further propagation algorithms can lead to a more precise analysis. In particular, we believe string analysis, based on the work by Wassermann and Su [302], can help reduce the set of possible variables in an alias set.

This chapter is based on original research by Biggar [28], containing more discussion on our alias analysis, our form of SSA construction, and the problem of static analysis of dynamic scripting languages.

References

1. Aho, A. V., & Johnson, S. C. (1976). Optimal code generation for expression trees. *Journal of the ACM, 23*(3), 488–501.
2. Aho, A. V., Sethi, R., & Ullman, J. D. (1986). *Compilers: Principles, techniques, and tools. Addison-Wesley series in computer science/World student series edition.*
3. Allen, F. E. (1970). Control flow analysis. *Sigplan Notices, 5*(7), 1–19 (1970).
4. Allen, F. E., & Cocke, J. (1976). A program data flow analysis procedure. *Communications of the ACM, 19*(3), 137–146.
5. Allen, J. R., et al. (1983). Conversion of control dependence to data dependence. In *Proceedings of the Symposium on Principles of Programming Languages. POPL '83* (pp. 177–189).
6. Alpern, B., Wegman, M. N., & Kenneth Zadeck, F. (1988). Detecting equality of variables in programs. In *Proceedings of the Symposium on Principles of Programming Languages. POPL '88*, pp. 1–11.
7. Amaral, J. N., et al. (2001). Using the SGI Pro64 open source compiler infra-structure for teaching and research. In *Symposium on Computer Architecture and High Performance Computing* (pp. 206–213).
8. Ananian, S. (1999). *The Static Single Information Form*. Master's Thesis. MIT.
9. Appel, A. W. (1992). *Compiling with continuations*. Cambridge: Cambridge University Press.
10. Appel, A. W. (1998). *Modern compiler implementation in {C,Java,ML}*. Cambridge: Cambridge University Press.
11. Appel, A. W. (1998). SSA is functional programming. *Sigplan Notices, 33*(4), 17–20 (1998).
12. Ashenden, P. J. (2001). *The designer's guide to VHDL*. Burlington: Morgan Kaufmann Publishers Inc.
13. August, D. I., et al. (1998). Integrated predicated and speculative execution in the IMPACT EPIC architecture. In *Proceedings of the International Symposium on Computer Architecture. ISCA '98* (pp. 227–237).
14. Aycock, J., & Horspool, N., (2000). Simple generation of static single assignment form. In *International Conference on Compiler Construction. CC '00* (pp. 110–125).
15. Bachmann, O., Wang, P. S., & Zima, E. V. (1994). Chains of recurrences—A method to expedite the evaluation of closed-form functions. In *Proceedings of the International Symposium on Symbolic and Algebraic Computation* (pp. 242–249).
16. Banerjee, U. (1988). *Dependence analysis for supercomputing*. Alphen aan den Rijn: Kluwer Academic Publishers.
17. Barik, R. (2010). *Efficient Optimization of Memory Accesses in Parallel Programs*. PhD Thesis. Rice University.

18. Barik, R., & Sarkar, V. (2009). Interprocedural load elimination for dynamic optimization of parallel programs. In *Proceedings of the International Conference on Parallel Architectures and Compilation Techniques. PACT '09* (pp. 41–52).

19. Barthou, D., Collard, J.-F., & Feautrier, P. (1997). Fuzzy array dataflow analysis. *Journal of Parallel and Distributed Computing, 40*(2), 210–226.

20. Bates, S., & Horwitz, S. (1993). Incremental program testing using program dependence graphs. In *Proceedings of the Symposium on Principles of Programming Languages. POPL '93* (pp. 384–396).

21. Baxter, W. T., & Bauer III, H. R. (1989). The program dependence graph and vectorization. *Proceedings of the Symposium on Principles of Programming Languages. POPL '89* (pp. 1–11).

22. Belady, L. A. (1966). A study of replacement of algorithms for a virtual storage computer. In *IBM Systems Journal, 5*, 78–101.

23. Bender, M., & Farach-Colton, M. (2000). The LCA problem revisited. In *Latin 2000: Theoretical informatics. Lecture notes in computer science* (pp. 88–94).

24. Benton, N., Kennedy, A., & Russell, G. (1998). Compiling standard ML to Java bytecodes. In *Proceedings of the International Conference on Functional Programming. ICFP '98. SIGPLAN Notices, 34*(1), 129–140.

25. Beringer, L. (2007). Functional elimination of phi-instructions. In *Electronic Notes in Theoretical Computer Science, 176*(3), 3–20.

26. Beringer, L., MacKenzie, K., & Stark, I. (2003). Grail: A functional form for imperative mobile code. In *Electronic Notes in Theoretical Computer Science, 85*(1), 3–23.

27. Berson, D. A., Gupta, R., & Soffa, M. L. (1999). Integrated instruction scheduling and register allocation techniques. In *Proceedings of the International Workshop on Languages and Compilers for Parallel Computing. LCPC '98* (pp. 247–262).

28. Biggar, P. (2010). *Design and Implementation of an Ahead-of-Time Compiler for PHP*. PhD Thesis. Trinity College Dublin.

29. Bilardi, G., & Pingali, K. (1996). A framework for generalized control dependence. *Sigplan Notices, 31*(5), 291–300.

30. Blech, J. O., et al. (2005). Optimizing code generation from SSA form: A comparison between two formal correctness proofs in Isabelle/HOL. In *Electronic Notes in Theoretical Computer Science, 141*(2), 33–51.

31. Blickstein, D. S., et al. (1992). The GEM optimizing compiler system. *Digital Technical Journal, 4*(4), 121–136.

32. Bodík, R., Gupta, R., & Sarkar, V. (2000). ABCD: Eliminating array bounds checks on demand. In *International Conference on Programming Languages Design and Implementation. PLDI '00* (pp. 321–333).

33. Böhm, W., et al. (2002). Mapping a single assignment programming language to reconfigurable systems. *The Journal of Supercomputing, 21*(2), 117–130.

34. Boissinot, B., et al. (2008). Fast liveness checking for SSA-form programs. In *Proceedings of the International Symposium on Code Generation and Optimization. CGO '08* (pp. 35–44).

35. Boissinot, B., et al. (2009). Revisiting out-of-SSA translation for correctness, code quality and efficiency. In *Proceedings of the International Symposium on Code Generation and Optimization. CGO '09* (pp. 114–125).

36. Boissinot, B., et al. (2011). A non-iterative data-flow algorithm for computing liveness sets in strict SSA programs. In *Asian Symposium on Programming Languages and Systems. APLAS '11* (pp. 137–154).

37. Boissinot, B., et al. (2012). SSI properties revisited. In *ACM Transactions on Embedded Computing Systems*. Special Issue on Software and Compilers for Embedded Systems.

38. Bouchez, F. (2009). *A Study of Spilling and Coalescing in Register Allocation as Two Separate Phases*. PhD Thesis. École normale supérieure de Lyon, France.

39. Bouchez, F., et al. (2006). Register allocation: What does the NP-completeness proof of chaitin et al. really prove? Or revisiting register allocation: Why and how. In *Proceedings of the International Workshop on Languages and Compilers for Parallel Computing. LCPC '06* (pp. 283–298).

40. Bouchez, F., Darte, A., & Rastello, F. (2007). On the complexity of register coalescing. In *Proceedings of the International Symposium on Code Generation and Optimization. CGO '07* (pp. 102–114).

41. Bouchez, F., Darte, A., & Rastello, F. (2007). On the complexity of spill everywhere under SSA form. In *Proceedings of the International Conference on Languages, Compilers, and Tools for Embedded Systems. LCTES '07* (pp. 103–112).

42. Bouchez, F., Darte, A., & Rastello, F. (2008). Advanced conservative and optimistic register coalescing. In *Proceedings of the International Conference on Compilers, Architecture, and Synthesis for Embedded Systems. CASES '08* (pp. 147–156).

43. Bouchez, F., et al. (2010). Parallel copy motion. In *Proceedings of the International Workshop on Software & Compilers for Embedded Systems. SCOPES '10* (pp. 1:1–1:10).

44. Brandis, M. M., & Mössenböck, H. (1994). Single-pass generation of static single assignment form for structured languages. *ACM Transactions on Programming Language and Systems, 16*(6), 1684–1698.

45. Braun, M., & Hack, S. (2009). Register spilling and live-range splitting for SSA-form programs. In *International Conference on Compiler Construction. CC '09* (pp. 174–189).

46. Braun, M., Mallon, C., & Hack, S. (2010). Preference-guided register assignment. In *International Conference on Compiler Construction. CC '10* (pp. 205–223). New York: Springer.

47. Braun, M., et al. (2013). Simple and efficient construction of static single assignment form. In *International Conference on Compiler Construction. CC '13* (pp. 102–122).

48. Briggs, P., Cooper, K. D., & Torczon, L. (1992). Rematerialization. In *International Conference on Programming Languages Design and Implementation. PLDI '92* (pp. 311–321).

49. Briggs, P., Cooper, K. D., & Taylor Simpson, L. (1997). Value numbering. In *Software Practice and Experience, 27*(6), 701–724.

50. Briggs, P., et al. (1998). Practical improvements to the construction and destruction of static single assignment form. In *Software—Practice and Experience, 28*(8), 859–881.

51. Brisk, P., et al. (2005). Polynomial-time graph coloring register allocation. In *International Workshop on Logic and Synthesis. IWLS '05*.

52. Bruel, C. (2006). If-conversion SSA framework for partially predicated VLIW architectures. In *Odes 4. SIGPLAN* (pp. 5–13).

53. Budimlić, Z., et al. (2002). Fast copy coalescing and live range identification. In *International Conference on Programming Languages Design and Implementation. PLDI '02* (pp. 25–32).

54. Budiu, M., & Goldstein, S. C. (2002). Compiling application-specific hardware. In *International Conference on Field Programmable Logic and Applications. FPL '02* (pp. 853–863).

55. Callahan, T. J., Hauser, J. R., & Wawrzynek, J. (2000). The Garp architecture and C compiler. *Computer, 33*(4), 62–69 (2000).

56. Canis, A., et al. (2011). High-level synthesis for FPGA-based processor/accelerator systems. In *Proceedings of the 19th ACM/SIGDA International Symposium on Field Programmable Gate Arrays. FPGA '11* (pp. 33–36).

57. Cardoso, J. M. P., & Diniz, P. C. (2008). *Compilation techniques for reconfigurable architectures*. New York: Springer.

58. Carter, L., et al. (1999). Predicated static single assignment. In *Proceedings of the International Conference on Parallel Architectures and Compilation Techniques. PACT '99* (p. 245).

59. Chaitin, G. J. (1982). Register allocation & spilling via graph coloring. In *Proceedings of the Symposium on Compiler Construction. SIGPLAN '82* (pp. 98–105).

60. Chaitin, G. J., et al. (1981). Register allocation via coloring. In *Computer Languages, 6,* 47–57.

61. Chaitin, G. J., et al. (1981). Register allocation via graph coloring. *Journal of Computer Languages, 6,* 45–57 (1981).

62. Chakravarty, M. M. T., Keller, G., & Zadarnowski, P. (2003). A functional perspective on SSA optimisation algorithms. *Electronic Notes in Theoretical Computer Science, 82*(2), 15.

63. Chambers, C., & Ungar, D. (1989). Customization: Optimizing compiler technology for SELF, a dynamically-typed object-oriented programming language. *Sigplan Notices, 24*(7), 146–160.

64. Chan, S., et al. (2008). Open64 compiler infrastructure for emerging multicore/manycore architecture. Tutorial at *the International Symposium on Parallel and Distributed Processing. SPDP '08.*

65. Chapman, B., Eachempati, D., & Hernandez, O. (2013). Experiences developing the OpenUH compiler and runtime infrastructure. *International Journal of Parallel Programming, 41*(6), 825–854.

66. Chen, C.-H. (1988). *Signal processing handbook* (Vol. 51). Boca Raton: CRC Press.

67. Choi, J.-D., Cytron, R. K., & Ferrante, J. (1991). Automatic construction of sparse data flow evaluation graphs. In *Proceedings of the Symposium on Principles of Programming Languages. POPL '91* (pp. 55–66).

68. Choi, J.-D., Cytron, R. K., & Ferrante, J. (1994). On the efficient engineering of ambitious program analysis. *IEEE Transactions on Software Engineering, 20*, 105–114.

69. Choi, J.-D., Sarkar, V., & Schonberg, E. (1996). Incremental computation of static single assignment form. In: *International Conference on Compiler Construction. CC '96* (pp. 223–237).

70. Chow, F. (1988). Minimizing register usage penalty at procedure calls. In *International Conference on Programming Languages Design and Implementation. PLDI '88* (pp. 85–94).

71. Chow, F., & Hennessy, J. L. (1990). The priority-based coloring approach to register allocation. *ACM Transactions on Programming Language and Systems, 12*(4), 501–536.

72. Chow, F., et al. (1996). Effective representation of aliases and indirect memory operations in SSA form. In *International Conference on Compiler Construction. CC '96* (pp. 253–267).

73. Chow, F., et al. (1997). A new algorithm for partial redundancy elimination based on SSA form. In *International Conference on Programming Languages Design and Implementation. PLDI '97* (pp. 273–286).

74. Chuang, W., Calder, B., & Ferrante, J. (2003). Phi-predication for light-weight if-conversion. In *Proceedings of the International Symposium on Code Generation and Optimization. CGO '03* (pp. 179–190).

75. Click, C. (1995). *Combining Analyses, Combining Optimizations*. PhD Thesis. Rice University.

76. Click, C. (1995). Global code motion/global value numbering. In *International Conference on Programming Languages Design and Implementation. PLDI '95* (pp. 246–257).

77. Cocke, J. W., & Schwartz, J. T. (1970). *Programming languages and their compilers*. New York: New York University.

78. Codina, J. M., Sánchez, J., & González, A. (2001). A unified modulo scheduling and register allocation technique for clustered processors. In *Proceedings of the International Conference on Parallel Architectures and Compilation Techniques. PACT '01* (pp. 175–184).

79. Colombet, Q. (2012). *Decoupled (SSA-Based) Register Allocators: From Theory to Practice, Coping with Just-in-Time Compilation and Embedded Processors Constraints*. PhD Thesis. École normale supérieure de Lyon, France, 2012.

80. Colombet, Q., et al. (2011). Graph-coloring and register allocation using repairing. In *Proceedings of the International Conference on Compilers, Architecture, and Synthesis for Embedded Systems. CASES '04* (pp. 45–54).

81. Colwell, R. P., et al. (1987). A VLIW architecture for a trace scheduling compiler. In *Proceedings of the International Conference on Architectual Support for Programming Languages and Operating Systems. ASPLOS-II* (pp. 180–192).

82. Cooper, K. D., & Taylor Simpson, L. (1998). Live range splitting in a graph coloring register allocator. In *International Conference on Compiler Construction. CC '98* (pp. 174–187). New York: Springer.

83. Cooper, K. D., & Torczon, L. (2004). *Engineering a compiler*. Burlington: Morgan Kaufmann.

84. Cooper, K. D., Taylor Simpson, L., & Vick, C. A. (2001). Operator strength reduction. *ACM Transactions on Programming Language and Systems, 23*(5), 603–625.

85. Cooper, K. D., Harvey, T. J., & Kennedy, K. W. (2006). An empirical study of iterative dataflow analysis. In *International Conference on Computing. ICC '06* (pp. 266–276).

86. Cousot, P., & Halbwachs, N. (1978). Automatic discovery of linear restraints among variables of a program. In *Proceedings of the Symposium on Principles of Programming Languages. POPL '78* (pp. 84–96).

87. Cytron, R. K., & Ferrante, J. (1987). What's in a name? or the value of renaming for parallelism detection and storage allocation. In *Proceedings of the 1987 International Conference on Parallel Processing* (pp. 19–27).

88. Cytron, R. K., & Gershbein, R. (1993). Efficient accommodation of may-alias information in SSA form. In *International Conference on Programming Languages Design and Implementation. PLDI '93* (pp. 36–45).

89. Cytron, R. K., et al. (1989). An efficient method of computing static single assignment form. In *Proceedings of the Symposium on Principles of Programming Languages. POPL '89* (pp. 25–35).

90. Cytron, R. K., et al. (1991). Efficiently computing static single assignment form and the control dependence graph. *ACM Transactions on Programming Language and Systems, 13*(4), 451–490.

91. Danvy, O., & Schultz, U. P. (2000). Lambda-dropping: Transforming recursive equations into programs with block structure. *Theoretical Computer Science, 248*(1–2), 243–287.

92. Danvy, O., Millikin, K., & Nielsen, L. R. (2007). On one-pass CPS transformations. *Journal of Functional Programming, 17*(6), 793–812.

93. Darte, A., Robert, Y., & Vivien, F. (2000). *Scheduling and Automatic Parallelization*, 1st ed. Boston: Birkhauser.

94. Das, D., & Ramakrishna, U. (2005). A practical and fast iterative algorithm for Φ-function computation using DJ graphs. *ACM Transactions on Programming Language and Systems, 27*, 426–440.

95. Dean, J., & Ghemawat, S. (2008). MapReduce: Simplified data processing on large clusters. *Communications of the ACM, 51*(1), 107–113.

96. Dennis, J. B. (1974). First version of a data flow procedure language. In *Programming Symposium, Proceedings Colloque sur la Programmation* (pp. 362–376).

97. Dennis, J. B. (1980). Data flow supercomputers. *Computer, 13*(11), 48–56 (1980).

98. Dhamdhere, D. M. (2002). E-path_PRE: Partial redundancy made easy. *Sigplan Notices, 37*(8), 53–65.

99. de Dinechin, B. D. (1999). Extending modulo scheduling with memory reference merging. In *International Conference on Compiler Construction. CC '99* (pp. 274–287).

100. de Dinechin, B. D. (2007). Time-indexed formulations and a large neighborhood search for the resource-constrained modulo scheduling problem. In *Multidisciplinary International Scheduling Conference: Theory and Applications. MISTA*.

101. de Dinechin, B. D. (2008). Inter-block scoreboard scheduling in a JIT compiler for VLIW processors. In *European Conference on Parallel Processing* (pp. 370–381).

102. de Dinechin, B. D., et al. (2000). Code generator optimizations for the ST120 DSP-MCU core. In *Proceedings of the International Conference on Compilers, Architecture, and Synthesis for Embedded Systems. CASES '00* (pp. 93–102).

103. de Dinechin, B. D., et al. (2000). DSP-MCU processor optimization for portable applications. *Microelectronic Engineering, 54*(1–2), 123–132.

104. de Ferrière, F. (2007). Improvements to the Psi-SSA representation. In *Proceedings of the International Workshop on Software & Compilers for Embedded Systems. SCOPES '07* (pp. 111–121).

105. Diniz, P. C., et al. (2005). Automatic mapping of C to FPGAs with the DEFACTO compilation and synthesis system. *Microprocessors and Microsystems, 29*(2–3), 51–62

106. Drechsler, K.-H., & Stadel, M. P. (1993). A variation of knoop, rüthing and steffen's lazy code motion. *Sigplan Notices, 28*(5), 29–38.
107. Duesterwald, E., Gupta, R., & Soffa, M. L. (1994). Reducing the cost of data flow analysis by congruence partitioning. In *International Conference on Compiler Construction. CC '94* (pp. 357–373).
108. Ebner, D., et al. (2008). Generalized instruction selection using SSA-graphs. *Proceedings of the International Conference on Languages, Compilers, and Tools for Embedded Systems. LCTES '08* (pp. 31–40).
109. Eckstein, E., König, O., & Scholz, B. (2003). Code instruction selection based on SSA-graphs. In: *Proceedings of the International Workshop on Software & Compilers for Embedded Systems. SCOPES '03* (pp. 49–65).
110. Emami, M., Ghiya, R., & Hendren, L. J. (1994). Context-sensitive interprocedural points-to analysis in the presence of function pointers. In *International Conference on Programming Languages Design and Implementation. PLDI '94* (pp. 242–256).
111. Fabri, J. (1979). Automatic storage optimization. In *SIGPLAN '79.*
112. Fang, J. Z. (1997). Compiler algorithms on if-conversion, speculative predicates assignment and predicated code optimizations. In *Lcpc '97* (pp. 135–153).
113. Faraboschi, P., et al. (2000). Lx: A technology platform for customizable VLIW embedded processing. *SIGARCH Computer Architecture News, 28*(2), 203–213.
114. Farach-Colton, M., & Liberatore, V. (1998). On local register allocation. In *Proceedings of the Symposium on Discrete Algorithms. SODA '98* (pp. 564–573).
115. Feautrier, P. (1988). Parametric integer programming. *Rairo recherche opérationnelle, 22*, 243–268.
116. Feautrier, P., & Lengauer, C. (2011). Polyhedron model. In *Encyclopedia of Parallel Computing* (pp. 1581–1592).
117. Ferrante, J., & Mace, M. (1985). On linearizing parallel code. In *Proceedings of the Symposium on Principles of Programming Languages. POPL '85* (pp. 179–190).
118. Ferrante, J., Ottenstein, K. J., & Warren, J. D. (1987). The program dependence graph and its use in optimization. *ACM Transactions on Programming Language and Systems, 9*(3), 319–349.
119. Ferrante, J., Mace, M., & Simons, B. (1988). Generating sequential code from parallel code. In *Proceedings of the International Conference on Supercomputing. ICS '88* (pp. 582–592).
120. Fink, S., Knobe, K., & Sarkar, V. (2000). Unified analysis of array and object references in strongly typed languages. In *International Static Analysis Symposium. SAS '00* (pp. 155–174).
121. Flanagan, C., et al. (1993). The essence of compiling with continuations. In *International Conference on Programming Languages Design and Implementation. PLDI '93* (pp. 237–247).
122. Gavril, F. (1974). The intersection graphs of subtrees in trees are exactly the chordal graphs. *Journal of Combinatorial Theory, Series B, 16*(1), 47–56.
123. Gawlitza, T., et al. (2009). Polynomial precise interval analysis revisited. *Efficient Algorithms, 1*, 422–437.
124. George, L., & Appel, A. W. (1996). Iterated register coalescing. *ACM Transactions on Programming Language and Systems, 18*, 300–324 (1996).
125. George, L., & Matthias, B. (2003). Taming the IXP network processor. In *International Conference on Programming Languages Design and Implementation. PLDI '03* (pp. 26–37).
126. Gerlek, M. P., Wolfe, M., & Stoltz, E. (1994). *A reference chain approach for live variables.* Technical Report CSE 94-029. Oregon Graduate Institute of Science & Technology.
127. Gerlek, M. P., Stoltz, E., & Wolfe, M. (1995). Beyond induction variables: detecting and classifying sequences using a demand-driven SSA form. In *ACM Transactions on Programming Language and Systems, 17*(1), 85–122.
128. Gillies, D. M., et al. (1996). Global predicate analysis and its application to register allocation. In *Micro 29* (pp. 114–125).

129. Glesner, S. (2004). An ASM semantics for SSA intermediate representations. In *Abstract State Machines 2004. Advances in Theory and Practice, 11th International Workshop (ASM 2004), Proceedings. Lecture Notes in Computer Science* (pp. 144–160).

130. Gonzalez, R. E. (2000). Xtensa: a configurable and extensible processor. *IEEE Micro, 20*(2), 60–70.

131. Goodman, J. R., & Hsu, W. (1988). Code scheduling and register allocation in large basic blocks. In: *Proceedings of the International Conference on Supercomputing. ICS '88* (pp. 442–452).

132. Grund, D., & Hack, S. (2007). A fast cutting-plane algorithm for optimal coalescing. In *International Conference on Compiler Construction. CC '07* (pp. 111—125).

133. Guo, Z., et al. (2008). A compiler intermediate representation for reconfigurable fabrics. *International Journal of Parallel Programming, 36*(5), 493–520.

134. Gurevich, Y. (2000). Sequential abstract-state machines capture sequential algorithms. In: *ACM Transactions on Computational Logic (TOCL), 1*(1), 77–111.

135. Hack, S. (2007). *Register Allocation for Programs in SSA Form*. PhD Thesis. Universität Karlsruhe.

136. Hack, S., Grund, D., & Goos, G. (2005). *Towards Register Allocation for Programs in SSA Form*. Technical Report 2005–27. University of Karlsruhe.

137. Hagiescu, A., et al. (2009). A computing origami: Folding streams in FPGAs. In *Proceedings of the Design Automation Conference. DAC '09* (pp. 282–287).

138. Hardekopf, B., & Lin, C. (2011). Flow-sensitive pointer analysis for millions of lines of code. In *Proceedings of the International Symposium on Code Generation and Optimization. CGO '11*. New York: IEEE Computer Society (pp. 289–298).

139. Hasti, R., & Horwitz, S. (1998). Using static single assignment form to improve flow-insensitive pointer analysis. In *International Conference on Programming Languages Design and Implementation. PLDI '98* (pp. 97–105).

140. Havanki, W. A., Banerjia, S., & Conte, T. M. (1998). Treegion scheduling for wide issue processors. In: *International Symposium on High-Performance Computer Architecture*, 266.

141. Havlak, P. (1993). Construction of thinned gated single-assignment form. In *Proceedings of the International Workshop on Languages and Compilers for Parallel Computing. LCPC '93* (pp. 477–499).

142. Havlak, P. (1997). Nesting of reducible and irreducible loops. *ACM Transactions on Programming Language and Systems, 19*, 557–567.

143. Hecht, M. S. (1977). *Flow Analysis of Computer Programs*. New York: Elsevier Science Inc.

144. Hecht, M. S., & Ullman, J. D. (1973). Analysis of a simple algorithm for global data flow problems. In *Proceedings of the Symposium on Principles of Programming Languages. POPL '73* (pp. 207–217).

145. Hormati, A., et al. (2008). Optimus: Efficient realization of streaming applications on FPGAs. In *Proceedings of the International Conference on Compilers, Architecture, and Synthesis for Embedded Systems. CASES '08*, pp. 41–50.

146. Hwu, W. -M. W., et al. (1993). The superblock: An effective technique for VLIW and super-scalar compilation. *The Journal of Supercomputing, 7*(1–2), 229–248.

147. Intel Corp. *Arria10 device overview*. 2017. https://www.altera.com/products/fpga/arria-series/arria-10/features.html (visited on 03/15/2017).

148. Jacome, M. F., De Veciana, G., & Pillai, S. (2001). Clustered VLIW architectures with predicated switching. In *Proceedings of the Design Automation Conference. DAC '01* (pp. 696–701).

149. Janssen, J., & Corporaal, H. (1997). Making graphs reducible with controlled node splitting. *ACM Transactions on Programming Language and Systems, 19*, 1031–1052.

150. Jensen, S. H., Møller, A., & Thiemann, P. (2009). Type analysis for JavaScript. In *International Static Analysis Symposium. SAS '09* (pp. 238–255).

151. Johnson, N. E. (2004). *Code Size Optimization for Embedded Processors*. Technical Report UCAM-CL-TR-607. University of Cambridge, Computer Laboratory.

152. Johnson, N. E., & Mycroft, A. (2003). Combined code motion and register allocation using the value state dependence graph. In: *International Conference on Compiler Construction. CC '03* (pp. 1–16).

153. Johnson, R., & Pingali, K. (1993). Dependence-based program analysis. In *International Conference on Programming Languages Design and Implementation. PLDI '93* (pp. 78–89).

154. Johnson, R., Pearson, D., & Pingali, K. (1994). The program tree structure. In *International Conference on Programming Languages Design and Implementation. PLDI '94* (pp. 171–185).

155. Johnsson, T. (1985). Lambda lifting: Transforming programs to recursive equations. In *Conference on Functional Programming Languages and Computer Architecture. Lecture Notes in Computer Science* (pp. 190–203).

156. Jovanovic, N., Kruegel, C., & Kirda, E. (2006). Pixy: A static analysis tool for detecting web application vulnerabilities (short paper). In *Symposium on Security and Privacy* (pp. 258–263).

157. Kam, J. B., & Ullman, J. D. (1976). Global data flow analysis and iterative algorithms. *Journal of the ACM, 23*(1), 158–171.

158. Kam, J. B., & Ullman, J. D. (1977). Monotone data flow analysis frameworks. *Acta Informatica, 7*(3), 305–317.

159. Kästner, D., & Winkel, S. (2001). ILP-based instruction scheduling for IA-64. In *Proceedings of the International Conference on Languages, Compilers, and Tools for Embedded Systems. LCTES '01* (pp. 145–154).

160. Kelsey, R. (1995). A correspondence between continuation passing style and static single assignment form. In *Intermediate Representations Workshop* (pp. 13–23).

161. Kennedy, A. (2007). Compiling with continuations, continued. In *Proceedings of the International Conference on Functional Programming. ICFP '07* (pp. 177–190).

162. Kennedy, K. W. (1975). Node listings applied to data flow analysis. In *Proceedings of the Symposium on Principles of Programming Languages. POPL '75* (pp. 10–21).

163. Kennedy, R., et al. (1998). Strength reduction via SSAPRE. In *International Conference on Compiler Construction. CC '98*.

164. Kennedy, R., et al. (1999). Partial redundancy elimination in SSA form. *ACM Transactions on Programming Language and Systems, 21*(3), 627–676 (1999).

165. Khedker, U. P., & Dhamdhere, D. M. (1999). Bidirectional data flow analysis: Myths and reality. In *SIGPLAN Notices, 34*(6), 47–57.

166. Kildall, G. A. (1973). A unified approach to global program optimization. In *Proceedings of the Symposium on Principles of Programming Languages. POPL '73* (pp. 194–206).

167. Kislenkov, V., Mitrofanov, V., & Zima, E. V. (1998). Multidimensional chains of recurrences. In *Proceedings of the International Symposium on Symbolic and Algebraic Computation* (pp. 199–206).

168. Knobe, K., & Sarkar, V. (1998). Array SSA form and its use in parallelization. In *Proceedings of the Symposium on Principles of Programming Languages. POPL '98*.

169. Knobe, K., & Sarkar, V. (1998). Conditional constant propagation of scalar and array references using array SSA form. In *International Static Analysis Symposium. Lecture Notes in Computer Science* (pp. 33–56).

170. Knoop, J., Rüthing, O., & Steffen, B. (1992). Lazy code motion. In *International Conference on Programming Languages Design and Implementation. PLDI '92* (pp. 224–234).

171. Knoop, J., Rüthing, O., & Steffen, B. (1993). Lazy strength reduction. *Journal of Programming Languages, 1*(1), 71–91.

172. Knoop, J., Rüthing, O., & Steffen, B. (1994). Optimal code motion: Theory and practice. *ACM Transactions on Programming Language and Systems, 16*(4), 1117–1155.

173. Kronos Group (2018). *OpenCL overview.* https://www.khronos.org/opencl/resources

174. Landin, P. (1965). *A Generalization of Jumps and Labels.* Technical Report. Reprinted in *Higher Order and Symbolic Computation, 11*(2), 125–143, 1998, with a foreword by Hayo Thielecke. UNIVAC Systems Programming Research, 1965.

175. Lapkowski, C., & Hendren, L. J. (1996). Extended SSA numbering: Introducing SSA properties to languages with multi-level pointers. In *Proceedings of the Conference of the Centre for Advanced Studies on Collaborative Research. CASCON '96* (pp. 23–34).

176. Lattner, C., & Adve, V. S. (2004). LLVM: a compilation framework for lifelong program analysis & transformation. In *Proceedings of the International Symposium on Code Generation and Optimization. CGO '04* (pp. 75–88).

177. Laud, P., Uustalu, T., & Vene, V. (2006). Type systems equivalent to dataflow analyses for imperative languages. *Theoretical Computer Science, 364*(3), 292–310.

178. Lawrence, A. C. (2007). *Optimizing Compilation with the Value State Dependence Graph.* Technical Report UCAM-CL-TR-705. University of Cambridge, Computer Laboratory.

179. Lee, E. A., & Messerschmitt, D. G. (1987). Synchronous data flow. *Proceedings of the IEEE, 75*(9), 1235–1245.

180. Lee, J.-Y., & Park, I.-C. (2003). Address code generation for DSP instruction-set architectures. *ACM Transactions on Design Automation of Electronic Systems, 8*(3), 384–395.

181. Lenart, A., Sadler, C., & Gupta, S. K. S. (2000). SSA-based flow-sensitive type analysis: Combining constant and type propagation. In *Proceedings of the Symposium on Applied Computing. SAC '00* (pp. 813–817).

182. Leung, A., & George, L. (1999). Static single assignment form for machine code. In *International Conference on Programming Languages Design and Implementation. PLDI '99* (pp. 204–214).

183. Leupers, R. (1997). *Retargetable code generation for digital signal processors.* Amsterdam: Kluwer Academic Publishers.

184. Leupers, R. (1999). Exploiting conditional instructions in code generation for embedded VLIW processors. In *Proceedings of the Conference on Design, Automation and Test in Europe. DATE '99.*

185. Liu, S.-M., Lo, R., & Chow, F. (1996). Loop induction variable canonicalization in parallelizing compilers. In *PACT '96* (p. 228).

186. *LLVM website.* http://llvm.cs.uiuc.edu.

187. Lo, R., et al. (1998). Register promotion by sparse partial redundancy elimination of loads and stores. In *International Conference on Programming Languages Design and Implementation. PLDI '98* (pp. 26–37).

188. Logozzo, F., & Fähndric, M. (2010). Pentagons: A weakly relational abstract domain for the efficient validation of array accesses. *Science of Computer Programming, 75*, 796–807.

189. Lowney, P. G., et al. (1993). The multiflow trace scheduling compiler. *The Journal of Supercomputing, 7*(1–2), 51–142 (1993).

190. Lueh, G.-Y., Gross, T., & Adl-Tabatabai, A.-R. (2000). Fusion-based register allocation. *ACM Transactions on Programming Language and Systems, 22*(3), 431–470 (2000).

191. Mahlke, S. A., et al. (1992). Effective compiler support for predicated execution using the hyperblock. In *Proceedings of the International Symposium on Microarchitecture. MICRO 25* (pp. 45–54).

192. Mahlke, S. A., et al. (1992). Sentinel scheduling for VLIW and superscalar processors. In *Proceedings of the International Conference on Architectural Support for Programming Languages and Operating Systems. ASPLOS-V* (pp. 238–247).

193. Mahlke, S. A., et al. (1995). A comparison of full and partial predicated execution support for ILP processors. In *Proceedings of the International Symposium on Computer Architecture. ISCA '95* (pp. 138–150)

194. Mahlke, S. A., et al. (2001). Bitwidth cognizant architecture synthesis of custom hardware accelerators. *IEEE Transactions on Computer-Aided Design of Integrated Circuits and Systems, 20*(11), 1355–1371.

195. Matsuno, Y., & Ohori, A. (2006). A type system equivalent to static single assignment. In *Proceedings of the International Conference on Principles and Practice of Declarative Programming. PPDP '06* (pp. 249–260).

196. May, C. (1989). The parallel assignment problem redefined. In *IEEE Transactions on Software Engineering, 15*(6), 821–824.

197. McAllester, D. (2002). On the complexity analysis of static analyses. *Journal of the ACM, 49*, 512–537.
198. Metzgen, P., & Nancekievill, D. (2005). Multiplexer restructuring for FPGA implementation cost reduction. In *Proceedings of the Design Automation Conference. DAC '05* (pp. 421–426).
199. Miné, A. (2006). The octagon abstract domain. *Higher-Order and Symbolic Computation, 19*, 31–100.
200. Moggi, E. (1991). Notions of computation and monads. *Information and Computation, 93*(1), 55–92.
201. Morel, E., & Renvoise, C. (1979). Global optimization by suppression of partial redundancies. *Communications of the ACM, 22*(2), 96–103.
202. Morgan, R. (1998). *Building an optimizing compiler*. Oxford: Butterworth-Heinemann.
203. Mössenböck, H., & Pfeiffer, M. (2002). Linear scan register allocation in the context of SSA form and register constraints. In *International Conference on Compiler Construction. CC '02* (pp. 229–246).
204. Motwani, R., et al. (1995). *Combining Register Allocation and Instruction Scheduling*. Technical Report. Stanford University.
205. Muchnick, S. S. (1997). *Advanced compiler design and implementation*. Burlington: Morgan Kaufmann.
206. Murphy, B. R., et al. (2008). Fault-safe code motion for type-safe languages. In *Proceedings of the International Symposium on Code Generation and Optimization. CGO '08* (pp. 144–154).
207. Nanda, M. G., & Sinha, S. (2009). Accurate interprocedural null-dereference analysis for Java. In *International Conference on Software Engineering* (pp. 133–143).
208. Nielson, F., Nielson, H. R., & Hankin, C. (2005). *Principles of program analysis*. New York: Springer.
209. Norris, C., & Pollock, L. L. (1993). A scheduler-sensitive global register allocator. In *Proceedings of the International Conference on Supercomputing. ICS '93* (pp. 804–813). New York: IEEE.
210. Novillo, D. (2005). A propagation engine for GCC. In *Proceedings of the GCC Developers Summit* (pp. 175–184).
211. Novillo, D. (2007). Memory SSA — a unified approach for sparsely representing memory operations. In *Proceedings of the GCC Developers Summit*.
212. Nuzman, D., & Henderson, R. (2006). Multi-platform auto-vectorization. In *Proceedings of the International Symposium on Code Generation and Optimization. CGO '06*.
213. O'Donnell, C. (1994). *High Level Compiling for Low Level Machines*. PhD Thesis. Ecole Nationale Superieure des Telecommunications.
214. Ottenstein, K. J., & Ottenstein, L. M. (1984). The program dependence graph in a software development environment. In *ACM SIGSOFT Software Engineering Notes, 99*(3), 177–184.
215. Ottenstein, K. J., Ballance, R. A., & MacCabe, A. B. (1990). The program dependence web: A representation supporting control-, data-, and demand-driven interpretation of imperative languages. In *International Conference on Programming Languages Design and Implementation. PLDI '90* (pp. 257–271).
216. D. A. Padua (Ed.) (2011). *Encyclopedia of parallel computing*. New York: Springer.
217. Paleri, V. K., Srikant, Y. N., & Shankar, P. (2003). Partial redundancy elimination: a simple, pragmatic and provably correct algorithm. *Science of Programming Programming, 48*(1), 1–20.
218. Panda, P. R. (2001). SystemC: A modeling platform supporting multiple design abstractions. In *Proceedings of International Symposium on Systems Synthesis. ISSS '01* (pp. 75–80).
219. Park, J., & Moon, S.-M. (2004). Optimistic register coalescing. *ACM Transactions on Programming Language and Systems, 26*(4), 735–765.
220. Park, J. C. H., & Schlansker, M. S. (1991). *On Predicated Execution*. Technical Report HPL-91-58. Hewlett Packard Laboratories.
221. Pereira, F. M. Q., & Palsberg, J. (2005). Register allocation via coloring of chordal graphs. In *Asian Symposium on Programming Languages and Systems. APLAS '05* (pp. 315–329).

222. Pereira, F. M. Q., & Palsberg, J. (2008). Register allocation by puzzle solving. In *International Conference on Programming Languages Design and Implementation. PLDI '08* (pp. 216–226).

223. Peyton Jones, S., et al. (1998). Bridging the gulf: A common intermediate language for ML and Haskell. In *Proceedings of the Symposium on Principles of Programming Languages. POPL '98* (pp. 49–61).

224. Pingali, K., & Bilardi, G. (1995). APT: A data structure for optimal control dependence computation. In *International Conference on Programming Languages Design and Implementation. PLDI '95* (pp. 211–222).

225. Pingali, K., & Bilardi, G. (1997). Optimal control dependence computation and the roman chariots problem. In *ACM Transactions on Programming Language and Systems* (pp. 462–491) (1997).

226. Pinter, S. S. (1993). Register allocation with instruction scheduling. In *International Conference on Programming Languages Design and Implementation. PLDI '93* (pp. 248–257).

227. Pioli, A., & Hind, M. (1999). *Combining Interprocedural Pointer Analysis and Conditional Constant Propagation*. Technical Report. IBM T. J. Watson Research Center.

228. Plevyak, J. B. (1996). Optimization of Object-Oriented and Concurrent Programs. PhD Thesis. University of Illinois at Urbana-Champaign.

229. Plotkin, G. D. (1975). Call-by-name, call-by-value and the lambda-calculus. *Theoretical Computer Science, 1*(2), 125–159.

230. Poletto, M., & Sarkar, V. (1999). Linear scan register allocation. In *ACM Transactions on Programming Language and Systems, 21*(5), 895–913.

231. Pop, S. (2006). *The SSA Representation Framework: Semantics, Analyses and GCC Implementation*. PhD Thesis. Research center for computer science (CRI) of the Ecole des mines de Paris.

232. Pop, S., Cohen, A., & Silber, G.-A. (2005). Induction variable analysis with delayed abstractions. In *Proceedings of the First International Conference on High Performance Embedded Architectures and Compilers. HiPEAC'05* (pp. 218–232).

233. Pop, S., Jouvelot, P., & Silber, G.-A. (2007). *In and Out of SSA: A Denotational Specification*. Technical Report. Research center for computer science (CRI) of the Ecole des mines de Paris.

234. Prosser, R. T. (1959). Applications of boolean matrices to the analysis of flow diagrams. In *Eastern Joint IRE-AIEE-ACM Computer Conference* (pp. 133–138).

235. Pugh, W. (1991). Uniform techniques for loop optimization. In *Proceedings of the International Conference on Supercomputing. ICS '91* (pp. 341–352).

236. Ramalingam, G. (2002). On loops, dominators, and dominance frontiers. *ACM Transactions on Programming Language and Systems, 24*(5), 455–490.

237. Ramalingam, G. (2002). On sparse evaluation representations. *Theoretical Computer Science, 277*(1–2), 119–147.

238. Ramalingam, G., & Reps, T. (1994). An incremental algorithm for maintaining the dominator tree of a reducible flowgraph. In *Proceedings of the Symposium on Principles of Programming Languages. POPL '94* (pp. 287–296).

239. Rastello, F. (2012). *On sparse intermediate representations: some structural properties and applications to just in time compilation*. Habilitation à diriger des recherches. École normale supérieure de Lyon, France.

240. Rastello, F., de Ferrière, F., & Guillon, C. (2004). Optimizing translation out of SSA using renaming constraints. In *Proceedings of the International Symposium on Code Generation and Optimization. CGO '04* (pp. 265–278).

241. Rau, B. R. (1996). Iterative modulo scheduling. *International Journal of Parallel Programming, 24*(1), 3–65.

242. Rawat, P. S., et al. (2018). Register optimizations for stencils on GPUs. In *Proceedings of the Symposium on Principles and Practice of Parallel Programming. PPoPP '18* (pp. 168–182).

243. Reppy, J. H. (2002). Optimizing nested loops using local CPS conversion. *Higher-Order and Symbolic Computation, 15*(2–3), 161–180.

244. Reynolds, J. C. (1972). Definitional interpreters for higher-order programming languages. In *Proceedings of the ACM National Conference*. Reprinted in *Higher-Order and Symbolic Computation, 11*(4), 363–397, 1998 (pp. 717–714).

245. Reynolds, J. C. (1974). On the relation between direct and continuation semantics. In *Proceedings of the Colloquium on Automata, Languages and Programming* (pp. 141–156).

246. Reynolds, J. C. (1993). The discoveries of continuations. *Lisp and Symbolic Computation, 6*(3–4), 233–248.

247. Rideau, L., Serpette, B. P., & Leroy, X. (2008). Tilting at windmills with Coq: Formal verification of a compilation algorithm for parallel moves. *Journal of Automated Reasoning, 40*(4), 307–326.

248. Rimsa, A. A., D'Amorim, M., & Pereira, F. M. Q. (2011). Tainted flow analysis on e-SSA-form programs. In *International Conference on Compiler Construction. CC '11* (pp. 124–143).

249. Rosen, B. K., Wegman, M. N., & Zadeck, F. K. (1988). Global value numbers and redundant computations. In *Proceedings of the Symposium on Principles of Programming Languages. POPL '88* (pp. 12–27).

250. Ruf, E. (1995). Optimizing sparse representations for dataflow analysis. *Sigplan Notices, 30*(3), 50–61.

251. Ryder, B. G., & Paull, M. C. (1986). Elimination algorithms for data flow analysis. In *ACM Computing Surveys, 18*(3), 277–316.

252. Sarkar, V., & Barik, R. (2007). Extended linear scan: an alternate foundation for global register allocation. In *International Conference on Compiler Construction. CC '07* (pp. 141–155). New York: Springer.

253. Sarkar, V., & Fink, S. (2001). Efficient dependence analysis for Java arrays. In *European Conference on Parallel Processing. Lecture Notes in Computer Science* (Vol. 2150, pp. 273–277).

254. Schäfer, S., & Scholz, B. (2007). Optimal chain rule placement for instruction selection based on SSA graphs. In *Proceedings of the International Workshop on Software & Compilers for Embedded Systems. SCOPES '07* (pp. 91–100).

255. Schlansker, M. S., Mahlke, S. A., & Johnson, R. (1999). Control CPR: A branch height reduction optimization for EPIC architectures. In *International Conference on Programming Languages Design and Implementation. PLDI '99* (pp. 155–168).

256. Seshan, N. (1998). High VelociTI processing. In *IEEE Signal Processing Magazine* (pp. 86–101).

257. Shanley, T (2010). *X86 instruction set architecture*. New York: Mindshare Press.

258. Shapiro, R. M., & Saint, H. (1969). *The Representation of Algorithms*. Technical Report RADC-TR-69-313. Rome Air Development Center.

259. Simons, B., Alpern, D., & Ferrante, J. (1990). A foundation for sequentializing parallel code. In *Proceedings of the Symposium on Parallel Algorithms and Architectures. SPAA '90* (pp. 350–359).

260. Singer, J. (2004). Sparse bidirectional data flow analysis as a basis for type inference. In *Web Proceedings of the Applied Semantics Workshop. APPSEM '04*.

261. Singer, J. (2005). Static Program Analysis Based on Virtual Register Renaming. PhD Thesis. University of Cambridge.

262. Smith, M. D., Ramsey, N., & Holloway, G. H. (2004). A generalized algorithm for graph-coloring register allocation. In *International Conference on Programming Languages Design and Implementation. PLDI '04* (pp. 277–288).

263. Sreedhar, V. C., & Gao, G. R. (1995). A linear time algorithm for placing ϕ-nodes. In *Proceedings of the Symposium on Principles of Programming Languages. POPL '95* (pp. 62–73).

264. Sreedhar, V. C., Gao, G. R., & Lee, Y.-F. (1995). Incremental computation of dominator trees. In *Sigplan Notices, 30*(3), 1–12.

265. Sreedhar, V. C., Gao, G. R., & Lee, Y.-F. (1996). A new framework for exhaustive and incremental data flow analysis using DJ graphs. In *International conference on programming languages design and implementation. PLDI '96* (pp. 278–290).

266. Sreedhar, V. C., Gao, G. R., & Lee, Y.-F. (1996). Identifying loops using DJ graphs. textit*ACM Transactions on Programming Language and Systems*, 18(6), 649–658.

267. Sreedhar, V. C., et al. (1999). Translating out of static single assignment form. In *International Static Analysis Symposium. SAS '99* (pp. 194–210).

268. Stallman, R. M., & GCC Dev. Community (2017). *GCC 7.0 GNU compiler collection internals*. Samurai Media Limited.

269. Stanier, J. (2011). *Removing and Restoring Control Flow with the Value State Dependence Graph*. PhD Thesis. University of Sussex, School of Informatics.

270. Stanier, J., & Watson, D. (2013). Intermediate representations in imperative compilers: A survey. In *ACM Computing Surveys, 45*(3), 26:1–26:27.

271. Steensgaard, B. (1993). *Sequentializing Program Dependence Graphs for Irreducible Programs*. Technical Report MSR-TR-93-14. Microsoft Research, Redmond, WA.

272. Steensgaard, B. (1995). Sparse functional stores for imperative programs. In *Workshop on Intermediate Representations*.

273. Stephenson, M., Babb, J., & Amarasinghe, S. (2000). Bidwidth analysis with application to silicon compilation. In *International Conference on Programming Languages Design and Implementation. PLDI '00* (pp. 108–120).

274. Stoltz, E., Gerlek, M. P., & Wolfe, M. (1994). Extended SSA with factored use-def chains to support optimization and parallelism. In *Proceedings of the Hawaii International Conference on System Sciences* (pp. 43–52).

275. Stoutchinin, A., & de Ferrière, F. (2001). Efficient static single assignment form for predication. In *Proceedings of the International Symposium on Microarchitecture. MICRO 34* (pp. 172–181).

276. Stoutchinin, A., & Gao, G. R. (2004). If-conversion in SSA form. In *European Conference on Parallel Processing. Lecture Notes in Computer Science* (pp. 336–345).

277. Su, Z., & Wagner, D. (2005). A class of polynomially solvable range constraints for interval analysis without widenings. *Theoretical Computeter Science, 345*(1), 122–138.

278. Surawski, M. J. (2016). *Loop Optimizations for MLton*. Master's Thesis. Department of Computer Science, Rochester Institute of Technology.

279. Sussman, G. J., & Steele, G. L. Jr. (1998). *Scheme: An Interpreter for Extended Lambda Calculus*. Technical Report AI Lab Memo AIM-349. Reprinted in *Higher-Order and Symbolic Computation, 11*(4), 405–439, 1998. MIT AI Lab, 1975.

280. Tarditi, D., et al. (1996). TIL: A type-directed optimizing compiler for ML. In *International Conference on Programming Languages Design and Implementation. PLDI '96* (pp. 181–192).

281. Tavares, A. L. C., et al. (2011). Decoupled graph-coloring register allocation with hierarchical aliasing. In *Proceedings of the International Workshop on Software & Compilers for Embedded Systems. SCOPES '11* (pp. 1–10).

282. Tavares, A. L. C., et al. (2014). Parameterized construction of program representations for sparse dataflow analyses. In *International Conference on Compiler Construction. CC '14* (pp. 18–39).

283. Thomas, D., & Moorby, P. (1998). *The verilog hardware description language*. Alphen aan den Rijn: Kluwer Academic Publishers.

284. Tobin-Hochstadt, S., & Felleisen, M. (2008). The design and implementation of typed scheme. In *POPL '08* (pp. 395–406).

285. Tolmach, A. P., & Oliva, D. (1998). From ML to Ada: Strongly-typed language interoperability via source translation. *Journal of Functional Programming, 8*(4), 367–412.

286. Touati, S., & Eisenbeis, C. (2004). Early periodic register allocation on ILP processors. *Parallel Processing Letters, 14*(2), 287–313.

287. Traub, O., Holloway, G. H., & Smith, M. D. (1998). Quality and speed in linear-scan register allocation. In *International Conference on Programming Languages Design and Implementation. PLDI '98* (pp. 142–151).
288. Tripp, J. L., Jackson, P. A., & Hutchings, B. L. (2002). Sea cucumber: A synthesizing compiler for FPGAs. In: *International Conference on Field Programmable Logic and Applications. FPL '02* (pp. 875–885).
289. Tu, P., & Padua, D. (1995). Efficient building and placing of gating functions. In *International Conference on Programming Languages Design and Implementation. PLDI '95* (pp. 47–55).
290. Tu, P., & Padua, D. (1995). Gated SSA-based demand-driven symbolic analysis for parallelizing compilers. In *Proceedings of the International Conference on Supercomputing. ICS '95* (pp. 414–423).
291. Upton, E. (2003). Optimal sequentialization of gated data dependence graphs is NP-complete. In *Proceedings of the International Conference on Parallel and Distributed Processing Techniques and Applications* (pp. 1767–1770).
292. Upton, E. (2006). *Compiling with Data Dependence Graphs*. PhD Thesis. University of Cambridge, Computer Laboratory.
293. van Engelen, R. A. (2001). Efficient symbolic analysis for optimizing compilers. In *International Conference on Compiler Construction. CC '01* (pp. 118–132).
294. van Wijngaarden, A. (1966). Recursive defnition of syntax and semantics. In *Formal Language Description Languages for Computer Programming* (pp. 13–24).
295. VanDrunen, T. J. (2004). *Partial Redundancy Elimination for Global Value Numbering*. PhD Thesis. Purdue University.
296. Vandrunen, T., & Hosking, A. L. (2004). Anticipation-based partial redundancy elimination for static single assignment form. In *Software—Practice and Experience, 34*(15), 1413–1439 (2004).
297. Vandrunen, T., & Hosking, A. L. (2004). Value-based partial redundancy elimination. In *International Conference on Compiler Construction. CC '04* (pp. 167–184).
298. Verma, A. K., Brisk, P., & Ienne, P. (2008). Dataflow transformations to maximize the use of carry-save representation in arithmetic circuits. *IEEE Transactions on Computer-Aided Design of Integrated Circuits and Systems, 27*(10), 1761–1774.
299. Wadsworth, C. P. (2000). Continuations revisited. *Higher-Order and Symbolic Computation, 13*(1/2), 131–133.
300. Wand, M. (1983). Loops in combinator-based compilers. *Information and Control, 57*(2/3), 148–164.
301. Wang, J., et al. (1994). Software pipelining with register allocation and spilling. In *Proceedings of the International Symposium on Microarchitecture. MICRO 27* (pp. 95–99).
302. Wassermann, G., & Su, Z. (2007). Sound and precise analysis of web applications for injection vulnerabilities. In *International Conference on Programming Languages Design and Implementation. PLDI '07* (pp. 32–41).
303. Wegman, M. N., & Kenneth Zadeck, F. (1991). Constant propagation with conditional branches. *ACM Transactions on Programming Language and Systems, 13*(2), 181–210.
304. Weise, D., et al. (1994). Value dependence graphs: Representation without taxation. In *Proceedings of the Symposium on Principles of Programming Languages. POPL '94* (pp. 297–310).
305. Wimmer, C., & Mössenböck, H. (2005). Optimized interval splitting in a linear scan register allocator. In *Proceedings of the 1st ACM/USENIX International Conference on Virtual Execution Environments* (pp. 132–141). Chicago: ACM.
306. Wolfe, M. (1992). Beyond induction variables. In *International Conference on Programming Languages Design and Implementation. PLDI '92* (pp. 162–174).
307. Wolfe, M. (1994). "J+ =j". *Sigplan Notices, 29*(7), 51–53.
308. Wolfe, M. (1996). *High performance compilers for parallel computing*. Reading, MA: Addison-Wesley.

309. Wu, P., Cohen, A., & Padua, D. (2001). Induction variable analysis without idiom recognition: Beyond monotonicity. In *Proceedings of the International Workshop on Languages and Compilers for Parallel Computing. LCPC '01* (p. 2624).

310. Xilinx Inc. (2014). *Vivado high-level synthesis.* http://www.xilinx.com/products/design-%20tools/vivado/integration/esl-design/index.htm (visited on 06/2014).

311. Xilinx Inc. (2016). *Virtex ultrascale+ FPGA devices.* https://www.xilinx.com/products/silicon-devices/fpga/virtex-ultrascale-plus.html

312. Xue, J., & Cai, Q. (2006). A lifetime optimal algorithm for speculative PRE. *ACM Transactions on Architecture and Code Optimization, 3*(2), 115–155.

313. Xue, J., & Knoop, J. (2006). A fresh look at PRE as a maximum flow problem. In *International Conference on Compiler Construction. CC '06* (pp. 139–154).

314. Yang, B.-S., et al. (1999). LaTTe: A Java VM just-in-time compiler with fast and efficient register allocation. In *Proceedings of the International Conference on Parallel Architectures and Compilation Techniques. PACT '99* (pp. 128–138). New York: IEEE.

315. Yeung, J. H. C., et al. (2008). Map-reduce as a programming model for custom computing machines. In *Proceedings of the International Symposium on Field-Programmable Custom Computing Machines. FCCM '08* (pp. 149–159).

316. Zadeck, F. K. (1984). *Incremental Data Flow Analysis in a Structured Program Editor.* PhD Thesis. Rice University.

317. Zhou, H., Chen, W., & Chow, F. (2011). An SSA-based algorithm for optimal speculative code motion under an execution profile. In *International Conference on Programming Languages Design and Implementation. PLDI '11.*

318. Zima, E. V. (2001). On computational properties of chains of recurrences. In *Proceedings of the International Symposium on Symbolic and Algebraic Computation* (pp. 345–352).

Index

Page numbers are underlined in the index when they represent the definition or the main source of information about whatever is being indexed. A page number is given in *italics* when that page contains an instructive example, use, or discussion of the concept in question.

Printed in the United States
by Baker & Taylor Publisher Services